Deep Learning with MXNet Cookbook

Discover an extensive collection of recipes for creating and implementing AI models on MXNet

Andrés P. Torres

BIRMINGHAM—MUMBAI

Deep Learning with MXNet Cookbook

Group Product Manager: Ali Abidi
Publishing Product Manager: Sunith Shetty
Content Development Editor: Priyanka Soam
Technical Editor: Kavyashree K S
Copy Editor: Safis Editing
Project Coordinator: Shambhavi Mishra
Proofreader: Safis Editing
Indexer: Hemangini Bari
Production Designer: Ponraj Dhandapani
Marketing Coordinator: Vinishka Kalra

First published: Dec 2023

Production reference: 2201223

Published by Packt Publishing Ltd.
Grosvenor House
11 St Paul's Square
Birmingham
B3 1RB, UK.

ISBN 978-1-80056-960-7

`www.packtpub.com`

To my parents, thanks for giving me everything. To my sister Carmen and her family, for reminding me that life is not only work, above all Carlos, Adriana and Pablo, I love you madly.

I miss Madrid because of you all, but also because of my friends from there, Almu, Alba, Bea, @nosinmiscascos, Ducali, Cala, Rubens, Marco, Moi, Agus, Charlie, Puente, Johnny, Surcio, Chemo and Zaragoza, because I would like to see them so much more, and when I see them the cliché is true, it's indeed as if time hadn't passed. Specially, the last three, because you have supported me greatly with my latest changes and this book. To new friends, Julien, Santi, Paul, Laz, Ricciardo (and partners!), Karla, Carmen, Marina, Michele, Luis and Carlos. Without you, I wouldn't have adapted to the UK. Same with Valentina and Javi, thanks for those board game nights!. Biggest hug to the last three for their recent changes, we are all a team, and we are all with you in this.

Lastly, to Marta, thank you for always supporting me whatever my mood.

- Andrés P. Torres

Foreword

The best teachers speak from experience. The best teachers remember the learning journey they themselves undertook. The best teachers end up teaching the best topics from a perspective given by a profound understanding of the foundations.

This is nowhere more true than you can find here in this book. Andrés P. Torres has a wealth of practical hands-on experience in machine learning, built and enabled by his detailed and nuanced understanding of the fundamentals of the subject.

He has built it, he has deployed it, he has iterated and he has, like all the best seasoned practitioners, built an atlas on how to approach a flotilla of ML problems and challenges.

This book is a direct consequence of the author's professional journey, one that I have been privileged to share in as a colleague. It draws on his time at Amazon and his wider career to provide a compact, accessible and appropriately detailed practitioners guide to MXNet - a deep learning framework which is used widely in commercial and research.

If you are new to the MXNet framework or finding yourself in charge of maintaining and developing a code base built with MXNet then this book is, of course, for you. But more generally, this book allows any deep learning practitioner to gain deep insight into the foundations of techniques that transcend any particular framework. Indeed, understanding how several frameworks approach common aspects and problems of deep learning is a powerful way to understand concepts beyond and below any given API.

This book is full of worked examples which are carefully chosen to balance the common-case problems and topics you are likely to face, be appropriately pedagogical and also provide multiple jumping off points into the wider MXNet documentation and code base. You will enjoy it. Who is not deeply entertained by learning how to teach machines to learn (deeply) ?

Prof. Paul Newman CBE FREng FIEEE FIET

BP Chair of Information Engineering

University of Oxford, Oxford

CTO Oxa Autonomy

Contributors

About the author

Andrés P. Torres, is the Head of Perception at Oxa, a global leader in industrial autonomous vehicles, leading the design and development of State-Of The-Art algorithms for autonomous driving. Before, Andrés had a stint as an advisor and Head of AI at an early stage content generation startup, Maekersuite, where he developed several AI-based algorithms for mobile phones and the web. Prior to this, Andrés was a Software Development Manager at Amazon Prime Air, developing software to optimize operations for autonomous drones.

I want to specially thank Marta M. Civera for most of the illustrations in Chapter 5, another example of her wonderful skills, apart from being a fantastic Architect and, above all, partner.

About the reviewers

Tuhin Sharma is Sr. Principal Data Scientist at Redhat in the Corporate Development and Strategy group. Prior to that, he worked at Hypersonix as an AI Architect. He also co-founded and has been CEO of Binaize, a website conversion intelligence product for eCommerce SMBs. He received a master's degree from IIT Roorkee and a bachelor's degree from IIEST Shibpur in Computer Science. He loves to code and collaborate on open-source and research projects. He has 4 research papers and 5 patents in the field of AI and NLP. He is a reviewer of the IEEE MASS conference in the AI track. He writes deep learning articles for O'Reilly in collaboration with the AWS MXNET team. He is a regular speaker at prominent AI conferences like O'Reilly AI, ODSC, GIDS, etc.

Anshu Trivedi, as a data scientist at BharatPe with three years plus of experience, delves into the realms of data engineering while mentoring budding data science enthusiasts at Newton School. Alongside, She actively contributes to open-source projects and regularly shares technical insights through her blog. Her passion lies in exploring intricate concepts of Artificial Intelligence and disseminating knowledge to foster learning and innovation.

Table of Contents

Preface xiii

1

Up and Running with MXNet 1

Technical requirements 2
Installing MXNet, Gluon, GluonCV, and GluonNLP 2
Getting ready 2
How to do it... 3
How it works... 9
There's more… 10
NumPy and MXNet ND arrays 10
Getting ready 11
How to do it... 12
How it works... 19
There's more… 19

2

Working with MXNet and Visualizing Datasets – Gluon and DataLoader 21

Technical requirements 21
Understanding regression datasets – loading, managing, and visualizing the House Sales dataset 22
Getting ready 23
How to do it... 24
How it works... 33
There's more… 33
Understanding classification datasets – loading, managing, and visualizing the Iris dataset 33
Getting ready 34
How to do it... 34
How it works... 40
There's more… 40
Understanding image datasets – loading, managing, and visualizing the Fashion-MNIST dataset 40
Getting ready 41
How to do it... 42
How it works... 49
There's more… 49

Understanding text datasets – loading, managing, and visualizing the Enron Email dataset 50
Getting ready 50
How to do it... 51
How it works... 62
There's more... 62

3

Solving Regression Problems 65

Technical requirements 65
Understanding the math of regression models 66
Getting ready 66
How to do it... 66
How it works... 74
There's more... 74
Defining loss functions and evaluation metrics for regression 74
Getting ready 75
How to do it... 76
How it works... 79
There's more... 79
Training regression models 79
Getting ready 80
How to do it... 80
How it works... 92
There's more... 93
Evaluating regression models 94
Getting ready 94
How to do it... 94
How it works... 98
There's more... 99

4

Solving Classification Problems 101

Technical requirements 101
Understanding math for classification models 102
Getting ready 102
How to do it... 102
How it works... 107
There's more... 107
Defining loss functions and evaluation metrics for classification 107
Getting ready 108
How to do it... 108
How it works... 113
There's more... 113
Training for classification models 114
Getting ready 114
How to do it... 114
How it works... 119
There's more... 120
Evaluating classification models 120
Getting ready 120
How to do it... 120
How it works... 124
There's more... 124

5

Analyzing Images with Computer Vision 125

Technical requirements 125

Understanding convolutional neural networks 126

Getting ready 126

How to do it... 126

How it works... 132

There's more... 132

Classifying images with MXNet - GluonCV Model Zoo, AlexNet, and ResNet 133

Getting ready 133

How to do it... 134

How it works... 143

There's more... 144

Detecting objects with MXNet - Faster R-CNN and YOLO 144

Getting ready 145

How to do it... 145

How it works... 158

There's more... 159

Segmenting objects in images with MXNet - PSPNet and DeepLab-v3 160

Getting ready 160

How to do it... 160

How it works... 172

There's more... 172

6

Understanding Text with Natural Language Processing 175

Technical requirements 175

Introducing NLP networks 176

Getting ready 176

How to do it... 176

Introducing Recurrent Neural Networks (RNNs) 178

Improving RNNs with Long Short-Term Memory (LSTM) 181

Introducing GluonNLP Model Zoo 183

Paying attention with Transformers 183

How it works... 187

There's more... 188

Classifying news highlights with topic modeling 189

Getting ready 189

How to do it... 189

How it works... 195

There's more... 196

Analyzing sentiment in movie reviews 196

Getting ready 196

How to do it... 197

How it works... 204

There's more... 205

Translating text from Vietnamese to English 205

Getting ready 206

How to do it... 206

How it works... 213

There's more... 213

7

Optimizing Models with Transfer Learning and Fine-Tuning 215

Technical requirements 216

Understanding transfer learning and fine-tuning 216

Getting ready 216

How to do it... 217

How it works... 225

There's more... 225

Improving performance for classifying images 226

Getting ready 226

How to do it... 226

Revisiting the ImageNet-1k and Dogs vs Cats datasets 227

How it works... 238

There's more... 238

Improving performance for segmenting images 239

Getting ready 239

How to do it... 239

How it works... 250

There's more... 251

Improving performance for translating English to German 251

Getting ready 252

How to do it... 252

How it works... 261

There's more... 262

8

Improving Training Performance with MXNet 263

Technical requirements 264

Introducing training optimization features 264

Getting ready 264

How to do it... 264

How it works... 274

There's more... 275

Optimizing training for image segmentation 275

Getting ready 276

How to do it... 276

How it works... 284

There's more... 285

Optimizing training for translating text from English to German 285

Getting ready 286

How to do it... 286

How it works... 294

There's more... 294

9

Improving Inference Performance with MXNet 297

Technical requirements 298

Introducing inference optimization features 298

Getting ready 298

How to do it... 299

How it works... 308

There's more… 309

Optimizing inference for image segmentation 309

Getting ready 310

How to do it... 310

How it works... 323

There's more… 323

Optimizing inference when translating text from English to German 324

Getting ready 324

How to do it... 324

How it works... 333

There's more… 333

Index 335

Other Books You May Enjoy 348

Preface

MXNet is an open-source deep learning framework that allows you to train and deploy neural network models and implement **state-of-the-art (SOTA)** architectures in Computer Vision, Natural Language Processing, and more. With this cookbook, you will be able to construct fast, scalable deep learning solutions using Apache MXNet.

This book will start by showing you the different versions of MXNet and what version to choose before installing your library. You will learn to start using MXNet/Gluon libraries to solve classification and regression problems and get an idea on the inner workings of these libraries. This book will also show how to use MXNet to analyze toy datasets in the areas of numerical regression, data classification, image classification, and text classification. You'll also learn to build and train deep-learning neural network architectures from scratch, before moving on to complex concepts like transfer learning. You'll learn to construct and deploy neural network architectures including CNN, RNN, Transformers, and integrate these models into your applications. You will also learn to analyze the performance of these models, and fine-tune them for increased accuracy, scalability, and speed.

By the end of the book, you will be able to utilize the MXNet and Gluon libraries to create and train deep learning networks using GPUs and learn how to deploy them efficiently in different environments.

Who is this book for?

This book is ideal for Data Scientists, Machine Learning Engineers, and Developers who want to work with Apache MXNet for building fast, scalable deep learning solutions. The reader is expected to have a good understanding of Python programming and a working environment with Python 3.7+. A good theoretical understanding of mathematics for deep learning will be beneficial.

What this book covers

Chapter 1, Up and Running with MXNet, To start working with MXNet, we need to install the library. There are several different versions of MXNet available to be installed, and in this chapter, we will cover how to help you choose the right version. The most important parameter will be the available hardware we have. In order to optimize performance, it is always best to maximize the use of our available Hardware. We will compare the usage of a well-known linear algebra library, NumPy, and how MXNet provides similar operations. We will then compare the performance of the different MXNet versions vs. Numpy.

Chapter 2, Working with MXNet and Visualizing Datasets: Gluon and DataLoader, In this chapter, we will start using MXNet to analyze some toy datasets in the domains of numerical regression, data classification, image classification and text classification. To manage those tasks efficiently, we will see new MXNet libraries and functions such as Gluon and DataLoader.

Chapter 3, Solving Regression Problems, In this chapter, we will learn how to use MXNet and Gluon libraries to apply supervised learning to solve regression problems. We will explore and understand a house prices dataset and will learn how to predict the price of a house. To achieve this objective, we will train neural networks and study the effect of the different hyper-parameters.

Chapter 4, Solving Classification Problems, In this chapter, we will learn how to use MXNet and Gluon libraries to apply supervised learning to solve classification problems. We will explore and understand a flowers dataset and will learn how to predict the type of a flower given some metrics. To achieve this objective, we will train neural networks and study the effect of the different hyper-parameters.

Chapter 5, Analyzing Images with Computer Vision, In this chapter, the reader will understand the different architectures and operations available in MXNet/GluonCV to work with images. Furthermore, the readers will get introduced to classic Computer Vision problems: Image Classification, Object Detection and Semantic Segmentation. They will then learn how to leverage GluonCV Model Zoo to use pre-existing models to solve these problems.

Chapter 6, Understanding Text with Natural Language Processing, In this chapter, the reader will understand the different architectures and operations available in MXNet/GluonNLP to work with text datasets. Furthermore, the readers will get introduced to classic Natural Language Processing problems: Word Embeddings, Text Classification, Sentiment Analysis and Translation. They will then learn how to leverage GluonNLP Model Zoo to use pre-existing models to solve these problems.

Chapter 7, Optimizing Models with Transfer Learning and Fine-Tuning, In this chapter, the reader will understand how to optimize pre-trained models for specific tasks using Transfer Learning and Fine-Tuning techniques. Furthermore, the readers will compare the performance of these techniques against training a model from scratch and the trade-offs involved. The reader will apply these techniques to problems such as image classifcation, image segmentation and translating text from English to German.

Chapter 8, Improving Training Performance with MXNet, In this chapter, the reader will learn how to leverage MXNet and Gluon libraries to optimize deep learning training loops. The reader will learn how MXNet and Gluon can take advantage of computational paradigms such as Lazy Evaluation and Automatic Parallelization. Furthermore, the reader will also learn to optimize Gluon DataLoaders for CPU and GPU, to apply **Automatic Mixed Precision** (**AMP**) and to train with multiple GPUs.

Chapter 9, Improving Inference Performance with MXNet, In this chapter, the reader will learn how to leverage MXNet and Gluon libraries to optimize deep learning inference. The reader will learn how MXNet and Gluon can take advantage of hybridizing Machine Learning models (combining imperative and symbolic programming). Furthermore, the reader will also learn to optimize inference time by applying Float16 data type combined with AMP, quantizing their models and profiling to find out further gains.

To get the most out of this book

The reader is expected to have a good understanding of Python programming and a working environment with Python 3.7+. A good theoretical understanding of mathematics for deep learning will be beneficial. MXNet 1.9.1 and the supplementary GluonCV and GluonNLP libraries will need to be installed as well (versions 0.10). These MXNet/Gluon requirements are described in detail in Chapter 1 and can be followed along by the reader. All code examples have been tested with Ubuntu 20.04, Python 3.10.12, MXNet 1.9.1, GluonCV 0.10 and GluonNLP 0.10. However, they should work with future releases too.

Software/hardware covered in the book	Operating system requirements
Python3.7+	Linux (Ubuntu recommended)
MXNet 1.9.1	
GluonCV 0.10	
GluonNLP 0.10	

In order to reproduce similar results to those described in Chapter 8, the reader will need access to a machine with multiple GPUs installed.

If you are using the digital version of this book, we advise you to type the code yourself or access the code from the book's GitHub repository (a link is available in the next section). Doing so will help you avoid any potential errors related to the copying and pasting of code.

Download the example code files

You can download the example code files for this book from GitHub at `https://github.com/PacktPublishing/Deep-Learning-with-MXNet-Cookbook`. If there's an update to the code, it will be updated in the GitHub repository.

We also have other code bundles from our rich catalog of books and videos available at `https://github.com/PacktPublishing/`. Check them out!

Conventions used

There are a number of text conventions used throughout this book.

`Code in text`: Indicates code words in text, database table names, folder names, filenames, file extensions, pathnames, dummy URLs, user input, and Twitter handles. Here is an example: "We will store the computation time in five dictionaries, one for each compute profile (timings_np, timings_mx_cpu, and timings_mx_gpu)."

A block of code is set as follows:

```
import mxnet
mxnet.__version__
features = mxnet.runtime.Features()
print(features)
print(features.is_enabled('CUDA'))
print(features.is_enabled('CUDNN'))
print(features.is_enabled('MKLDNN'))
```

Any command-line input or output is written as follows:

```
!python3 -m pip install gluoncv gluonnlp
!python3 -m pip install gluoncv gluonnlp
```

Bold: Indicates a new term, an important word, or words that you see onscreen. For instance, words in menus or dialog boxes appear in **bold**. Here is an example: "For this step, we will use the **pyplot** module from a library called **Matplotlib**, which will allow us to create charts easily."

> **Tips or important notes**
> Appear like this.

Get in touch

Feedback from our readers is always welcome.

General feedback: If you have questions about any aspect of this book, email us at `customercare@packtpub.com` and mention the book title in the subject of your message.

Errata: Although we have taken every care to ensure the accuracy of our content, mistakes do happen. If you have found a mistake in this book, we would be grateful if you would report this to us. Please visit `www.packtpub.com/support/errata` and fill in the form.

Piracy: If you come across any illegal copies of our works in any form on the internet, we would be grateful if you would provide us with the location address or website name. Please contact us at `copyright@packt.com` with a link to the material.

If you are interested in becoming an author: If there is a topic that you have expertise in and you are interested in either writing or contributing to a book, please visit `authors.packtpub.com`

Share Your Thoughts

Once you've read *Deep Learning with MXNet Cookbook*, we'd love to hear your thoughts! Scan the QR code below to go straight to the Amazon review page for this book and share your feedback.

`https://packt.link/r/1-800-56960-2`

Your review is important to us and the tech community and will help us make sure we're delivering excellent quality content.

Download a free PDF copy of this book

Thanks for purchasing this book!

Do you like to read on the go but are unable to carry your print books everywhere?

Is your eBook purchase not compatible with the device of your choice?

Don't worry, now with every Packt book you get a DRM-free PDF version of that book at no cost.

Read anywhere, any place, on any device. Search, copy, and paste code from your favorite technical books directly into your application.

The perks don't stop there, you can get exclusive access to discounts, newsletters, and great free content in your inbox daily

Follow these simple steps to get the benefits:

1. Scan the QR code or visit the link below

`https://packt.link/free-ebook/9781800569607`

2. Submit your proof of purchase
3. That's it! We'll send your free PDF and other benefits to your email directly

1
Up and Running with MXNet

MXNet is one of the most used deep learning frameworks and is an Apache open source project. Before 2016, **Amazon Web Services** (**AWS**)'s research efforts did not use a preferred deep learning framework, allowing each team to research and develop according to their choices. Although some deep learning frameworks have thriving communities, sometimes AWS was not able to fix code bugs at the required speed (among other issues). To solve these issues, at the end of 2016, AWS announced MXNet as its deep learning framework of choice, investing in internal teams to develop it further. Research institutions that support MXNet are Intel, Baidu, Microsoft, Carnegie Mellon University, and MIT, among others. It was co-developed by Carlos Guestrin at Carnegie Mellon University and the University of Washington (along with GraphLab).

Some of its advantages are as follows:

- Imperative/symbolic programming and hybridization (which will be covered in *Chapters 1* and *9*)
- Support for multiple GPUs and distributed training (which will be covered in *Chapters 7* and *8*)
- Highly optimized for inference production systems (which will be covered in *Chapters 7* and *9*)
- A large number of pre-trained models on its Model Zoos in the fields of computer vision and natural language processing, among others (covered in *Chapters 6*, *7*, and *8*)

To start working with MXNet, we need to install the library. There are several different versions of MXNet available to be installed, and in this chapter, we will cover how to choose the right version. The most important parameter will be the available hardware we have. In order to optimize performance, it is always best to maximize the use of our available hardware. We will compare the usage of a well-known linear algebra library, NumPy, with similar operations in MXNet. We will then compare the performance of the different MXNet versions versus NumPy.

MXNet includes its own API for deep learning, Gluon, and moreover, Gluon provides different libraries for computer vision and natural language processing that include pre-trained models and utilities. These libraries are known as GluonCV and GluonNLP.

In this chapter, we will cover the following topics:

- Installing MXNet, Gluon, GluonCV, and GluonNLP
- NumPy and MXNet ND arrays – comparing their performance

Technical requirements

Apart from the technical requirements specified in the *Preface*, no other requirements apply to this chapter.

The code for this chapter can be found at the following GitHub URL: `https://github.com/PacktPublishing/Deep-Learning-with-MXNet-Cookbook/tree/main/ch01`.

Furthermore, you can access directly each recipe from Google Colab – for example, use the following for the first recipe of this chapter: `https://colab.research.google.com/github/PacktPublishing/Deep-Learning-with-MXNet-Cookbook/blob/main/ch01/1_1_Installing_MXNet.ipynb`.

Installing MXNet, Gluon, GluonCV, and GluonNLP

In order to get the maximum performance out of the available software (programming languages) and hardware (CPU and GPU), there are different MXNet library versions available to install. We shall learn how to install them in this recipe.

Getting ready

Before getting started with the MXNet installation, let us review the different versions available of the software packages that we will use, including MXNet. The reason we do that is that our hardware configuration must map to the chosen versions of our software packages in order to maximize performance:

- **Python**: MXNet is available for different programming languages – Python, Java, R, and C++, among others. We will use MXNet for Python, and Python 3.7+ is recommended.
- **Jupyter**: Jupyter is an open source web application that provides an easy-to-use interface to show Markdown text, working code, and data visualizations. It is very useful for understanding deep learning, as we can describe concepts, write the code to run through those concepts, and visualize the results (typically comparing them with the input data). Jupyter Core 4.5+ is recommended.

- **CPUs and GPUs**: MXNet can work with any hardware configuration – that is, any single CPU can run MXNet. However, there are several hardware components that MXNet can leverage to improve performance:
  - **Intel CPUs**: Intel developed a library known as **Math Kernel Library** (**MKL**) for optimized math operations. MXNet has support for this library, and using the optimized version can improve certain operations. Any modern version of Intel MKL is sufficient.
  - **NVIDIA GPUs**: NVIDIA developed a library known as **Compute Unified Device Architecture (CUDA)** for optimized parallel operations (such as matrix operations, which are very common in deep learning). MXNet has support for this library, and using the optimized version can dramatically improve large deep learning workloads, such as model training. CUDA 11.0+ is recommended.
- **MXNet version**: At the time of writing, MXNet 1.9.1 is the most up-to-date stable version that has been released. All the code throughout the book has been verified with this version. MXNet, and deep learning in general, can be considered a live ongoing project, and therefore, new versions will be released periodically. These new versions will have improved functionality and new features, but they might also contain breaking changes from previous APIs. If you are revisiting this book in a few months and a new version has been released with breaking changes, how to install MXNet version 1.8.0 specifically is also described here.

> **Tip**
>
> I have used Google Colab as the platform to run the code described in this book. At the time of writing, it provides Python 3.10.12, up-to-date Jupyter libraries, Intel CPUs (Xeon @ 2.3 GHz), and NVIDIA GPUs (which can vary: K80s, T4s, P4s, and P100s) with CUDA 11.8 pre-installed. Therefore, minimal steps are required to install MXNet and get it running.

How to do it...

Throughout the book, we will not only use code extensively but also clarify comments and headings in that code to provide structure, as well as several types of visual information such as images or generated graphs. For these reasons, we will use Jupyter as the supporting development environment. Moreover, in order to facilitate setup, installation, and experimentation, we will use Google Colab.

Google Colab is a hosted Jupyter Notebook service that requires no setup to use, while providing free access to computing resources, including GPUs. In order to set up Google Colab properly, this section is divided into two main points:

- Setting up the notebook
- Verifying and installing libraries

> **Important note**
>
> If you prefer, you can use any local environment that supports Python 3.7+, such as Anaconda, or any other Python distribution. This is highly encouraged if your hardware specifications are better than Google Colab's offering, as better hardware will reduce computation time.

Setting up the notebook

In this section, we will learn how to work with Google Colab and set up a new notebook, which we will use to verify our MXNet installation:

1. Open your favorite web browser. In my case, I have used Google Chrome as the web browser throughout the book. Visit `https://colab.research.google.com/` and click on **NEW NOTEBOOK**.

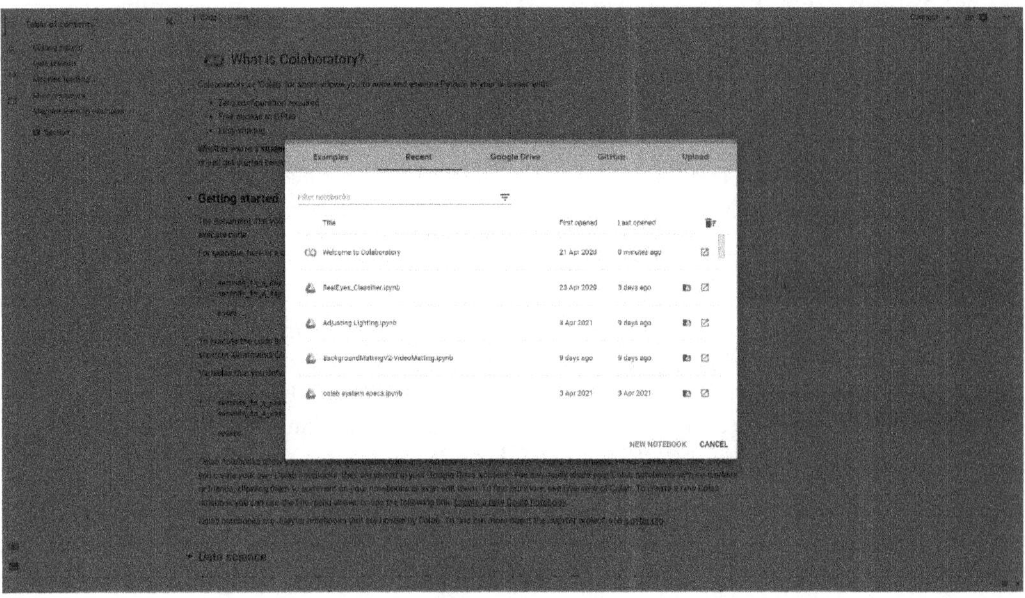

Figure 1.1 – The Google Colab start screen

2. Change the title of the notebook – for example, as you can see in the following screenshot, I have changed the title to `DL with MXNet Cookbook 1.1 Installing MXNet`.

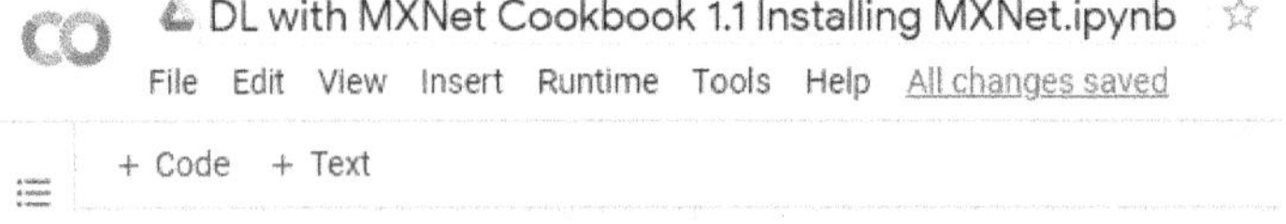

Figure 1.2 – A Google Colab notebook

3. Change your Google Colab runtime type to use a GPU:

   I. Select **Change runtime type** from the **Runtime** menu.

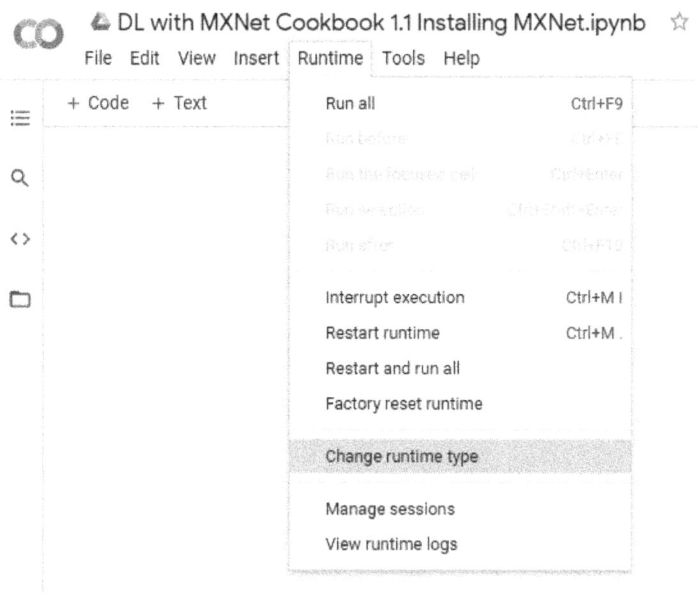

Figure 1.3 – Change runtime type

   II. In **Notebook settings**, select **GPU** as the **Hardware accelerator** option.

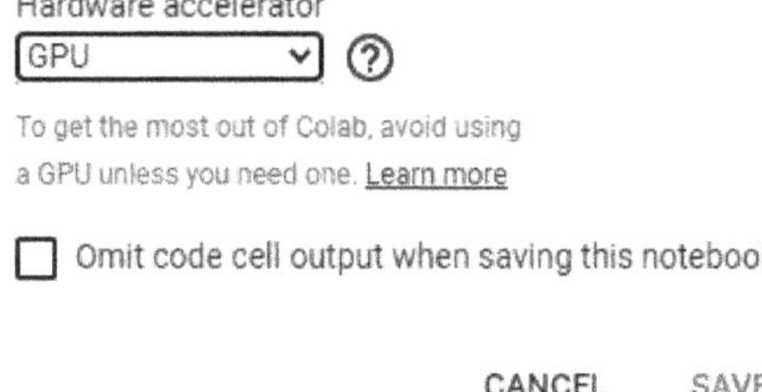

Figure 1.4 – Hardware accelerator | GPU

Verifying and installing libraries

In this section, go to the first cell (make sure it is a code cell) and type the following commands:

1. Verify the Python version by typing the following:

   ```
   import platform
   platform.python_version()
   ```

   This will yield an output as follows:

   ```
   3.7.10
   ```

   Check the version, and make sure that it is 3.7+.

> **Important note**
>
> In Google Colab, you can directly run commands as if you were in the Linux Terminal by adding the `!` character to the command. Feel free to try other commands such as `!ls`.

2. We now need to verify the Jupyter version (Jupyter Core 4.5.0 or above will suffice):

   ```
   !jupyter --version
   ```

   This is one potential output from the previous command:

   ```
   jupyter core : 4.5.0
   jupyter-notebook : 5.2.2
   qtconsole : 4.5.2
   ipython : 5.5.0
   ipykernel : 4.10.1
   jupyter client : 5.3.1
   jupyter lab : not installed
   nbconvert : 5.5.0
   ipywidgets : 7.5.0
   nbformat : 4.4.0
   traitlets : 4.3.2
   ```

> **Tip**
>
> Jupyter, an open source notebook application, is assumed to be installed, as is the case for Google Colab. For further instructions on how to install it, visit `https://jupyter.org/install`.

3. Verify whether an Intel CPU is present in the hardware:

   ```
   !lscpu | grep 'Model name'
   ```

This will yield a similar output to the following:

```
Model name: Intel(R) Xeon(R) CPU @ 2.20GHz
```

The more up to date the processor the better, but for the purposes of this book, the dependency is larger with the GPU than with the CPU.

4. Verify the NVIDIA GPU is present in the hardware (there are devices listed below) and that NVIDIA CUDA is installed:

```
!nvidia-smi
```

This will yield a similar output to the following:

```
+-----------------------------------------------------------------
---+
| NVIDIA-SMI 460.67 Driver Version: 460.32.03 CUDA Version:
11.2   |
|-----------------------------+--------------+-------------------
---+
|GPU Name        Persistence-M|Bus-Id  Disp.A| Volatile Uncorr.
ECC |
|Fan Temp Perf Pwr:Usage/Cap|  Memory-Usage|  GPU-Util Compute
M. |
|                             |              |                MIG
M. | |============================+==============+===============
========|
|   0 Tesla T4          Off |0:00:04.0 Off
|                          0 |
| N/A  37C   P8     9W / 70W |0MiB/15109MiB
|        0%      Default |
                 |              |                   N/A |
+-----------------------------+--------------+-------------------
---+
+-----------------------------------------------------------------
---+
| Proces
ses:                                                            |
|  GPU    GI   CI       PID  Type   Process  name            GPU
Memory |
        ID   ID                                           Usage      |
|=================================================================
===|
| No running processes
found                                             |
+-----------------------------------------------------------------
---+
```

> **Important note**
>
> CUDA 11.0 has known issues with the NVIDIA K80. If you have an NVIDIA K80 and are having issues with the examples described, uninstall CUDA 11.0 and install CUDA 10.2. Afterward, install MXNet for CUDA 10.2 following the steps described here.

5. Verify that the CUDA version is 11.0 or above:

```
!nvcc --version
```

This will yield a similar output to the following:

```
nvcc: NVIDIA (R) Cuda compiler driver
Copyright (c) 2005-2020 NVIDIA Corporation
Built on Wed_Jul_22_19:09:09_PDT_2020
Cuda compilation tools, release 11.0, V11.0.221
Build cuda_11.0_bu.TC445_37.28845127_0
```

6. Install MXNet, depending on your hardware configuration. The following are the different MXNet versions that you can install:

   - **Recommended/Google Colab**: The latest MXNet version (1.9.1) with GPU support:

   ```
   !python3 -m pip install mxnet-cu117
   ```

   - **No Intel CPU nor NVIDIA GPU**: Install MXNet with the following command:

   ```
   !python3 -m pip install mxnet
   ```

   - **Intel CPU without NVIDIA GPU**: Install MXNet with Intel MKL, with the following command:

   ```
   !python3 -m pip install mxnet-mkl
   ```

   - **No Intel CPU with NVIDIA GPU**: Install MXNet with NVIDIA CUDA 10.2, with the following command:

   ```
   !python3 -m pip install mxnet-cu102
   ```

   - **Intel CPU and NVIDIA GPU**: Install MXNet with Intel MKL and NVIDIA CUDA 11.0, with the following command:

   ```
   !python3 -m pip install mxnet-cu110
   ```

> **Tip**
>
> `pip3`, a Python 3 package manager, is assumed to be installed, as is the case for Google Colab. If a different installation method for MXNet is preferred, visit `https://mxnet.apache.org/versions/master/get_started` for instructions.

After version 1.6.0, MXNet is released by default with the Intel MKL library extension; therefore, there is no need to add the `mkl` suffix anymore when installing the most recent versions, as seen previously in the recommended installation.

7. Verify that the MXNet installation has been successful with the following two steps:

   I. The following commands must not return any error and must successfully display MXNet version 1.9.1:

   ```
   import mxnet
   mxnet.__version__
   ```

   II. The list of features that appear in the following contain the `CUDA`, `CUDNN`, and `MKLDNN` features:

   ```
   features = mxnet.runtime.Features()
   print(features)
    print(features.is_enabled('CUDA'))
    print(features.is_enabled('CUDNN'))
    print(features.is_enabled('MKLDNN'))
   ```

   The output will list all the features and `True` for each one.

8. Install GluonCV and GluonNLP:

   ```
   !python3 -m pip install gluoncv gluonnlp
   ```

This command will install the latest versions of GluonCV and GluonNLP, which at the time of writing were, respectively, 0.10 and 0.10.

How it works...

The training, inference, and evaluation of deep learning networks are highly complex operations, involving hardware and several layers of software, including drivers, low-level performance libraries such as MKL and CUDA, and high-level programming languages and libraries such as Python and MXNet.

Important note

MXNet is an actively developed project, part of the Apache Incubator program. Therefore, new versions are expected to be released, and they might contain breaking changes. The preceding command will install the latest stable version available. Throughout this book, the version of MXNet used is 1.9.1. If your code fails and it uses a different MXNet version, try installing MXNet version 1.9.1 by running the following:

```
!python3 -m pip install mxnet-cu117==1.9.1
```

By checking all the hardware and software components, we can install the most optimized version of MXNet. We can use Google Colab, which easily transfers to other local configurations such as the Anaconda distribution.

Moreover, we can identify the right combination of CUDA drivers and MXNet versions that will maximize performance and verify a successful installation.

There's more...

It is highly recommended to always use the latest versions of all the software components discussed. Deep learning is an evolving field and there are always improvements such as new functionalities added, changes in the APIs, and updates in the internal functions to increase performance, among other changes.

However, it is very important that all components (CPU, GPU, CUDA, and the MXNet version) are compatible. To match these components, it is highly recommended to visit `https://mxnet.apache.org/versions/master/get_started` and check for the latest CUDA and MXNet versions you can install to maximize your hardware performance.

As an example, for a Python 3-based Linux distribution, installed using `pip3`, these are the MXNet versions available (note with/without CPU acceleration and/or with GPU acceleration).

If you are interested in knowing more about Intel's **MKL**, the following link is a very good starting point: `https://software.intel.com/content/www/us/en/develop/articles/getting-started-with-intel-optimization-for-mxnet.html`.

NumPy and MXNet ND arrays

If you have worked with data previously in Python, chances are you have found yourself working with NumPy and its **N-dimensional arrays** (**ND arrays**). These are also known as tensors, and the 0D variants are called **scalars**, the 1D variants are called **vectors**, and the 2D variants are called **matrixes**.

MXNet provides its own ND array type, and there are two different ways to work with them. On one hand, there is the `nd` module, MXNet's native and optimized way to work with MXNet ND arrays. On the other hand, there is the `np` module, which has the same interfaces and syntax as the NumPy ND array type and has also been optimized, but it's limited due to the interface constraints. With MXNet ND arrays, we can leverage its underlying engine, with compute optimizations such as Intel MKL and/or NVIDIA CUDA, if our hardware configuration is compatible. This means we will be able to use almost the same syntax as when working with NumPy, but accelerated with the MXNet engine and our GPUs, not supported by NumPy.

Moreover, as we will see in the next chapters, a very common operation that we will execute on MXNet is automatic differentiation on these ND arrays. By using MXNet ND array libraries, this operation will also leverage our hardware for optimum performance. NumPy does not provide automatic differentiation out of the box.

Getting ready

If you have already installed MXNet, as described in the previous recipe, in terms of executing accelerated code, the only remaining steps before using MXNet ND arrays is importing their libraries:

```
import numpy as np
import mxnet as mx
```

However, it is worth noting here an important underlying difference between NumPy ND array operations and MXNet ND array operations. NumPy follows an eager evaluation strategy – that is, all operations are evaluated at the moment of execution. Conversely, MXNet uses a lazy evaluation strategy, more optimal for large compute loads, where the actual calculation is deferred until the values are actually needed.

Therefore, when comparing performances, we will need to force MXNet to finalize all calculations before computing the time needed for them. As we will see in the examples, this is achieved by calling the `wait_to_read()` function, Furthermore, when accessing the data with functions such as `print()` or `.asnumpy()`, execution is then completed before calling these functions, yielding the wrong impression that these functions are actually time-consuming:

1. Let's check a specific example and start by running it on the CPU:

   ```
   import time
   x_mx_cpu = mx.np.random.rand(1000, 1000, ctx = mx.cpu())
   start_time = time.time()
   mx.np.dot(x_mx_cpu, x_mx_cpu).wait_to_read()
   print("Time of the operation: ", time.time() - start_time)
   ```

   This will yield a similar output to the following:

   ```
   Time of the operation: 0.04673886299133301
   ```

2. However, let's see what happens if we measure the time without the call to `wait_to_read()`:

   ```
   x_mx_cpu = mx.np.random.rand(1000, 1000, ctx = mx.cpu())
   start_time = time.time()
   x_2 = mx.np.dot(x_mx_cpu, x_mx_cpu)
    print("(FAKE, MXNet has lazy evaluation)")
    print("Time of the operation : ", time.time() - start_time)
    start_time = time.time()
   print(x_2)
    print("(FAKE, MXNet has lazy evaluation)")
    print("Time to display: ", time.time() - start_time)
   ```

The following will be the output:

```
(FAKE, MXNet has lazy evaluation)
 Time of the operation : 0.00118255615234375
 [[256.59583 249.70404 249.48639 ... 251.97151 255.06744
255.60669]
 [255.22629 251.69475 245.7591 ... 252.78784 253.18878
247.78052]
 [257.54187 254.29262 251.76346 ... 261.0468 268.49127 258.2312
]
 ...
 [256.9957 253.9823 249.59073 ... 256.7088 261.14255 253.37457]
 [255.94278 248.73282 248.16641 ... 254.39209 252.4108
249.02774]
 [253.3464 254.55524 250.00716 ... 253.15712 258.53894
255.18658]]
 (FAKE, MXNet has lazy evaluation)
 Time to display: 0.042133331298828125
```

As we can see, the first experiment indicated that the computation took ~50 ms to complete; however, the second experiment indicated that the computation took ~1 ms (50 times less!), and the visualization was more than 40 ms. This is an incorrect result. This is because we measured our performance incorrectly in the second experiment. Refer to the first experiment and the call to `wait_to_read()` for a proper performance measurement.

How to do it...

In this section, we will compare performance in terms of computation time for two compute-intensive operations:

- Matrix creation
- Matrix multiplication

We will compare five different compute profiles for each operation:

- Using the NumPy library (no CPU or GPU acceleration)
- Using the MXNet `np` module with CPU acceleration but no GPU
- Using the MXNet `np` module with CPU acceleration and GPU acceleration
- Using the MXNet `nd` module with CPU acceleration but no GPU
- Using the MXNet `nd` module with CPU acceleration and GPU acceleration

To finalize, we will plot the results and draw some conclusions.

Timing data structures

We will store the computation time in five dictionaries, one for each compute profile (`timings_np`, `timings_mx_cpu`, and `timings_mx_gpu`). The initialization of the data structures is as follows:

```
timings_np = {}
timings_mx_np_cpu = {}
timings_mx_np_gpu = {}
timings_mx_nd_cpu = {}
timings_mx_nd_gpu = {}
```

We will run each operation (matrix generation and matrix multiplication) with matrixes in a different order, namely the following:

```
matrix_orders = [1, 5, 10, 50, 100, 500, 1000, 5000, 10000]
```

Matrix creation

We define three functions to generate matrixes; the first function will use the NumPy library to generate a matrix, and it will receive as an input parameter the matrix order. The second function will use the MXNet np module, and the third function will use the MXNet and module. For the second and third functions, as input parameters we will provide the context where the matrix needs to be created, apart from the matrix order. This context specifies whether the result (the created matrix in this case) must be computed in the CPU or the GPU (and which GPU if there are multiple devices available):

```
def create_matrix_np(n):
    """
    Given n, creates a squared n x n matrix,
    with each matrix value taken from a random
    uniform distribution between [0, 1].
    Returns the created matrix a.
    Uses NumPy.
    """
    a = np.random.rand(n, n)
    return a
def create_matrix_mx(n, ctx=mx.cpu()):
    """
    Given n, creates a squared n x n matrix,
    with each matrix value taken from a random
    uniform distribution between [0, 1].
    Returns the created matrix a.
    Uses MXNet NumPy syntax and context ctx
    """
    a = mx.np.random.rand(n, n, ctx=ctx)
    a.wait_to_read()
```

```
    return a
def create_matrix_mx_nd(n, ctx=mx.cpu()):
    """
    Given n, creates a squared n x n matrix,
    with each matrix value taken from a random
    uniform distribution between [0, 1].
    Returns the created matrix a.
    Uses MXNet ND native syntax and context ctx
    """
    a = mx.nd.random.uniform(shape=(n, n), ctx=ctx)
    a.wait_to_read()
    return a
```

To store necessary data for our performance comparison later, we use the structures created previously, with the following code:

```
timings_np["create"] = []
for n in matrix_orders:
    result = %timeit -o create_matrix_np(n)
    timings_np["create"].append(result.best)
timings_mx_np_cpu["create"] = []
for n in matrix_orders:
    result = %timeit -o create_matrix_mx_np(n)
    timings_mx_np_cpu["create"].append(result.best)
timings_mx_np_gpu["create"] = []
ctx = mx.gpu()
for n in matrix_orders:
    result = %timeit -o create_matrix_mx_np(n, ctx)
    timings_mx_np_gpu["create"].append(result.best)
timings_mx_nd_cpu["create"] = []
for n in matrix_orders:
    result = %timeit -o create_matrix_mx_nd(n)
    timings_mx_nd_cpu["create"].append(result.best)
timings_mx_nd_gpu["create"] = []
ctx = mx.gpu()
for n in matrix_orders:
    result = %timeit -o create_matrix_mx_nd(n, ctx)
    timings_mx_nd_gpu["create"].append(result.best)
```

Matrix multiplication

We define three functions to compute the matrixes multiplication; the first function will use the NumPy library and will receive as input parameters the matrixes to multiply. The second function will use the MXNet np module, and the third function will use the MXNet nd module. For the second and third functions, the same parameters are used. The context where the multiplication will happen is given by the context where the matrixes were created; no parameter needs to be added. Both matrixes need to have been created in the same context, or an error will be triggered:

```
def multiply_matrix_np(a, b):
    """
    Multiplies 2 squared matrixes a and b
    and returns the result c.
    Uses NumPy.
    """
    #c = np.matmul(a, b)
    c = np.dot(a, b)
    return c
def multiply_matrix_mx_np(a, b):
    """
    Multiplies 2 squared matrixes a and b
    and returns the result c.
    Uses MXNet NumPy syntax.
    """
    c = mx.np.dot(a, b)
    c.wait_to_read()
    return c

def multiply_matrix_mx_nd(a, b):
    """
    Multiplies 2 squared matrixes a and b
    and returns the result c.
    Uses MXNet ND native syntax.
    """
    c = mx.nd.dot(a, b)
    c.wait_to_read()
    return c
```

To store the necessary data for our performance comparison later, we will use the structures created previously, with the following code:

```
timings_np["multiply"] = []
for n in matrix_orders:
    a = create_matrix_np(n)
```

```
    b = create_matrix_np(n)
    result = %timeit -o multiply_matrix_np(a, b)
    timings_np["multiply"].append(result.best)
timings_mx_np_cpu["multiply"] = []
for n in matrix_orders:
    a = create_matrix_mx_np(n)
    b = create_matrix_mx_np(n)
    result = %timeit -o multiply_matrix_mx_np(a, b)
    timings_mx_np_cpu["multiply"].append(result.best)
timings_mx_np_gpu["multiply"] = []
ctx = mx.gpu()
for n in matrix_orders:
    a = create_matrix_mx_np(n, ctx)
    b = create_matrix_mx_np(n, ctx)
    result = %timeit -o multiply_matrix_mx_np(a, b)
    timings_mx_gpu["multiply"].append(result.best)
timings_mx_nd_cpu["multiply"] = []
for n in matrix_orders:
    a = create_matrix_mx_nd(n)
    b = create_matrix_mx_nd(n)
    result = %timeit -o multiply_matrix_mx_nd(a, b)
    timings_mx_nd_cpu["multiply"].append(result.best)
timings_mx_nd_gpu["multiply"] = []
ctx = mx.gpu()
for n in matrix_orders:
    a = create_matrix_mx_nd(n, ctx)
    b = create_matrix_mx_nd(n, ctx)
    result = %timeit -o multiply_matrix_mx_nd(a, b)
    timings_mx_nd_gpu["multiply"].append(result.best)
```

Drawing conclusions

The first step before making any assessments is to plot the data we have captured in the previous steps. For this step, we will use the `pyplot` module from a library called Matplotlib, which will allow us to create charts easily. The following code plots the runtime (in seconds) for the matrix generation and all the matrix orders computed:

```
import matplotlib.pyplot as plt

fig = plt.figure()
plt.plot(matrix_orders, timings_np["create"], color='red', marker='s')
plt.plot(matrix_orders, timings_mx_np_cpu["create"], color='blue',
marker='o')
plt.plot(matrix_orders, timings_mx_np_gpu["create"], color='green',
```

```
marker='^')
plt.plot(matrix_orders, timings_mx_nd_cpu["create"], color='yellow',
marker='p')
plt.plot(matrix_orders, timings_mx_nd_gpu["create"], color='orange',
marker='*')
plt.title("Matrix Creation Runtime", fontsize=14)
plt.xlabel("Matrix Order", fontsize=14)
plt.ylabel("Runtime (s)", fontsize=14)
plt.grid(True)
ax = fig.gca()
ax.set_xscale("log")
ax.set_yscale("log")
plt.legend(["NumPy", "MXNet NumPy (CPU)", "MXNet NumPy (GPU)", "MXNet
ND (CPU)", "MXNet ND (GPU)"])
plt.show()
```

Quite similarly as shown in the previous code block, the following code plots the runtime (in seconds) for the matrix multiplication and all the matrix orders computed:

```
import matplotlib.pyplot as plt
fig = plt.figure()
plt.plot(matrix_orders, timings_np["multiply"], color='red',
marker='s')
 plt.plot(matrix_orders, timings_mx_np_cpu["multiply"], color='blue',
marker='o')
 plt.plot(matrix_orders, timings_mx_np_gpu["multiply"], color='green',
marker='^')
 plt.plot(matrix_orders, timings_mx_nd_cpu["multiply"],
color='yellow', marker='p')
 plt.plot(matrix_orders, timings_mx_nd_gpu["multiply"],
color='orange', marker='*')
 plt.title("Matrix Multiplication Runtime", fontsize=14)
 plt.xlabel("Matrix Order", fontsize=14)
 plt.ylabel("Runtime (s)", fontsize=14)
 plt.grid(True)
 ax = fig.gca()
ax.set_xscale("log")
ax.set_yscale("log")
plt.legend(["NumPy", "MXNet NumPy (CPU)", "MXNet NumPy (GPU)", "MXNet
ND (CPU)", "MXNet ND (GPU)"])
 plt.show()
```

These are the plots displayed (the results will vary according to the hardware configuration):

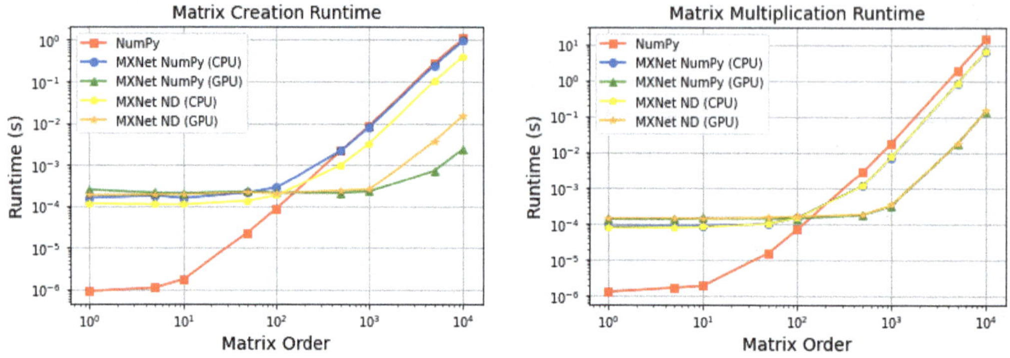

Figure 1.5 – Runtimes – a) Matrix creation, and b) Matrix multiplication

> **Important note**
>
> Note that the charts use a logarithmic scale for both axes, horizontal and vertical (the differences are larger than they seem). Furthermore, the actual values depend on the hardware architecture that the computations are run on; therefore, your specific results will vary.

There are several conclusions that can be drawn, both from each individual operation and collectively:

- For smaller matrix orders, using NumPy is much faster in both operations. This is because MXNet works in a different memory space, and the amount of time to move the data to this memory space is longer than the actual compute time.
- In matrix creation, for larger matrix orders, the difference between NumPy (remember, it's CPU only) and MXNet with the np module and CPU acceleration is negligible, but with the nd module and CPU, acceleration is ~2x faster. For matrix multiplication, and depending on your hardware, MXNet with CPU acceleration can be ~2x faster (regardless of the module). This is because MXNet uses Intel MKL to optimize CPU computations.
- In the ranges that are interesting for deep learning – that is, large computational loads involving matrix orders > 1,000 (which can represent data such as images composed of several megapixels or large language dictionaries), GPUs deliver typical gains of several orders of magnitude (~200x for creation, and ~40x for multiplication, exponentially growing with every increase of matrix order). This is by far the most compelling reason to work with GPUs when running deep learning experiments.

- When using the GPU, the MXNet np module is faster than the MXNet nd module in creation (~7x), but the difference is negligible in multiplication. Typically, deep learning algorithms are more similar to multiplications to terms of computational loads, and therefore, a priori, there is no significant advantage in using the np module or the nd module. However, *MXNet recommends using the native MXNet nd module* (and the author subscribes to this recommendation) because some operations on the np module are not supported by `autograd` (MXNet's auto-differentiation module). We will see in the upcoming chapters, when we train neural networks, how the `autograd` module is used and why it is critical.

How it works...

MXNet provides two optimized modules to work with ND arrays, including one that is an in-place substitute for NumPy. The advantages of operating with MXNet ND arrays are twofold:

- MXNet ND array operations support automatic differentiation. As we will see in the following chapters, automatic differentiation is a key feature that allows developers to concentrate on the forward pass of the models, letting the backward pass be automatically derived.
- Conversely, operations with MXNet ND arrays are optimized for the underlying hardware, yielding impressive results with GPU acceleration. We computed results for matrix creation and matrix multiplication to validate this conclusion experimentally.

There's more...

In this recipe, we have barely scratched the surface of MXNet operations with ND arrays. If you want to read more about MXNet and ND arrays, this is the link to the official MXNet API reference: `https://mxnet.apache.org/versions/1.0.0/api/python/ndarray/ndarray.html`.

Furthermore, a very interesting tutorial can be found in the official MXNet documentation: `https://gluon.mxnet.io/chapter01_crashcourse/ndarray.html`.

Moreover, we have taken a glimpse at how to measure performance on MXNet. We will revisit this topic in the following chapters; however, a good deep-dive into the topic is given in the official MXNet documentation: `https://mxnet.apache.org/versions/1.8.0/api/python/docs/tutorials/performance/backend/profiler.html`.

2
Working with MXNet and Visualizing Datasets – Gluon and DataLoader

In the previous chapter, we learned how to set up MXNet. We also verified how MXNet could leverage our hardware to provide maximum performance. Before applying **deep learning** (DL) to solve specific problems, we need to understand how to load, manage, and visualize the datasets we will be working with. In this chapter, we will start using MXNet to analyze some toy datasets in the domains of numerical regression, data classification, image classification, and text classification. To manage those tasks efficiently, we will see new MXNet libraries and functions such as Gluon (an API for DL) and DataLoader.

In this chapter, we will cover the following topics:

- Understanding regression datasets – loading, managing, and visualizing the *House Sales* dataset
- Understanding classification datasets – loading, managing, and visualizing the Iris dataset
- Understanding image datasets – loading, managing, and visualizing the Fashion-MNIST dataset
- Understanding text datasets – loading, managing, and visualizing the Enron Email dataset

Technical requirements

Apart from the technical requirements specified in the *Preface*, no other requirements apply to this chapter.

The code for this chapter can be found at the following GitHub URL: `https://github.com/PacktPublishing/Deep-Learning-with-MXNet-Cookbook/tree/main/ch02`

Furthermore, you can directly access each recipe from Google Colab; for example, for the first recipe of this chapter, visit `https://colab.research.google.com/github/PacktPublishing/Deep-Learning-with-MXNet-Cookbook/blob/main/ch02/2_1_Toy_Dataset_for_Regression_Load_Manage_and_Visualize_House_Sales_Dataset.ipynb`.

Understanding regression datasets – loading, managing, and visualizing the House Sales dataset

The training process of **machine learning** (**ML**) models can be divided into three main sub-groups:

- **Supervised learning** (**SL**): The expected outputs are known for at least some data
- **Unsupervised learning** (**UL**): The expected outputs are not known but the data has some features that could help with understanding its internal distribution
- **Reinforcement learning** (**RL**): An agent explores the environment and makes decisions based on the inputs acquired from the environment

There is also an approach that falls in between the first two sub-groups called **weakly SL**, where there are not enough known outputs to follow an SL approach for one of the following reasons:

- The outputs are inaccurate
- Only some of the output features are known (incomplete)
- They are not exactly the expected outputs but are connected/related to the task we intend to achieve (inexact)

With SL, one of the most common problem types is **regression**. In regression problems, we want to estimate numerical outputs given a variable number of input features. In this recipe, we will analyze a toy regression dataset from Kaggle: *House Sales in King County, USA*.

The House Sales dataset presents the problem of estimating the price of a house (in $) given the following 19 features:

- `Date` of the home sale
- `Number of bedrooms`
- `Number of bathrooms`, where `0.5` accounts for a room with a toilet but no shower
- `Sqft_living`: Square feet of the apartment's interior living space
- `Sqft_lot`: Square feet of the land space
- `Number of floors`
- `Waterfront` view or not

- An index from 0 to 4 of how good the view of the property is
- An index from 1 to 5 on the condition of the apartment
- `Grade`: An index from 1 to 13, with 1 being the worst and 13 the best
- `Sqft_above`: Square feet of the interior housing space that is above ground level
- `Sqft_basement`: Square feet of the interior housing space that is below ground level
- `Yr_built`: The year the house was initially built
- `Yr_renovated`: The year of the house's last renovation
- `Zipcode` area the house is in
- Latitude (`Lat`)
- Longitude (`Long`)
- `Sqft_living15`: Square feet of interior housing living space for the nearest 15 neighbors
- `Sqft_lot15`: Square feet of the land lots of the nearest 15 neighbors

These data features are provided for *21,613* houses along with the price (value to be estimated).

Getting ready

The following dataset is provided under the *CC0 Public Domain* license and can be downloaded from `https://www.kaggle.com/harlfoxem/housesalesprediction`.

To read the data, we are going to use a very well-known library to manage data, `pandas`, and we will use the most common data structure for the library, **DataFrames**. Moreover, in order to plot the data and several visualizations we will compute, we will use the `matplotlib`, `pyplot`, and `seaborn` libraries. Therefore, we must run the following code:

```
import matplotlib.pyplot as plt
import pandas as pd
import seaborn as sns
```

If you do not have these libraries installed, they can be easily installed with the following terminal commands:

```
!pip3 install matplotlib==3.7.1
!pip3 install pandas==1.5.3
 !pip3 install seaborn==0.12.2
```

Therefore, to load the data, we can simply retrieve the file containing the dataset (available in the GitHub repository for the book) and process it:

```
# Retrieve Dataset (House Sales Prices) from GitHub repository for
Deep Learning with MXNet Cookbook by Packt
!wget https://github.com/PacktPublishing/Deep-Learning-with-
MXNet-Cookbook/raw/main/ch02/kc_house_data.zip
# Uncompress kc_house_data.csv file
!unzip /content/kc_house_data.zip
house_df = pd.read_csv("kc_house_data.csv")
```

This is all we need to start working with our regression dataset.

How to do it...

In this section, we will run an **exploratory data analysis** (**EDA**) that will help us understand which features are important (and which are not) to predict the price of a house:

- Data structure
- Correlation study
- Living square feet analysis, square feet above ground level analysis, and neighbors' living square feet analysis
- Grade analysis
- Rooms (bedrooms and bathrooms) analysis
- Views analysis
- Year-built and year-renovated analysis
- Location analysis

Data structure

Let's analyze what our data looks like. For this, we will use common operations on `pandas` DataFrames:

```
house_df.info()
```

From the output, we can draw the following conclusions:

- The data is complete (all columns have 21,613 values, as expected).
- There are no `NULL` values (the data is clean!).
- Apart from the features described previously, there is a feature called `id`. This feature is not needed as the index already allows us to uniquely identify each property.

In order to grasp what the values look like, let's display the first five properties:

```
house_df.head()
```

So far, we have had a look at the features. Now, let's take a look at the price distribution:

```
house_df.hist(column = "price", bins = 24)
 plt.show()
```

These commands will display a histogram of prices that shows how many houses in the dataset have a certain price (column selected in the previous commands). Histograms work in ranges (also known as *buckets* or *bins*); in our case, we have chosen 24. As the maximum price is $8M, when applying 24 ranges, we have 3 ranges per million dollars, specifically (all values in millions of $): [0 – 0.33), [0.33 - 0.66), [0.66 - 1), ... until [7.66 - 8].

The following is the output:

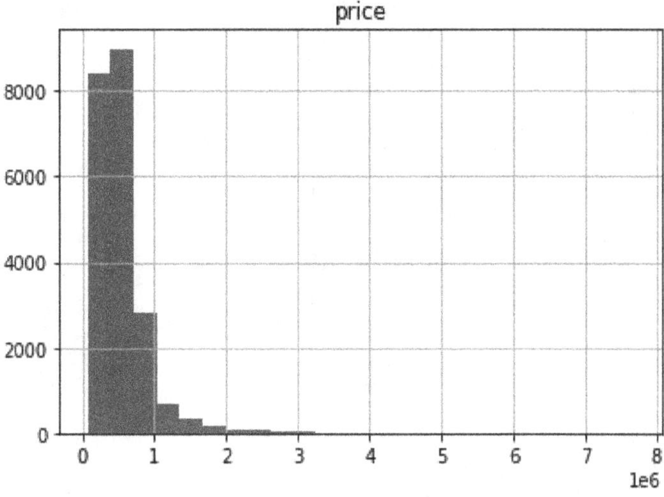

Figure 2.1 – Price distribution

Correlation study

Here, we will analyze how each feature correlates with each other and, most importantly, how each feature correlates with the price.

First, as previously discussed, we are going to remove the `id` feature:

```
house_df = house_df.drop(["id"], axis=1)
```

We can now compute the pairwise correlation diagram:

```
house_corr = house_df.corr()
```

To easily visualize the calculated correlations, we will plot a heatmap:

```
plt.figure(figsize=(20, 10))
colormap = sns.color_palette("rocket_r", as_cmap=True)
sns.heatmap(house_corr, annot=True, cmap=colormap)
plt.show()
```

These code statements yield the following result:

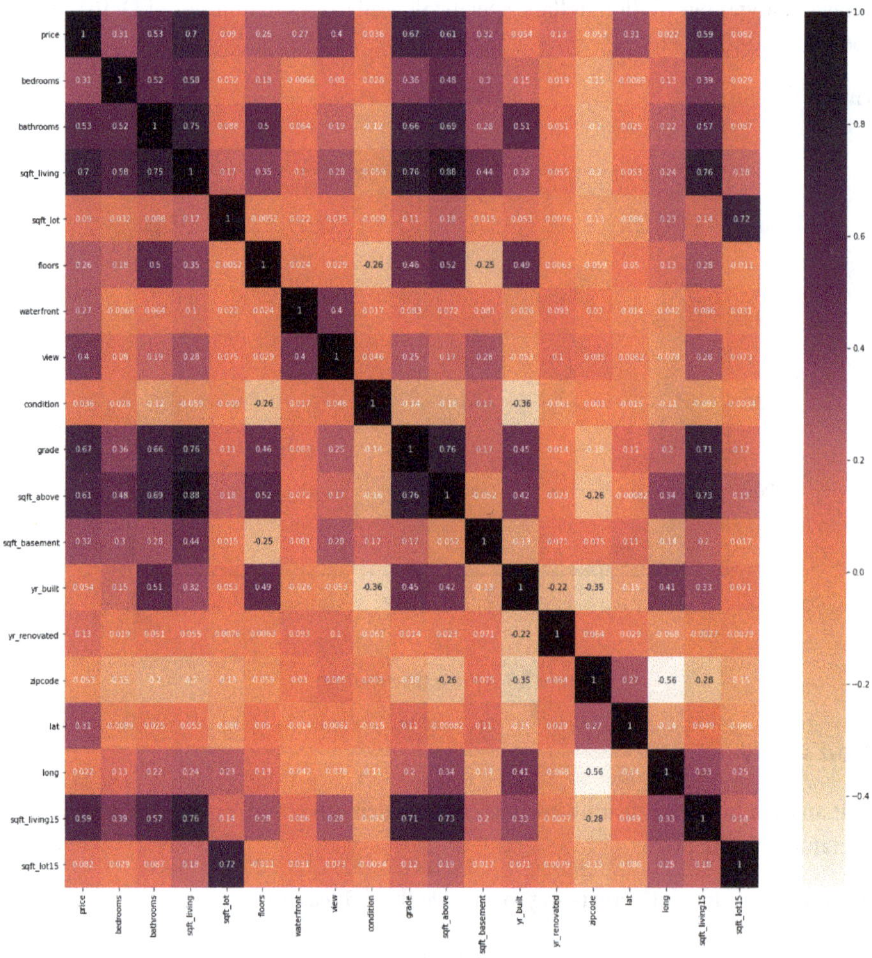

Figure 2.2 – House features correlation matrix

Note in *Figure 2.2* that the darker the cell is, the larger the correlation value.

To emphasize the first row (which is the most important as it shows the relationship between price and the input features), we will run the following code:

```
house_corr["price"].drop(["price"]).sort_values(ascending = False).
plot.bar(figsize=(5,5))
plt.show()
```

And we have the following result:

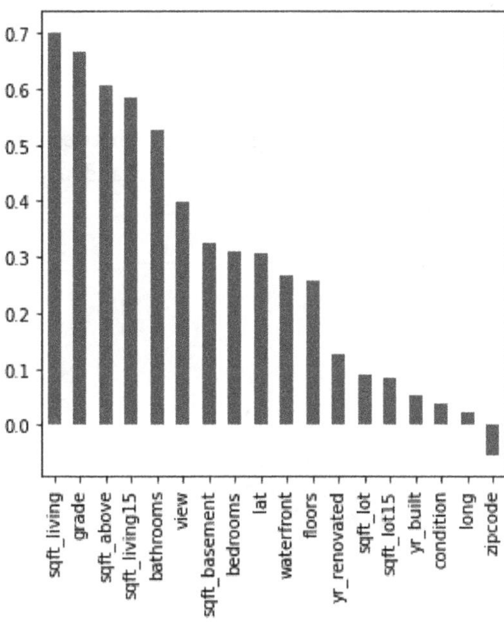

Figure 2.3 – House features: price correlation

The following conclusions can be drawn from *Figures 2.2* and *2.3*:

- `Living square feet` and `grade` are the most highly correlated features with price (0.7 and 0.67 respectively)
- `Above square feet` and `neighbors' living square feet` are very correlated with `living square feet` (0.88 and 0.76, respectively, which points to a degree of redundancy)
- The number of each type of room has the following correlation coefficients:
    - **Number of bathrooms:** 0.53
    - **Number of bedrooms:** 0.31
    - **Number of floors:** 0.26

- `View`, `waterfront`, and `renovation year` have some correlation with `price` (0.4, 0.27, and 0.13)
- `Location` is correlated with `price` as well, with `latitude` being the most important location feature (0.31)
- The rest of the features seem not to make a large contribution to the price of the property

Therefore, from an initial analysis, the most correlated features with *price* are, by order of importance: `living square feet`, `grade`, `number of bathrooms`, `view`, and `latitude`.

In the next sections, we will confirm these initial conclusions.

Square feet analysis

From the correlation diagram, we identified a strong correlation between *living square feet* and *price* (as expected), and a potential redundancy with *above square feet* and *neighbors' living square feet*. To analyze this in more detail, let's plot each variable versus price:

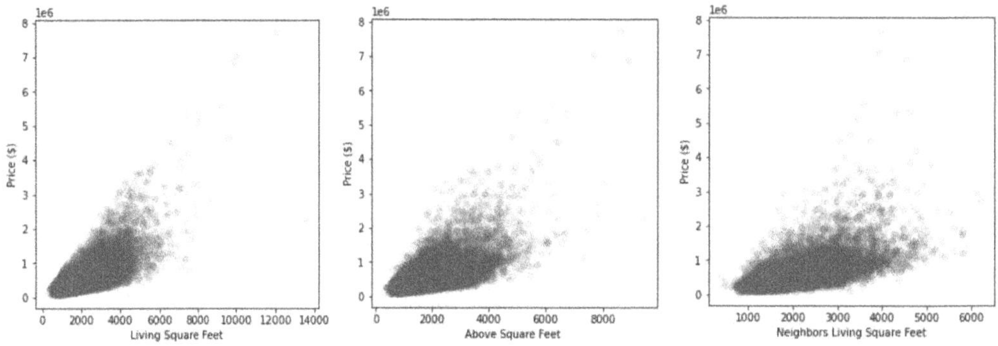

Figure 2.4 – Price compared with several features: a) living square feet, b) above square feet, c) neighbors' living square feet

As expected, the plots are very similar, which indicates a high correlation (and redundancy) among these variables. Furthermore, we can observe the largest density of data points occurs for prices less than $3M and less than 5,000 square feet. As most of our data lies in these areas, we can consider houses outside these ranges as outliers and remove them.

Grade analysis

Similarly, we can compare the *grade* feature against the price:

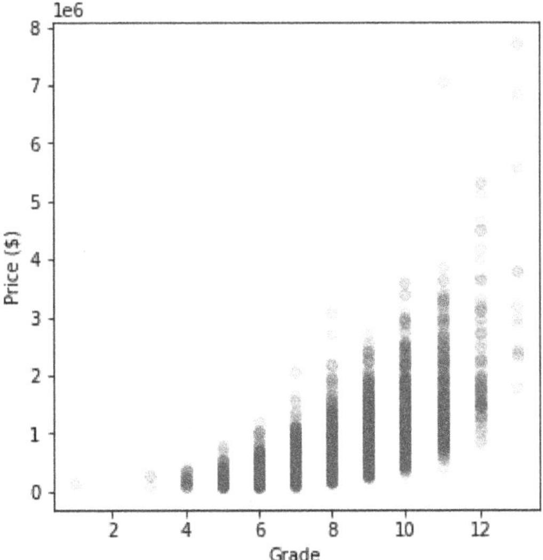

Figure 2.5 – House grade versus price

There is a clear direct correlation between the grade and the price; the higher the grade, the higher the price. It is also noteworthy that the highest values of grade are much less frequent.

Rooms analysis

Let's display in more detail the relationship between the price and the number of *floors*, *bedrooms*, and *bathrooms*:

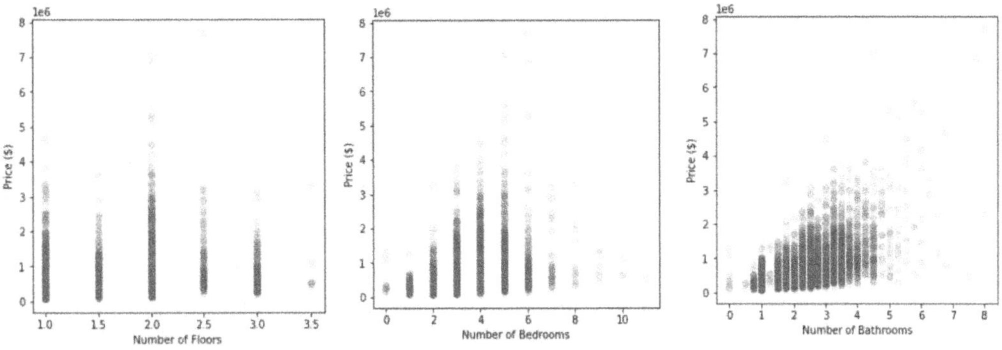

Figure 2.6 – Price compared with several features: a) number of floors, b) number of bedrooms, c) number of bathrooms

From the figure, you can observe the following:

- In *Figure 2.6 (a)*, we can see for a small number of floors (1-3), there is a direct correlation between the price of the house and this number. However, from the fourth floor, this correlation disappears, indicating a lack of data on this segment (houses with four or more floors are much less common).
- *Figure 2.6 (b)*, the comparison with the number of bedrooms, is a similar scenario to the previous chart comparing the number of floors. We can see how for a small number of bedrooms, there is a direct correlation between the price of the house and this number. However, from four bedrooms up, this correlation disappears, and other features need to be taken into account.

> **Important note**
>
> When looking carefully at the data, you will realize that in the row with index `15870`, there is an outlier; it is a house with 33 bedrooms. I do not know if this is the actual number of bedrooms of the house (I expect not!), but to properly analyze the dataset, this house, an outlier, was removed from it. See the code for details.

- In *Figure 2.6 (c)*, we can see there is a direct correlation between the number of bathrooms; nevertheless, there is also some uncertainty (the chart grows wider as we increase the number of bathrooms).

Views analysis

In this section, we will take a look in more detail at how *view* quality and a *waterfront view* (whether the house has this view or not) have a connection with the price:

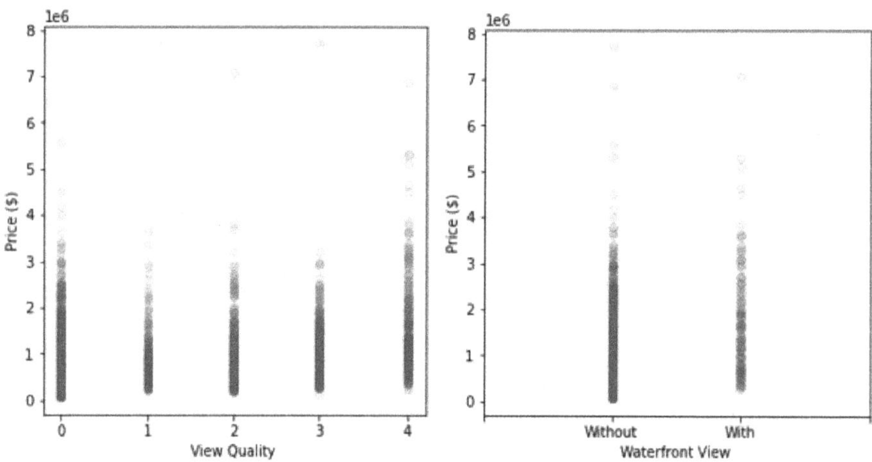

Figure 2.7 – View quality (a) and waterfront view (b) versus price

From these plots individually, it is a little bit more difficult to draw conclusions. Other variables seem to be needed to see a clear connection between the view quality and the price, and similarly with the waterfront view.

Year-built and year-renovated analysis

The following plots show how the features of which year a house was built and if and when a house was renovated are correlated with price:

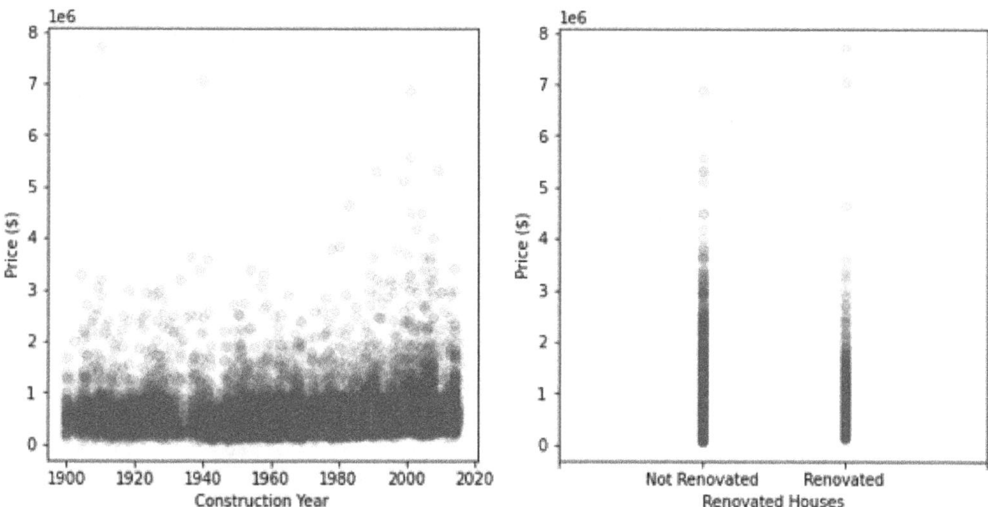

Figure 2.8 – Price compared with construction year (a) and renovation (b)

From the figure, you can observe the following:

- In *Figure 2.8 (a)*, we can see a slight linear increase in the price, suggesting that the more recently the house was built, the more expensive it is.
- For *Figure 2.8 (b)*, instead of analyzing the year, we split the dataset into two categories – houses that had been renovated and those that had not – and we plotted these two categories against price. Regardless, it is a little bit more difficult to draw conclusions. Other variables seem to be needed to see a clear connection between the renovation year and the price.

Location analysis

In this section, we will take a look in more detail at how latitude and longitude are connected to price:

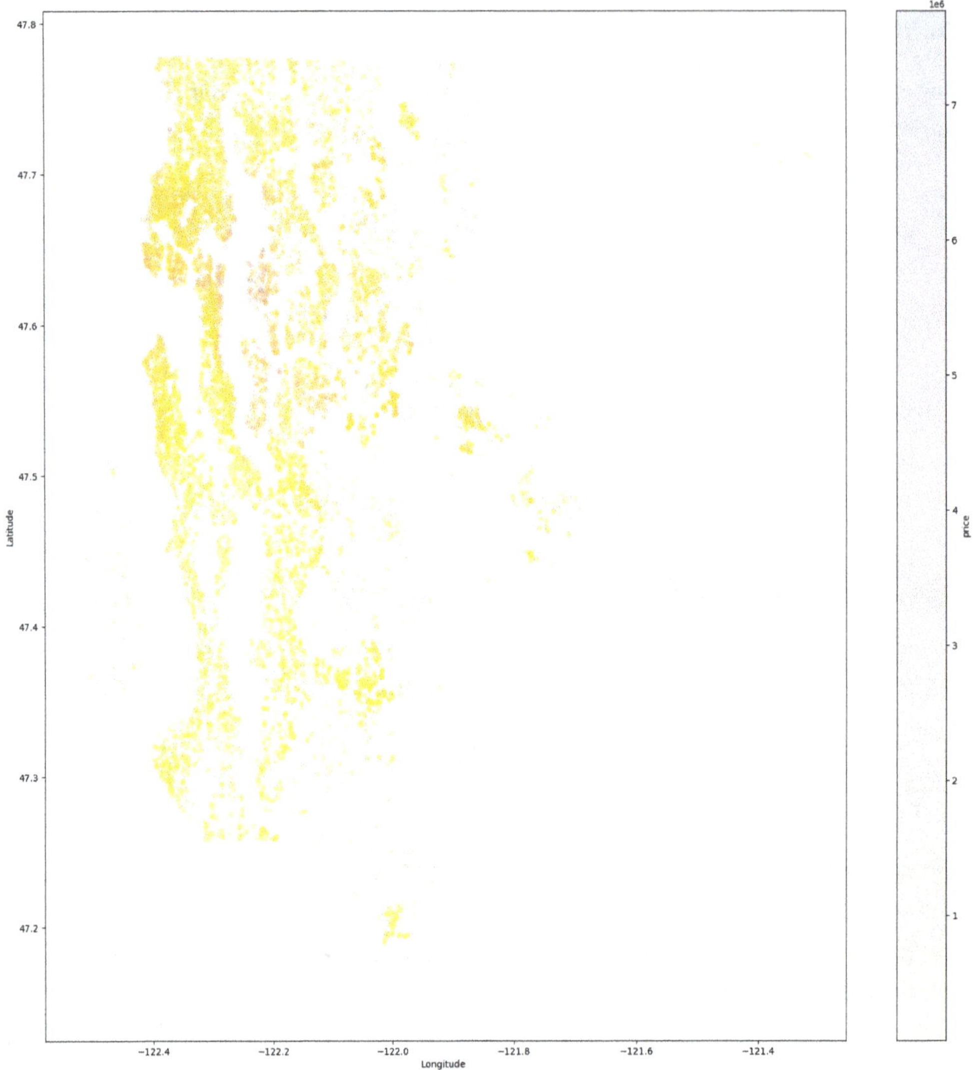

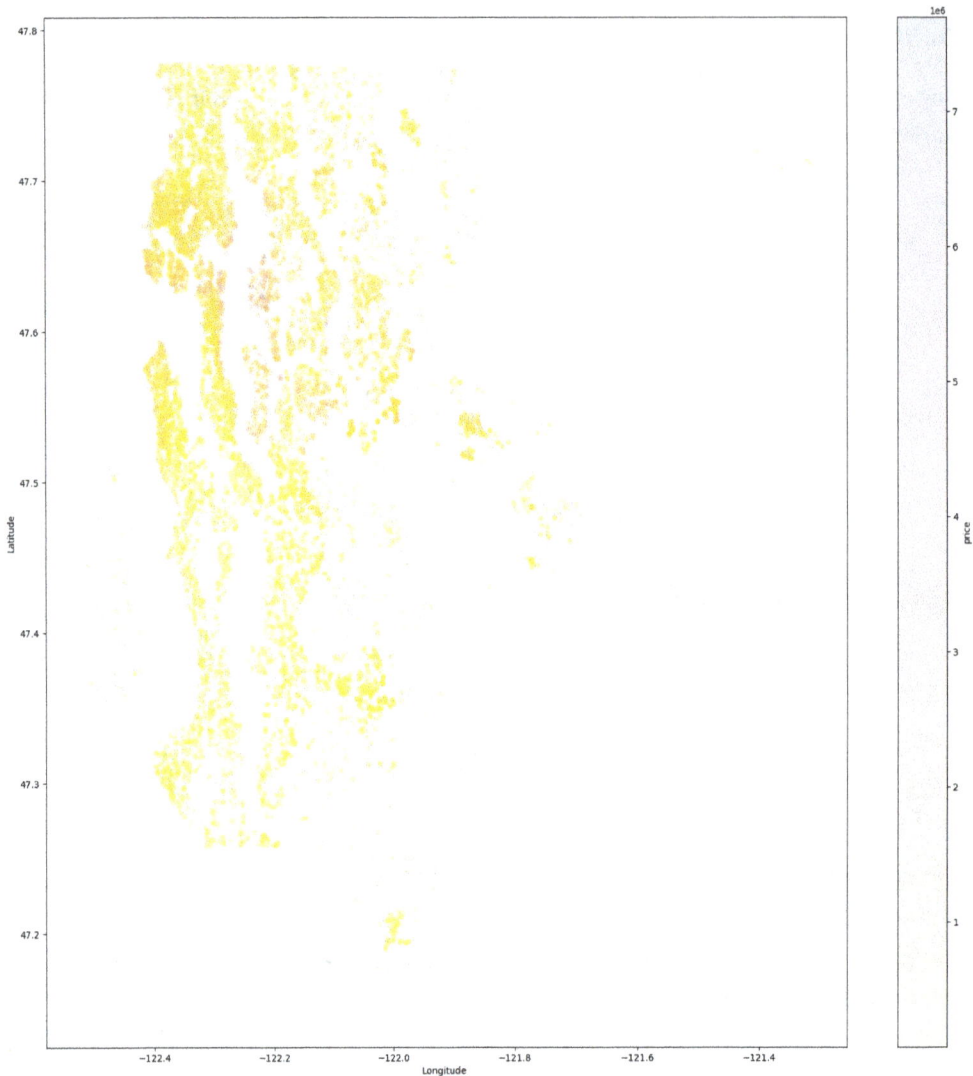

Figure 2.9 – Location versus price

From *Figure 2.9*, we can conclude that location plays an important role in the price of a house. Very clearly, the northern area of King County is more valued than the southern area. And there is a particular central region where houses are significantly more expensive than other nearby regions.

How it works...

Regression problems are one of the most common problems where SL approaches can be applied. By studying in depth a classic regression dataset, *King County House Price Prediction*, we can discover the most important connections between the input features (square feet, grade, and number of bathrooms) and the output feature (price). This analysis will help us build a model to predict the price in the next chapter.

There's more...

In this section, we focused on an individual analysis of each feature against price. However, some features are better understood when combined with others or preprocessed. We did a simple exploration of this topic, by combining the houses that have been renovated in one category and comparing this with the category of non-renovated houses. Furthermore, for the location analysis, we used a 2D map to plot the latitude and longitude to discover patterns.

However, there are plenty of relationships and analyses to be done, and I suggest you explore the dataset by yourself, create your own hypothese or hunches, and analyze the data to discover new insights.

Furthermore, there are many other regression datasets to play with; a small suggestion can be found here: `https://www.kaggle.com/rtatman/datasets-for-regression-analysis`.

Understanding classification datasets – loading, managing, and visualizing the Iris dataset

In the previous recipe, we studied one of the most common problem types in SL: regression. In this recipe, we will take a closer look at another of these problem types: **classification**.

In classification problems, we want to estimate a categorial output, a class, from a set of given classes, using a variable number of input features. In this recipe, we will analyze a toy classification dataset from Kaggle: the Iris dataset, one of the most renowned classification datasets.

The Iris dataset presents the problem of estimating the `iris` class of the flower of plants, from three classes (iris setosa, iris versicolor, and iris virginica) with the help of the following four features:

- Sepal length (in cm)
- Sepal width (in cm)
- Petal length (in cm)
- Petal width (in cm)

These data features are provided for 150 flowers, with 50 instances for each of the 3 classes (making it a balanced dataset).

Getting ready

This dataset is provided under the *CC0 Public Domain* license and can be downloaded from `https://www.kaggle.com/uciml/iris`.

To read, manage, and visualize the data, we are going to follow a similar approach to the toy regression dataset in the previous recipe. We will use `pandas` to manage the data, and we will use the most common data structure for the library, DataFrames. Moreover, in order to plot the data and several visualizations we will compute, we will use the `matplotlib`, `pyplot`, and `seaborn` libraries. Therefore, we must run the following code:

```
import matplotlib.pyplot as plt
import pandas as pd
import seaborn as sns
```

To load the data, we will introduce a new library that is very useful for managing datasets, called `scikit-learn`. This library comes pre-installed with a set of datasets, including the Iris dataset:

```
from sklearn import datasets
```

If you do not have the previously mentioned libraries installed, they can be easily installed with the following terminal commands:

```
!pip3 install matplotlib
!pip3 install pandas
!pip3 install seaborn
!pip3 install scikit-learn
```

Therefore, to load the data, we can simply read the dataset by making use of `scikit-learn` library functions:

```
iris = datasets.load_iris()
iris_df = pd.DataFrame(iris.data, columns = iris.feature_names)
 iris_df.insert(0, "class", iris.target)
```

This is all we need to start working with our classification dataset.

How to do it...

In this section, we will run an EDA that will help us understand which features are important (and which are not) to predict the iris class of a flower by completing the following:

- Data structure
- Correlation study

- One-versus-one comparison (pair plots)
- Violin plot

Data structure

Let's analyze what our data looks like. For this, we will use common operations on pandas DataFrames:

```
iris_df.info()
```

From the output, we can draw the following conclusions:

- The data is complete (all columns have 150 values, as expected)
- There are no NULL values (the data is clean!)

In order to grasp what the values look like, let's display the first five properties:

```
iris_df.head()
```

So far, we have looked at what the features look like. Now, let's take a look at what the iris class distribution looks like:

```
iris.target_names
```

This results in the following output:

```
array(['setosa', 'versicolor', 'virginica'], dtype='<U10')
```

If we want to confirm that there are 50 instances per class, we can run the following:

```
iris_df.groupby("class").size()
```

This yields the following output:

```
Class
0 50
1 50
2 50
dtype: int64
```

Here, 0 corresponds to setosa, 1 to versicolor, and 2 to virginica.

Correlation study

Here, we will analyze how each feature correlates with each other and, most importantly, how each feature correlates with the iris class.

We can compute a pairwise correlation diagram:

```
iris_corr = iris_df.corr()
```

To easily visualize the calculated correlations, we will plot a heatmap:

```
plt.figure(figsize=(10, 10))
colormap = sns.color_palette("rocket_r", as_cmap=True)
sns.heatmap(iris_corr, annot=True, cmap=colormap)
plt.show()
```

These code statements yield the following result:

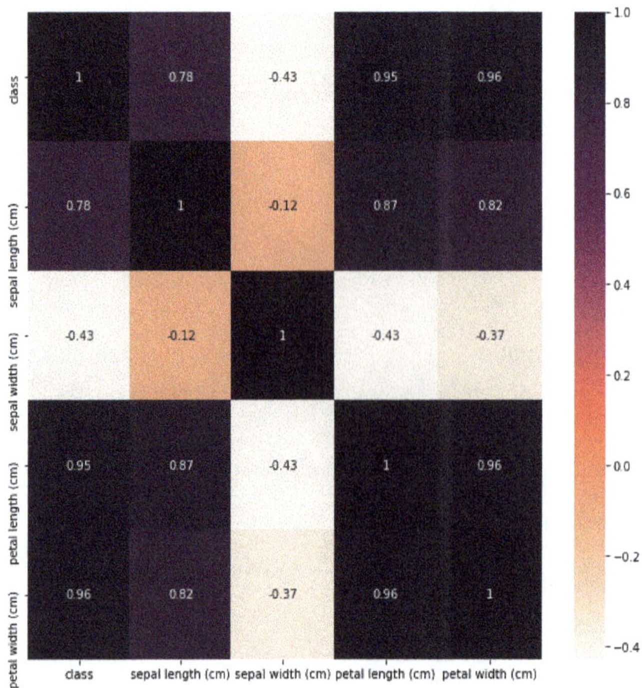

Figure 2.10 – Flower features correlation matrix

Let's note in *Figure 2.10* that the darker the cell is, the larger the correlation value.

To emphasize the first row (most important as it shows the relationship between the iris class and the input features), we will run the following code:

```
iris_corr["class"].drop(["class"]).sort_values(
    ascending = False).plot.bar(figsize=(5,5))
plt.show()
```

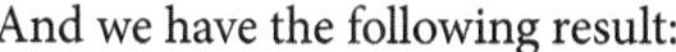

And we have the following result:

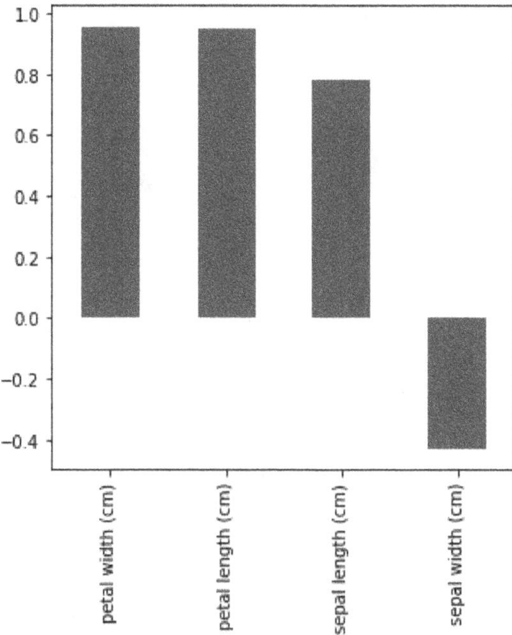

Figure 2.11 – Flower features: iris class correlation

The following conclusions can be drawn from *Figures 2.10* and *2.11*:

- Petal measurements (length and width) are highly correlated; analyzing and training both might not yield any additional information.
- Petal measurements are the most highly correlated features with the iris class.
- Sepal length and width are highly correlated as well but in opposite ways (sepal length is positively correlated while sepal width is negatively correlated).

One-versus-one comparison (pair plots)

In classification problems, hue/brightness can be used to indicate the class of a plot. Moreover, as in this dataset we are able to work with a limited set of features (four), a pair plot diagram will be very useful to compare all features in a single plot. The code to plot this diagram is shown here:

```
g = sns.pairplot(iris_df, hue="class", height=2, palette="rocket_r")
handles = g._legend_data.values()
labels = list(iris.target_names)
 g._legend.remove()
g.fig.legend(handles=handles, labels=labels, loc='upper left', ncol=3)
```

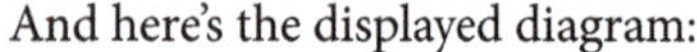

And here's the displayed diagram:

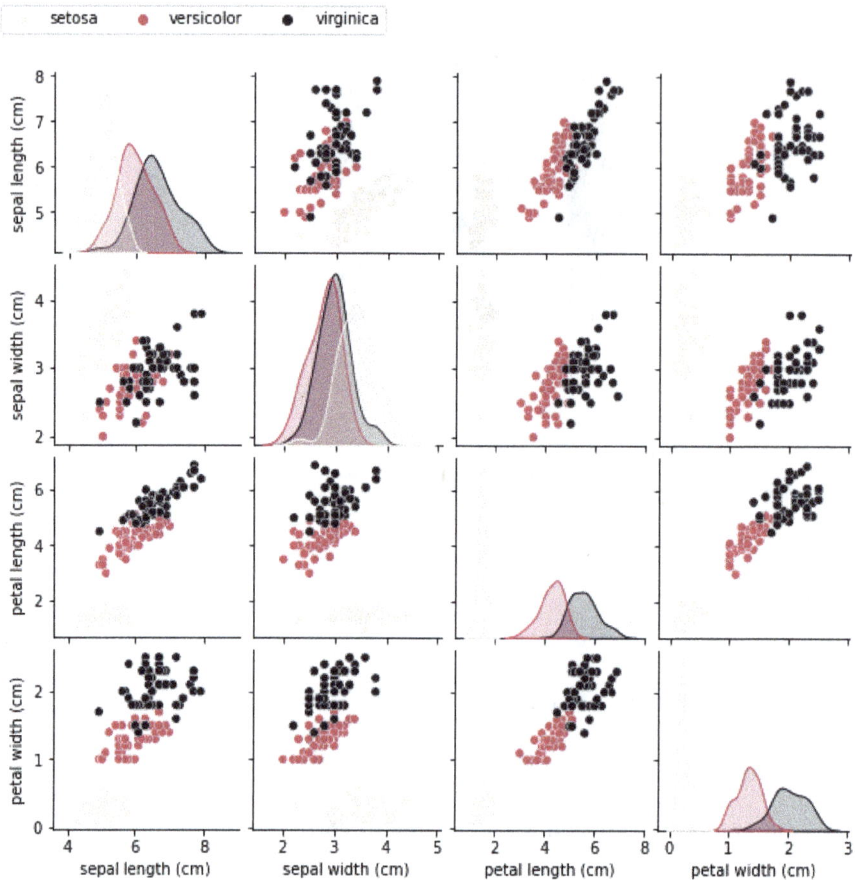

Figure 2.12 – Flower features pair plot

From this set of plots, we can draw the following conclusions:

- Setosa iris is easily differentiated using any of the features
- Sepal features overlap among the different iris classes
- Petal features have a direct relationship with the iris class; that is, the smallest numbers point to **setosa**, medium numbers point to **versicolor**, and the largest numbers point to **virginica**
- There is a region on the boundaries of **versicolor** and **virginica** where both groups overlap, for a petal length larger than ~5 cm and a petal width larger than ~1.5 cm

Violin plot

Another plot that might help with understanding the relationships between features and the iris class is a violin plot. The code to generate this plot is the following:

```
fig, axs = plt.subplots(2, 1)
 sns.violinplot(x="class", y="petal length (cm)", data=iris_df,
size=5, palette='rocket_r', ax = axs[0])
 sns.violinplot(x="class", y="petal width (cm)", data=iris_df, size=5,
palette='rocket_r', ax = axs[1])
```

And here's the displayed diagram:

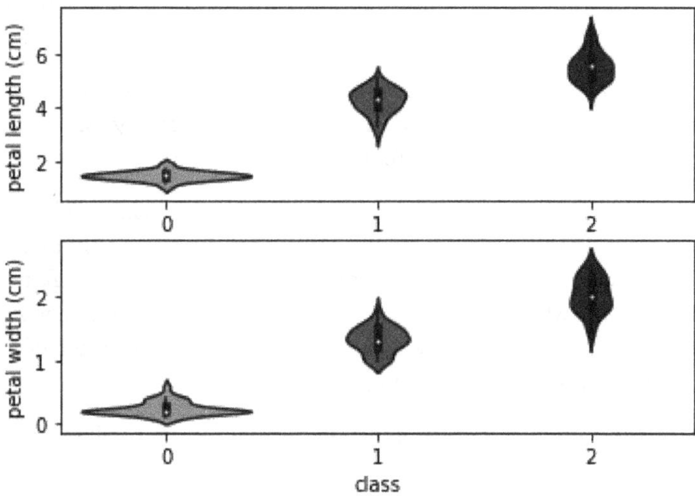

Figure 2.13 – Flower features violin plot

In these plots, the conclusions we found are even clearer, with the setosa (0) iris class clearly separable and where there is substantial overlap for versicolor (1) and virginica (2).

The violin plot also provides an indication of the distribution of the values in our case (starting with 0 to match the indexes of the classes in the code):

- **Setosa**: Values are more likely to be found around the mean (~1.5 cm for petal length and ~0.25 cm for petal width).
- **Versicolor**: Normal distribution with mean values ~4.25 cm and ~1.3 cm, and standard distribution values ~0.5 cm and ~0.2 cm (for petal length and petal width respectively).
- **Virginica**: Uniform distribution between [~5.1, ~5.9] cm and [~1.8, ~2.3] cm (for petal length and petal width respectively).

How it works...

Classification problems are one of the most common problems where **supervised ML (SML)** approaches can be applied. By studying a classic classification dataset, Iris Class in depth we can discover the connections between the input features (petal length and petal width) and the output feature (iris class). This analysis will help us build a model to predict the class in the next chapter.

There's more...

In this section, we focused on an individual analysis of each feature against the iris class. This was similar in principle to the analysis done for regression datasets, with the added information about the hue/brightness for each plot. A similar suggestion to the reader follows, to continue and deepen the analysis by themselves to discover new insights.

We mentioned the Iris dataset is one of the classical classification datasets; nonetheless, it dates back to 1936! Original reference: `https://onlinelibrary.wiley.com/doi/pdf/10.1111/j.1469-1809.1936.tb02137.x`.

Furthermore, as we will explore in the next chapter, classification problems can be seen as a special case of regression problems. In the regression recipe, we studied the prices of houses, and presented them in a way that would allow a buyer to determine which are affordable, by taking the price and comparing it with a threshold, which could be, for example, our budget limit. Therefore, we can use that threshold to classify houses that are below it as affordable and above it as not affordable. We will explore this connection in depth in the next chapter.

Furthermore, there are many other classification datasets to play with; a small selection can be found here: `https://www.kaggle.com/search?q=classification+tags%3Aclassification`.

Understanding image datasets – loading, managing, and visualizing the Fashion-MNIST dataset

One of the fields that has grown considerably in DL in the last years has been **computer vision (CV)**. Since the AlexNet revolution in 2012, CV has expanded from lab research to surpassing human performance in real-world datasets (known as "in the wild").

In this recipe, we will explore the simplest CV task: **image classification**. Given a set of images, our task is to correctly classify that image among a given set of labels (classes).

One of the most classic image classification datasets is the **MNIST** (which stands for the **Modified National Institute of Standards and Technology**) database. Similarly sized, but more suited for current CV analysis, is the *Fashion-MNIST dataset*. This dataset is a multi-label image classification dataset, with a training set of 60k examples and a test set of 10k examples, with each example belonging to 1 of these 10 categories (starting with 0 to match the indexes of the classes in the code):

- T-shirt/top
- Trouser
- Pullover
- Dress
- Coat
- Sandal
- Shirt
- Sneaker
- Bag
- Ankle boot

Each image is grayscale with 28x28 pixel dimensions. This can be seen as each data point having 784 features. The dataset is composed of 6k images per class in the training set and 1k images per class in the test set (balanced dataset).

Getting ready

This dataset is provided under the *MIT* license and can be downloaded from the following URL: `https://github.com/zalandoresearch/fashion-mnist`

This dataset is directly available from MXNet Gluon, and therefore we will use this library to access it. Moreover, as this dataset is significantly larger than the others we have explored so far, to handle the data efficiently, we will use the Gluon DataLoader functionality:

```
from mxnet import gluon
training_data_raw = gluon.data.vision.FashionMNIST(train=True)
test_data_raw = gluon.data.vision.FashionMNIST(train=False)
```

> **Tip**
> Gluon is installed with MXNet; no further steps are required.

This is all we need to start working with the Fashion-MNIST dataset.

> **Important note**
> Sometimes, data needs to be modified (transformed) for some operations. This can be done by defining a `transform` function and passing it as a parameter (`transform=<function_name>`).

How to do it...

In this section, we will run an EDA that will help us understand which features are important (and which are not) to predict the category of a garment, with the help of the following steps:

1. Identifying the data structure
2. Describing examples per class
3. Understanding dimensionality reduction techniques
4. Visualizing **Principal Component Analysis** (**PCA**)
5. Visualizing **t-distributed Stochastic Neighbor Embedding** (**t-SNE**)
6. Visualizing **Uniform Manifold Approximation and Projection** (**UMAP**)
7. Visualizing **Python Minimum-Distortion Embedding** (**PyMDE**)

Identifying the data structure

In order to optimize memory usage to work with large-scale datasets, instead of loading the full dataset in memory, datasets are usually accessed through **batches**, which are smaller packets of data.

Gluon has its own way of generating batches, while also applying **shuffling**, to increase robustness during training: on each epoch, batches are shown in a random manner to the network. For testing, as we want repeatable results, this is avoided. We select a **batch size** of `128`:

```
batch_size = 128
training_data_aux = gluon.data.DataLoader(
    training_data_raw, batch_size= batch_size, shuffle=True)
 test_data_aux = gluon.data.DataLoader(
    test_data_raw, batch_size= batch_size, shuffle=False)
```

> **Important note**
>
> DataLoader does not return a data structure but an iterator. Therefore, to access the data we need to iterate upon it, we use constructs such as `for` loops.

Let's verify the data structure is as expected:

```
training_data_size = 0
for X_batch, y_batch in training_data_aux:
    if not training_data_size:
        print("X_batch has shape {}, and y_batch has shape
{}"         .format(X_batch.shape, y_batch.shape))
    training_data_size += X_batch.shape[0]
 print("Training Dataset Samples: {}".format(training_data_size))
 test_data_size = 0
```

```
for X_batch, y_batch in test_data_aux:
    test_data_size += X_batch.shape[0]
print("Test Dataset Samples: {}".format(test_data_size))
```

We obtain the expected output:

```
X_batch has shape (128, 28, 28, 1), and y_batch has shape (128,)
 Training Dataset Samples: 60000
Test Dataset Samples: 10000
```

> **Important note**
>
> Gluon loads grayscale images as images with one channel, and the dimension for each batch is (batch size, height, width, number of channels); in our example, (128, 28, 28, 1).

Describing examples per class

The Fashion-MNIST dataset is a balanced dataset with 6k examples per class:

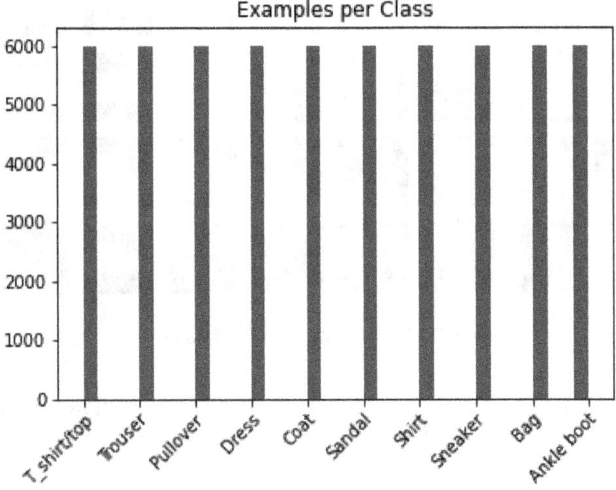

Figure 2.14 – Fashion-MNIST dataset labels

Let's take a look at what each class looks like. In order to do this, we can plot 10 examples per class:

T_shirt/top Trouser Pullover Dress Coat Sandal Shirt Sneaker Bag Ankle boot

Figure 2.15 – Fashion-MNIST dataset

As we can see in *Figure 2.15*, all instances can be differentiated fairly well by a human, except for the `T-shirt/top`, `Pullover`, `Coat`, and `Shirt` classes.

Understanding dimensionality reduction techniques

Apart from the large number of data points that our dataset contains, the number of features in the image (that is, the number of pixels per image) is very high. In our toy dataset, each image has 784 features, which can be seen as 1 point in 784-dimensional space. In this space, it is extremely difficult to analyze relationships among features (for example, correlation, as we explored in previous datasets). Furthermore, it is not rare to work with higher-quality images, with resolutions over 1 MP (more than 1 million features). For a 4k image, the number of features is ~8 million.

Therefore, in this subsection and the next ones (on *PCA*, *t-SNE*, *UMAP*, and *PyMDE*), we will work with techniques known as **dimensionality reduction** techniques. The idea behind these techniques is to be able to visualize high-dimensional features easily, typically in 2D or 3D, which are the kinds of visualizations humans are used to working with. These embeddings have two or three components that can be plotted in 2D or 3D. These representations are dataset-dependent; they are *learned* representations.

Each technique described has a different way of achieving this result. In this book, we will not deepen our knowledge of how each technique works, but the interested reader can find more information in the *There's more...* section.

Please also note that although each technique is different, all of them require a vector as an input (feature vector). This means that there is some spatial information that is lost. In our example, from 28x28 images, we will input 784 feature vectors.

Visualizing PCA

As expected, we can see some large clusters (**Sneaker** and **Ankle boot**), and others are mostly overlapping (**T-shirt**, **Pullover** and **Coat**):

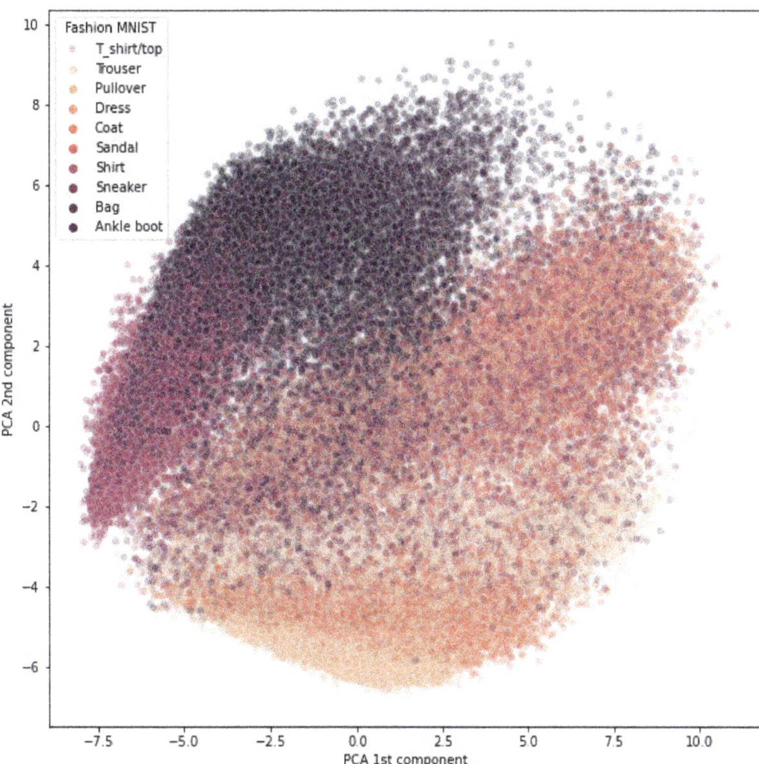

Figure 2.16 – Fashion-MNIST 2D PCA

Visualizing t-SNE

Another technique for dimensionality reduction is t-SNE. This technique is based on the idea of computing a probability distribution that represents similarities among neighbors. A recommended preliminary step is to compute PCA for 50 features and then pass these 50 feature vectors to the t-SNE algorithm. This is what we did to generate the following graph:

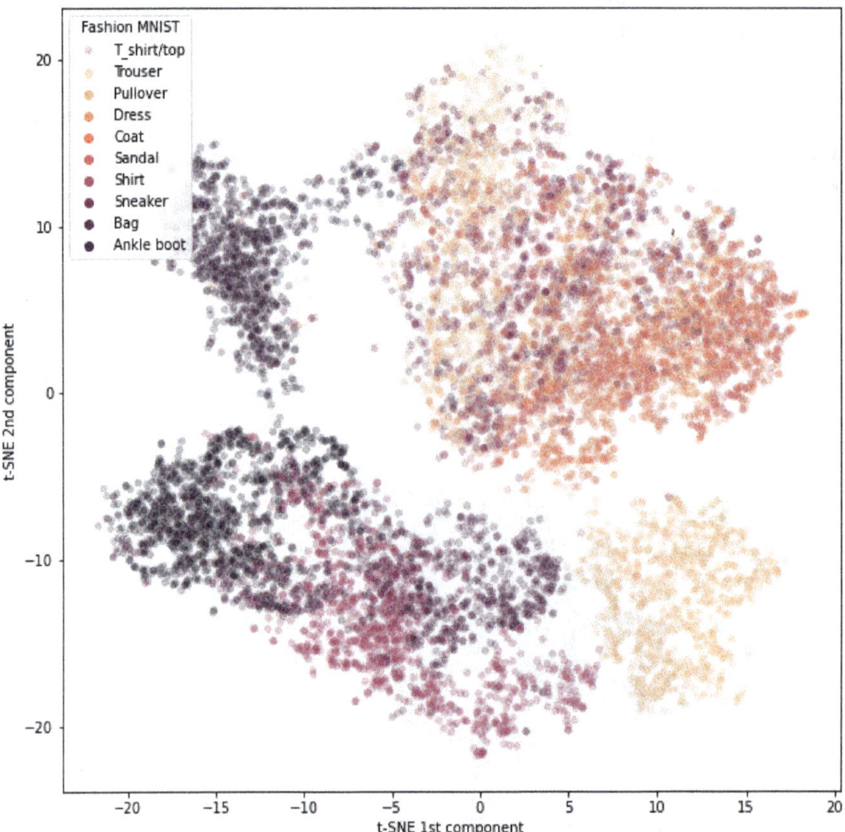

Figure 2.17 – Fashion-MNIST 2D t-SNE

In this plot, we can see in a clearer way how easily distinguishable objects are clustered in isolation (**Trouser**, on the lower right, and **Bag**, on the upper left).

> **Important note**
>
> For PCA and t-SNE, we can choose three components instead of two, which will yield a 3D plot. For the code, visit the GitHub repository of the book: `https://github.com/PacktPublishing/Deep-Learning-with-MXNet-Cookbook`.

Visualizing UMAP

Another method for dimensionality reduction is **UMAP**. UMAP allows us to play with different parameters, such as the *number of neighbors*, which helps the visualization of how to balance a local versus global structure. For an example with five neighbors, this is the visualization:

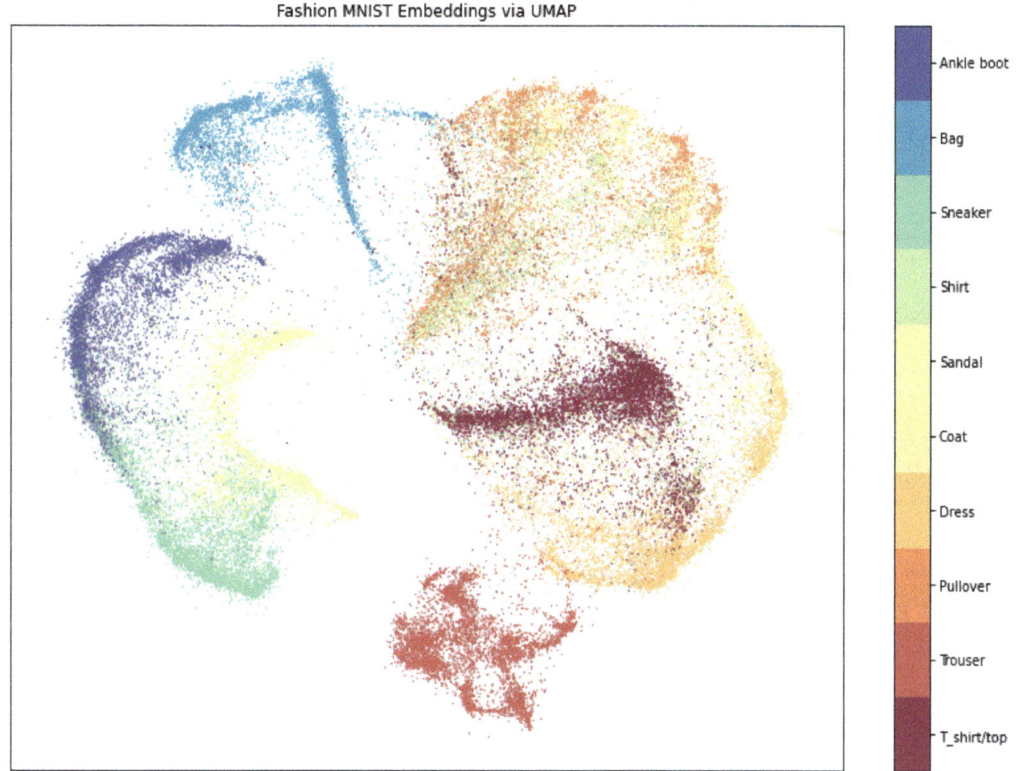

Figure 2.18 – Fashion-MNIST UMA

In this visualization, we can observe similar trends as in previous plots; that is, **Bag** is clustered in the upper-center region, and **Trouser** in the lower-center region. However, in this visualization, we can also note that there is a cluster on the left that contains data for **Ankle boot**, **Sneaker**, and **Sandal**, and another important cluster on the right for **Shirt**, **Coat**, **Dress**, and **T-shirt/top**, and we can see how these clusters overlap with each other.

To install UMAP, please run this command:

```
!pip3 install umap-learn
```

Visualizing PyMDE

Another popular technique that provides insightful visualizations is **PyMDE**. PyMDE allows two main approaches, to preserve neighbors (local structure of the data is preserved) and to preserve distances. This preserves relationship attributes such as pairwise distances in the data. The approach to preserve neighbors is similar to the plots we are seeing:

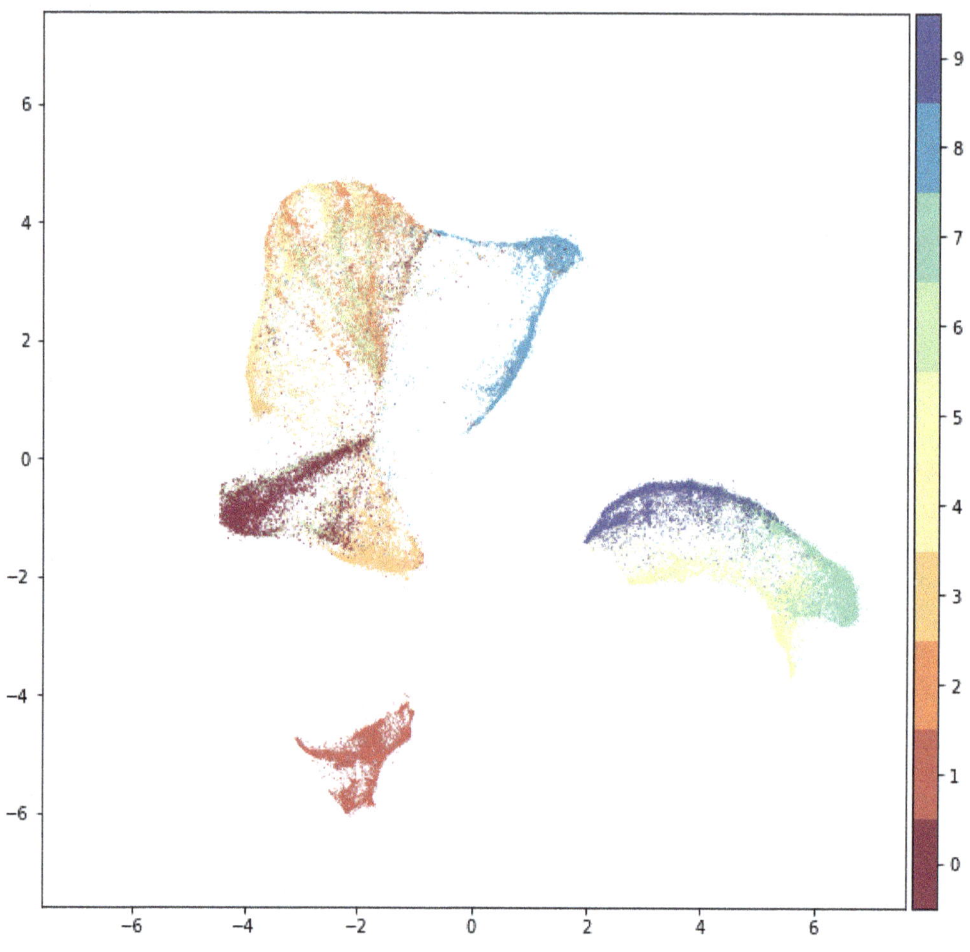

Figure 2.19 – Fashion-MNIST PyMDE

As we can see in *Figure 2.19*, very similar conclusions to UMAP can be drawn from a PyMDE visualization.

To install UMAP, please run the following command:

```
!pip3 install pymde
```

How it works...

To understand an image dataset, we need to understand the underlying connections among images in that dataset. One useful method to achieve this is with different visualizations.

In this recipe, we have learned how to discover patterns in our image datasets. We selected a well-researched dataset, Fashion-MNIST, and learned one of the most important approaches for working with large-scale datasets: **batching**.

We analyzed our dataset by taking a look at its internal structure and what the actual images looked like and tried to foresee where potential classification algorithms could have issues (such as similarities between coats and shirts and ankle boots and sneakers).

Every pixel is a dimension/feature of each image, and, therefore, to work with them, we learned about some dimensionality reduction techniques: PCA, t-SNE, UMAP, and PyMDE. With these visualizations, we were able to verify and extend our knowledge of the dataset.

There's more...

There are many resources for MNIST and Fashion-MNIST, as these are well-researched datasets. I personally recommend the following:

- **MNIST database:** `https://en.wikipedia.org/wiki/MNIST_database`
- **zalandoresearch/fashion-mnist:** `https://github.com/zalandoresearch/fashion-mnist`

We introduced some dimensionality reduction techniques, but we did not deepen our knowledge of them. If you want to understand better how each of these techniques works, I suggest the following resources:

- **PCA (from Caltech)**: `http://web.ipac.caltech.edu/staff/fmasci/home/astro_refs/PrincipalComponentAnalysis.pdf`
- **t-SNE**: `https://lvdmaaten.github.io/tsne/`
- **UMAP**: `https://umap-learn.readthedocs.io/`
- **PyMDE**: `https://pymde.org/`

In the code, you can find how to obtain the visualizations included. Furthermore, for PCA and t-SNE, as the number of components is a variable, 3D plots for both of them are included.

Finally, for readers interested in learning more about DL and its history, I recommend the following link: `https://www.skynettoday.com/overviews/neural-net-history`.

Understanding text datasets – loading, managing, and visualizing the Enron Email dataset

Another field that has grown considerably in DL in recent years is **natural language processing** (**NLP**). Similarly to CV, this field aims to surpass human performance in real-world datasets.

In this recipe, we will explore one of the simplest NLP tasks: **text classification**. Given a set of sentences and paragraphs, our task is to correctly classify that text among a given set of labels (classes).

One of the most classic text classification tasks is to distinguish whether received email is spam or not (ham). These datasets are binary text classification datasets (only two labels to assign, `0` and `1`, or `ham` and `spam`).

In our specific scenario, we will use a real-world email dataset. This set of emails was made public during the investigation of the Enron scandal in the early 2000s by the US Government. This dataset was first published in 2004 and is composed of emails from ~150 users, mostly senior management at Enron. Only a subset (known as `enron1`) is used in this section.

It contains 5,171 emails, with no training/test split (labels are provided for all examples). Being a real-world dataset, emails vary heavily with respect to subjects, content length, word count, and word length, and out of the box, the dataset only contains two features:

- **Label/class**: `0` corresponds to `Ham` and `1` to `Spam`
- **Text**: Includes the subject and the body of the email

The dataset is composed of 3,672 examples of ham email (~70%) and 1,499 examples of spam email (~30%); it is a highly imbalanced dataset.

Getting ready

This dataset is provided under the *CC0 Public Domain* license and can be downloaded from `https://www.kaggle.com/venky73/spam-mails-dataset`.

To read the data, we are going to follow a similar approach as seen in the recipe for regression tasks. We are going to load the data from a CSV file, and we are going to work with the data using very well-known Python libraries: `pandas`, `pyplot`, and `seaborn`. Therefore, we must run the following code:

```
import matplotlib.pyplot as plt
import pandas as pd
import seaborn as sns
```

Therefore, to load the data, we can simply read the file containing it (the file can be found in the book's GitHub repository):

```
emails_df = pd.read_csv("spam_ham_dataset.csv")
```

This is all we need to start working with our spam email dataset.

How to do it...

In this section, we will run an EDA that will help us understand which features are important (and which are not) to predict whether an email is spam or not, the following are not worded as steps. please either reword this to a more suitable lead-in or reword the following as steps. if the latter, please change the circular bullets to numbering:

1. Data structure
2. Examples per class
3. Content analysis
4. Data cleaning
5. N-grams
6. Word processing (tokenizing, stop words, stemming, and lemmatization)
7. Word clouds
8. Word embeddings (word2vec and **Global Vectors for Word Representation** (**GloVe**))
9. PCA and t-SNE

Data structure

The first step we will carry out will be to reformat the dataset for our purposes:

```
# Removing Unnecessary column
emails_df.drop("Unnamed: 0", axis=1, inplace=True)
 # Changing column names
emails_df.columns = ["label", "text", "class"]
```

After these modifications, the shape of our email DataFrame is as follows:

```
(5171, 3)
```

Examples per class

We will now take a look at each class distribution:

```
Label
ham 3672
spam 1499
dtype: int64
```

The following is the output:

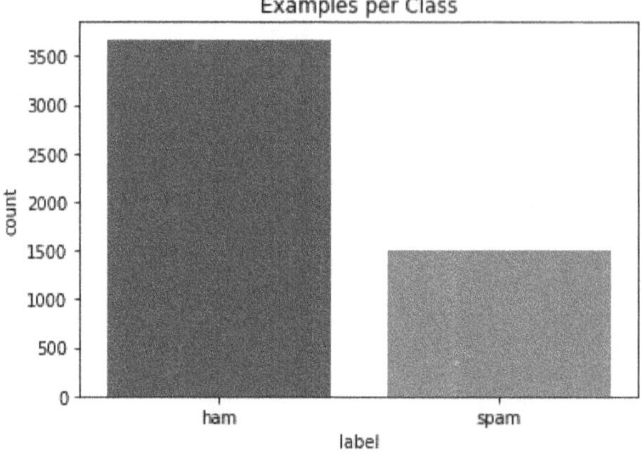

Figure 2.20 – Spam emails dataset

As we can see, the dataset is highly imbalanced.

Content analysis

In this section, we are going to analyze the emails' length and their distribution:

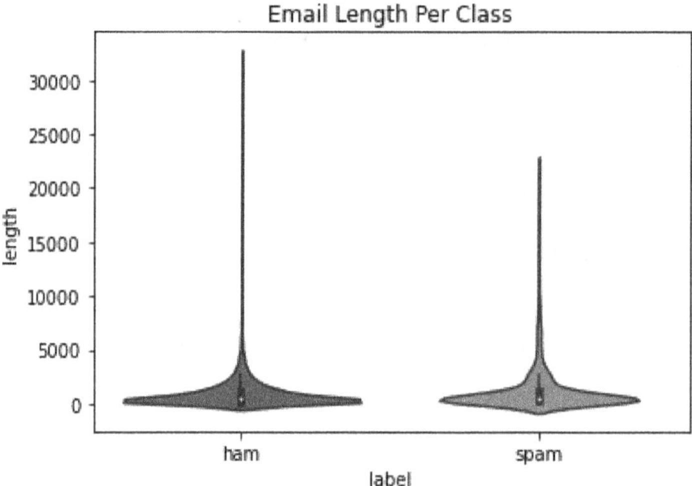

Figure 2.21 – Emails' length

There is a large outlier set corresponding to emails with more than 5,000 characters. Let's zoom in to the area where most of the emails lie and graph the length and the word count:

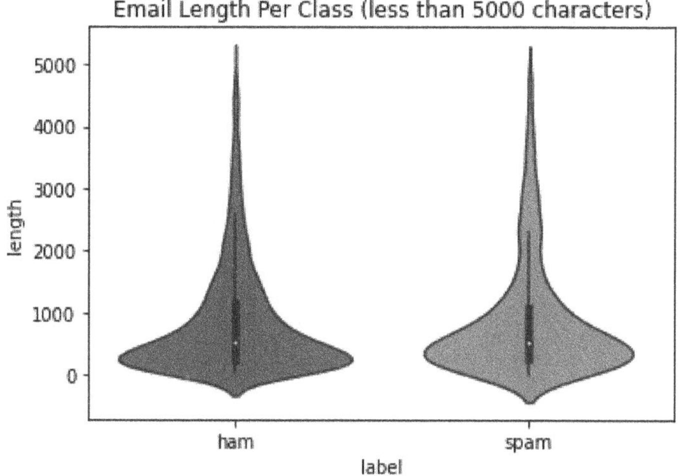

Figure 2.22 – Emails' length (detailed)

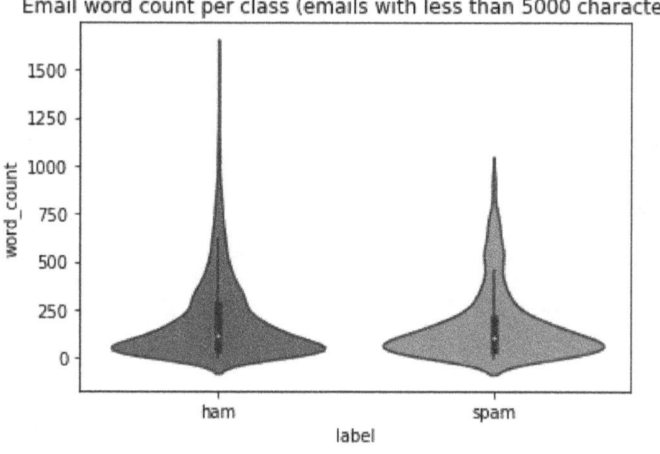

Figure 2.23 – Emails' word count

> **Important note**
>
> In *Figure 2.23*, we have defined a word without any semantic or dictionary approach, simply by specifying that each space-separated entity constitutes a word. This approach has disadvantages that we will analyze further in this recipe and in *Chapter 5*.

By looking at this graph, we can conclude that in terms of emails' length and word count, there is no significant difference between spam and legitimate emails. We will need to understand more about the words, their meaning, and their relationships to improve our analysis.

Therefore, let's start by looking at which words are most frequent in the dataset:

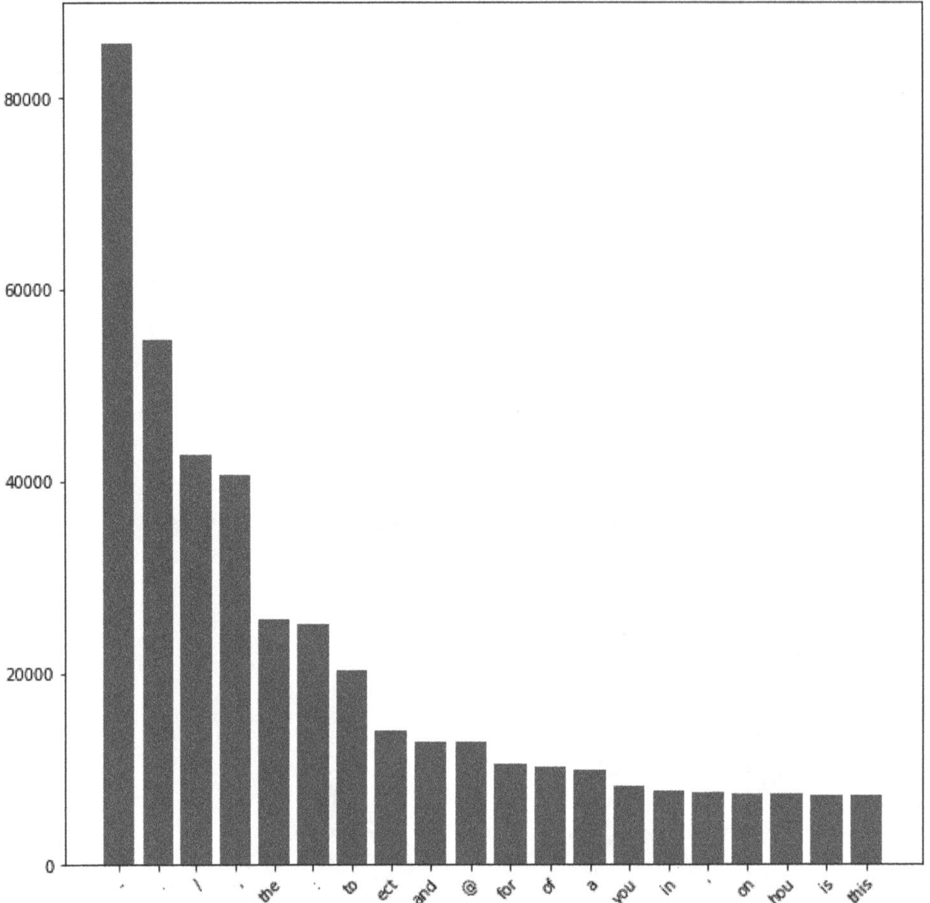

Figure 2.24 – Most frequent words

The first and most important conclusion after seeing *Figure 2.24* is that our initial space-separation approach to differentiate words was not enough when dealing with real-world datasets. Punctuation errors and typos are very common, and furthermore, as expected, very common words such as "the" and "to" yield no real benefit in understanding the differences between spam and legitimate emails.

Data cleaning

Let's get rid of some common issues when working with real-world text datasets:

- Punctuation
- Trailing characters
- Clarifications" (text between square brackets)
- Words containing numbers and links
- The word *subject* (specific to our email dataset structure)

After processing our corpus (text data specific to our problem) through our cleaning function, the results are more similar to our expectations:

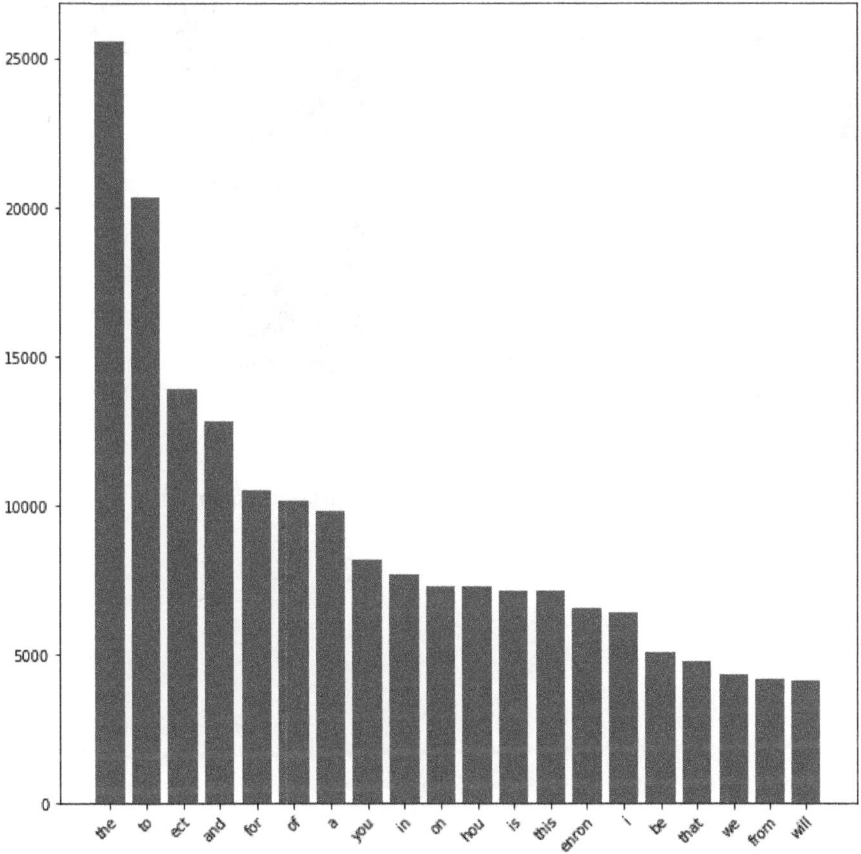

Figure 2.25 – Most frequent words (clean)

In *Figure 2.25*, we can see how the new word corpus we are analyzing contains real words. However, it is very clear that the most frequent words don't help to distinguish between spam and legitimate emails; words such as "the" and "to" are too common in the English language to be used properly for this classification.

N-grams

N-grams from a corpus in NLP are a set of *N* co-occurring words in the corpus. Typically in NLP, the most common N-grams are *unigrams* (one word), *bigrams* (two words), and *trigrams* (three words). Plotting the most frequent N-grams helps us understand relationships among words and classes (spam or not). A unigram is simply the most frequent word graph, as plotted in the previous section in *Figure 2.25*. For bigrams (per class), see the following:

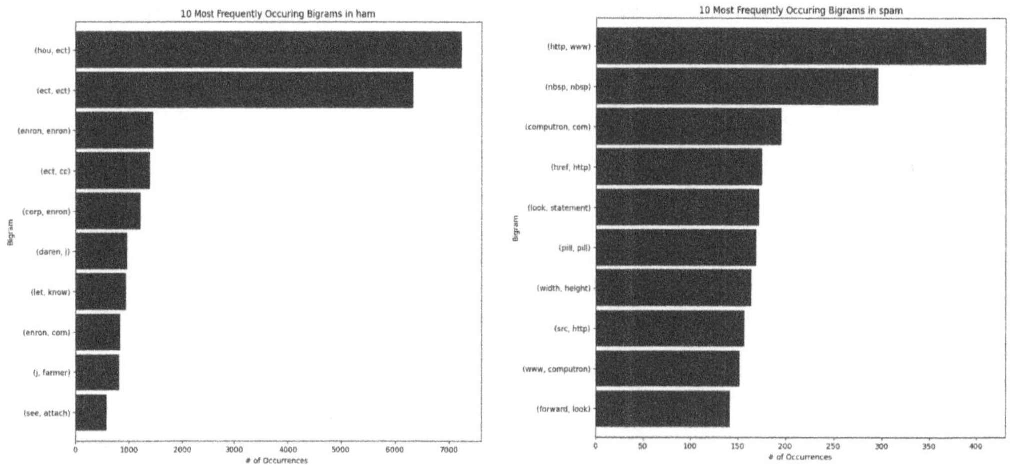

Figure 2.26 – Most frequent bigrams in ham (a) and spam (b)

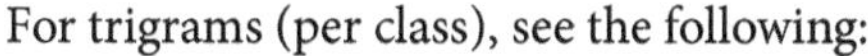

For trigrams (per class), see the following:

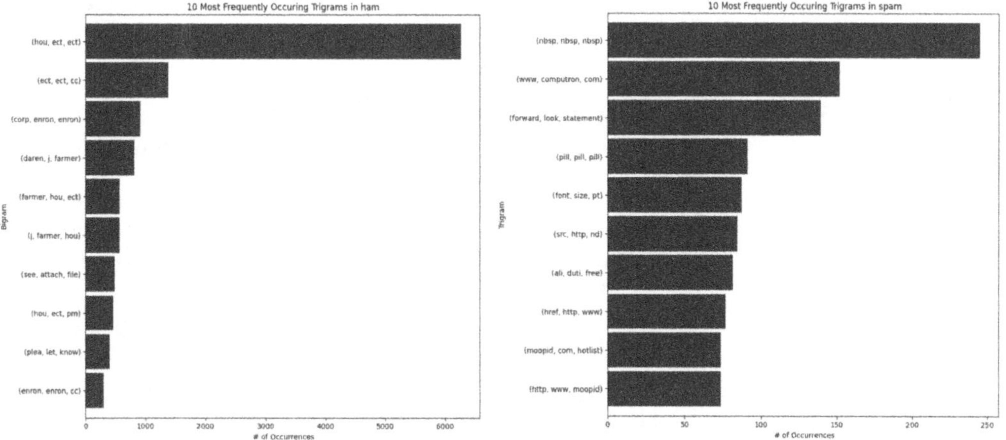

Figure 2.27 – Most frequent trigrams in ham (a) and spam (b)

In these graphs, we can start to grasp the underlying differences between the two classes:

- If there is a mention of *Enron Corp*, it is very likely to be a legitimate email
- If there is a polite call to action ("please let me know"), it is very likely to be a legitimate email
- If there is a link, it is very likely to be spam
- If there are typos (*hou* instead of *how*, *ect* instead of *etc*, and so on), it is very likely to be a legitimate email
- If *pills* are mentioned, it is very likely to be spam (plus a bonus for repetitiveness)

We have also discovered that there are some nuances related to how the emails have been coded: `nbsp` (for non-breaking space). It is very likely that the email parser found some kind of unstructured spaces in the text and has redacted them with the `nbsp` keyword. Coincidentally, these types of parsing nuances are much more common in spam emails than in legitimate ones, which will help us in our analysis.

Word processing

Processing words in text is typically composed of four steps:

1. Tokenizing
2. Stop-words filtering
3. Stemming
4. Lemmatization

These steps have their own individual complexity, and therefore we will use libraries available to run these steps, such as the **Natural Language Toolkit** (**NLTK**). To install it, run the following command:

```
!pip3 install nltk
```

Tokenizing is the step that processes a text and returns a list of **tokens**. Each word is a token, but if there are also punctuation marks, these would become separate tokens. Nevertheless, for our corpus, these have been removed in a previous step.

Please note that in this step, we have moved from a list of sentences and paragraphs per email, the corpus, to what is known as **bag of words** or **BOW**, which is directly connected to the vocabulary used in the corpus.

After we have each word as an entity, we can remove those common words we had already identified such as "the" or "to." These are known as stop words, and NLTK contains a set of these stop words for several languages. We will use this available set to filter our corpus.

Stemming is the process of reducing derived (and inflected, if we want to be formal) words to their root, known as the stem.

Lemmatization is the process of grouping together several different forms that can be analyzed as a single item, identified by the word's lemma, or dictionary form:

Figure 2.28 – Stemming and lemmatization

After processing our BOW through these steps, we have reduced the number of words we are working with to ~10% of the corpus:

```
Raw Corpus (Ham): 3133632
Processed Corpus (Ham): 317496 (~10%)
Raw Corpus (Spam): 1712737
Processed Corpus (Spam): 177780 (~10%)
```

Word clouds

With our postprocessed BOW, we can generate one of the most impactful and popular visualizations of a text corpus, a word cloud:

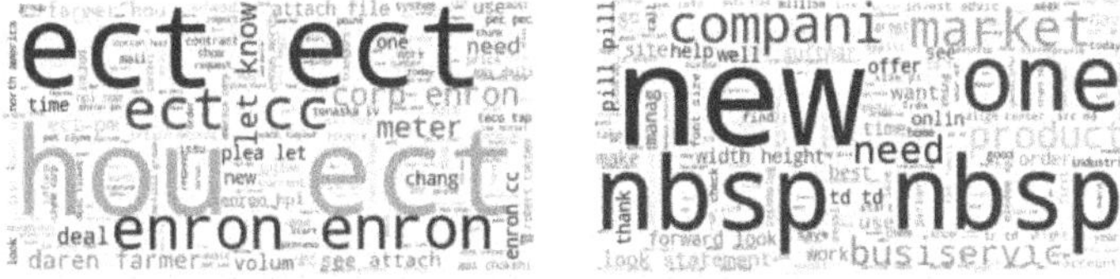

Figure 2.29 – Word clouds for (a) ham and (b) spam

In these visualizations we can clearly see how *Enron*, *please*, and *let know* are relevant for legitimate emails, whereas *new*, *nbsp*, *compani*, *market*, and *product* are typically connected to spam emails.

Word embeddings

So far, we have taken a look at how individual words behave (frequency and length) and their connections to other words in terms of the most frequent N-grams (bigrams and trigrams). However, we have not connected the meaning of the words among them. For example, one would expect that *Enron*, *corp*, and *company* (*compani*, its stemmed counterpart) are close from a semantic point of view. Therefore, we would like to have a representation where words with a similar meaning would have a similar representation. Furthermore, if that representation had a constant number of dimensions, we could easily make comparisons (find similarities) among words. These are word embeddings, and the representation is a vector.

There are infinite ways of generating vectors from words; for example, the most naive way to accomplish this is to generate as many dimensions (features of our vector, columns in a matrix representation) as our vocabulary, and then in each email (a row in the matrix representation), for each word that it contains, we can put a *1* (checkmark) in the column for that word in the vocabulary, resulting in vectors of the form [0, 0, 0, 0 ......, 1, 0, 0, 0,...., 1.....]. This representation is called one-hot encoding, but it is very inefficient as the number of features is the number of distinct words in the corpus, the length of the vocabulary, which is typically very high (~500k with our reduced vocabulary).

Therefore, we will take a look at more optimal ways to represent our words: word2vec and GloVe:

- **word2vec**: This algorithm was developed by Google in 2013 and is available pre-trained on the *Google News* corpus. It has a corpus of 3 billion words and a vocabulary of 3 million distinct words, each represented with 300 features. The intuition behind this algorithm is to calculate the probability of a given word by taking into account its context (surrounding words). The window size (how many words are being looked at the same time) is a parameter of the model and is a constant, which makes the model rely solely on the local context of each word.
- **GloVe**: This algorithm was developed at Stanford University in 2014 and is available pre-trained in several datasets. We will use Wikipedia articles (as of 2014) and a news dataset corpus called *Gigaword*. This dataset has a corpus of 6 billion words and a vocabulary of 400k distinct words, each represented with 50, 100, 200, or 300 features. We will use the 50-features model. This algorithm merges the locality of the approach from Google in `word2vec` combined with word co-occurrence (global statistics) to provide a more complete representation.

With these representations, we can now compute operations in words in their new vector representations:

- **Example 1**: `stronger` is to `strong` what `weaker` is to `weak`:

```
math_weaker = w2v["stronger"] - w2v["strong"] + w2v["weak"]
np.linalg.norm(math_weaker - w2v["weaker"])
```

  This generates an output of ~`1.9`, which is close.

- **Example 2**: Most similar words to `king`:

```
[('kings', 0.7138046026229858), ('queen', 0.6510956883430481),
('monarch', 0.6413194537162781), ('crown_prince',
0.6204220056533813), ('prince', 0.6159993410110474), ('sultan',
0.5864822864532471), ('ruler', 0.5797567367553711), ('princes',
0.5646552443504333), ('Prince_Paras', 0.543294370174408),
('throne', 0.5422104597091675)]
```

The observant reader will have realized that word embeddings are similar to the techniques we saw in the previous recipe for dimensionality reduction, as those were learned representations as well. However, in this case, we are actually increasing the dimensionality to obtain new advantages (constant number of features and similar meaning representation).

Important note

Word embeddings are typically *learned* representations; that is, the representations are trained to minimize the distance among words that have a similar meaning or, in our case, are classified with the same label. In this recipe, we will use pre-trained representations for `word2vec` and GloVe, and in *Chapter 5*, we will take a look at training.

PCA and t-SNE

As discussed in "the subsection", our current embeddings have either 300 features (`word2vec`) or 50 (GloVe). For proper visualizations, we need to apply dimensionality reduction techniques, as we saw in the previous recipe for CV.

For this dataset, we can apply PCA:

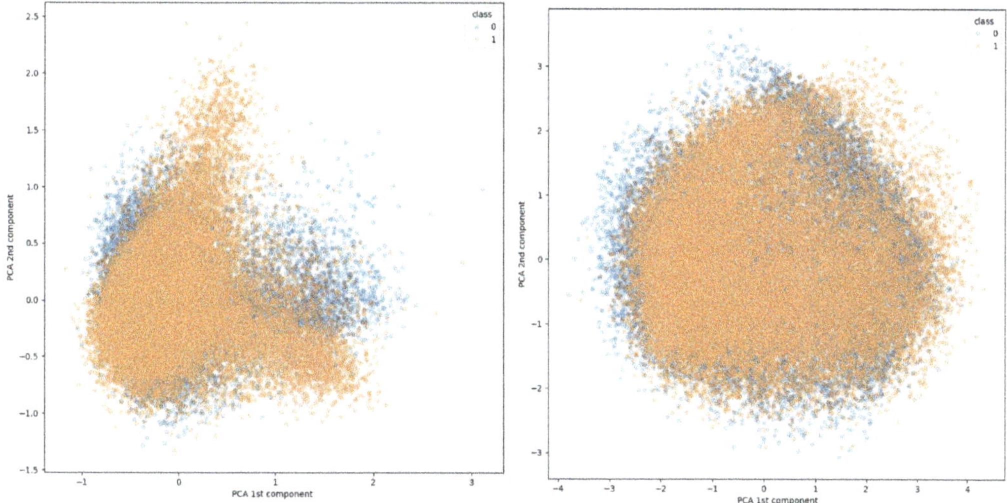

Figure 2.30 – PCA for (a) word2vec and (b) GloVe embeddings

Moreover, we can apply t-SNE as well:

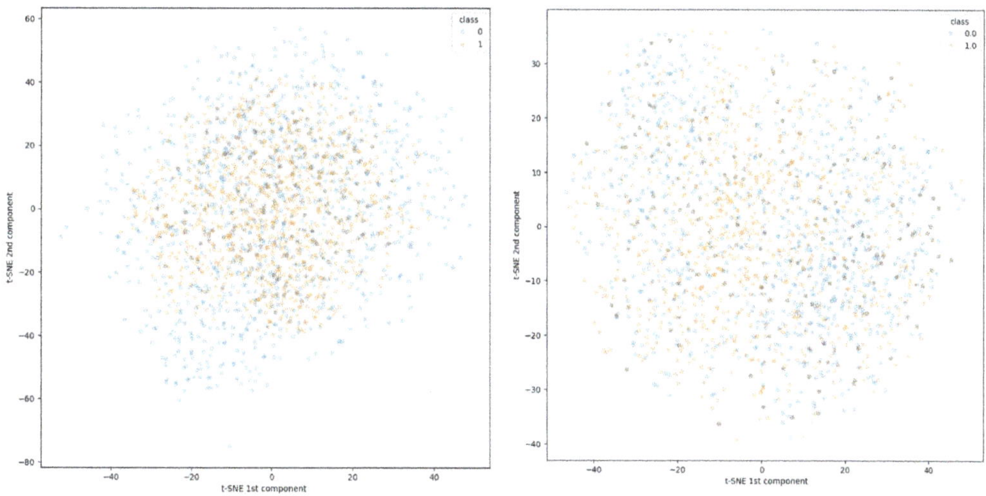

Figure 2.31 – t-SNE for (a) word2vec and (b) GloVe embeddings

From the preceding plots, we can see how the ham and spam words in our embedding space are very close to each other, making it very difficult to separate the clusters. This is due to the fact that we are using pre-trained embeddings, from the news and Wikipedia datasets. These datasets and the corresponding embeddings are not suited to our task at hand. We will see how to train word embeddings to achieve better results in *Chapter 5*.

> **Important note**
>
> For PCA and t-SNE, we can choose three components instead of two, which will yield a 3D plot. For the code, visit the GitHub repository of the book: `https://github.com/PacktPublishing/Deep-Learning-with-MXNet-Cookbook`.

How it works...

To understand the corpus of a text dataset, we need to understand the underlying connections among words in that corpus. One useful method to achieve this is with different visualizations of the corpus.

In this recipe, we have learned how to discover patterns in our text datasets. We selected an imbalanced dataset, the *Enron Email* dataset, and we learned how to deal with binary classification datasets.

We analyzed our dataset by taking a look at its internal structure and what the class imbalance looked like and checked the most common words in search of patterns and errors. We cleaned the dataset by removing punctuation marks, and we graphed the most frequent **bigrams** and **trigrams** and noticed several keywords that would help us classify our emails correctly.

We learned how to generate some cool visualizations such as **word clouds**, and we understood why **word embeddings** are important and plotted them using the **dimensionality reduction techniques** we learned previously.

There's more...

If you want to know more about the Enron Email dataset and the Enron scandal, the following links will help:

- **Enron Email dataset**: `http://www.cs.cmu.edu/~enron/`
- **Enron scandal**: `https://en.wikipedia.org/wiki/Enron_scandal`

We had a brief overview of several important concepts I invite you to learn more about:

- **BOW**: `https://machinelearningmastery.com/gentle-introduction-bag-words-model/`
- **N-grams**: `https://web.stanford.edu/~jurafsky/slp3/3.pdf`
- **Word clouds**: `https://amueller.github.io/word_cloud/`

Furthermore, we barely scratched the surface of what word embeddings can offer:

- **word2vec**: `https://code.google.com/archive/p/word2vec/`
- **GloVe**: `https://nlp.stanford.edu/projects/glove/`

3

Solving Regression Problems

In the previous chapters, we learned how to set up and run MXNet, work with Gluon and DataLoaders, and visualize datasets for regression, classification, image, and text problems. We also discussed the different learning methodologies (supervised learning, unsupervised learning, and reinforcement learning). In this chapter, we are going to focus on supervised learning, where the expected outputs are known for at least some examples. Depending on the given type of these outputs, supervised learning can be decomposed into regression and classification. Regression outputs are numbers from a continuous distribution (such as predicting the stock price of a public company), whereas classification outputs are defined from a known set (for example, identifying whether an image corresponds to a mouse, a cat, or a dog).

Classification problems can be seen as a subset of regression problems, and therefore, in this chapter, we will start working with the latter ones. We will learn why these problems are suitable for deep learning models with an overview of the equations that define these problems. We will learn how to create suitable models and how to train them, emphasizing the choice of hyperparameters. We will end each section by evaluating the models according to our data, as expected in supervised learning, and we will see the different evaluation criteria for regression problems.

We will cover the following recipes in this chapter:

- Understanding the math of regression models
- Defining loss functions and evaluation metrics for regression
- Training regression models
- Evaluating regression models

Technical requirements

Apart from the technical requirements specified in the *Preface*, the following are some of the additional requirements needed for this chapter:

- Ensure that you have completed *Recipe, Installing MXNet, Gluon, GluonCV and GluonNLP* from *Chapter 1 Up and Running with MXNet*.

- Ensure that you have completed *Recipe 1, Toy dataset for regression – load, manage, and visualize a house sales dataset* from *Chapter 2, Working with MXNet and Visualizing Datasets: Gluon and DataLoader*.

The code for this chapter can be found at the following GitHub URL: `https://github.com/PacktPublishing/Deep-Learning-with-MXNet-Cookbook/tree/main/ch03`.

Furthermore, you can access each recipe directly from Google Colab, for example, for the first recipe of this chapter: `https://colab.research.google.com/github/PacktPublishing/Deep-Learning-with-MXNet-Cookbook/blob/main/ch03/3_1_Understanding_Maths_for_Regression_Models.ipynb`.

Understanding the math of regression models

As we saw in the previous chapter, **regression** problems are a type of **supervised learning** problem whose output is a number from a continuous distribution, such as the price of a house or the predicted value of a company stock price.

The simplest model we can use for a regression problem is a **linear regression** model. However, these models are extremely powerful for simple problems, as their parameters can be trained and are very fast and explainable, given the small number of parameters involved. As we will see, this number of parameters is completely dependent on the number of features we use.

Another interesting property of linear regression models is that they can be represented by neural networks, and as neural networks will be the basis for most models that we will be using throughout the book, this is the linear regression model based on neural networks that we will be using.

The simplest neural network model is known as the **Perceptron**, and this is the building block we will be working on not only in this recipe, but throughout the whole chapter.

Getting ready

Before jumping in to understand our model, let me mention that for the math part of this recipe, we will encounter a little bit of matrix operations and linear algebra, but it will not be hard at all.

How to do it...

In this recipe, we will work through the following steps:

1. Modeling a biological neuron mathematically
2. Defining a regression model
3. Describing basic activation functions

4. Defining the features
5. Initializing the model
6. Evaluating the model

Modeling a biological neuron mathematically

The Perceptron was first introduced by American psychologist Frank Rosenblatt in 1958 at Cornell Aeronautical Laboratory, and it was an initial attempt at replicating how information is processed by the neurons in our brain.

Rosenblatt analyzed a biological neuron and developed a mathematical model that behaved similarly. To make a comparison between these architectures, we will start with a very simple model of a neuron.

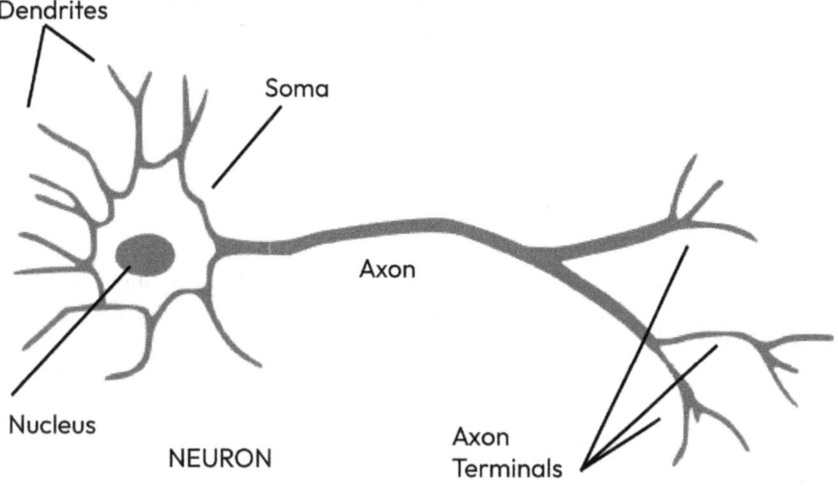

Figure 3.1 – Biological neuron

As we can see in *Figure 3.1*, the neuron is composed of three main parts:

- **Dendrites**: Where the neuron receives inputs from other neurons. Depending on how strong the connection is, an input will be increased or decreased in the dendrites.
- **Cell Body** or **Soma**: Contains the nucleus, which is the structure that receives all the inputs from the dendrites and processes them. The nucleus might trigger an electrical message to be communicated to other neurons.
- **Axon/Axon Terminals**: This is the output structure, which communicates messages with other neurons.

Rosenblatt took the preceding simplified model of the neuron and assigned to it certain mathematical properties:

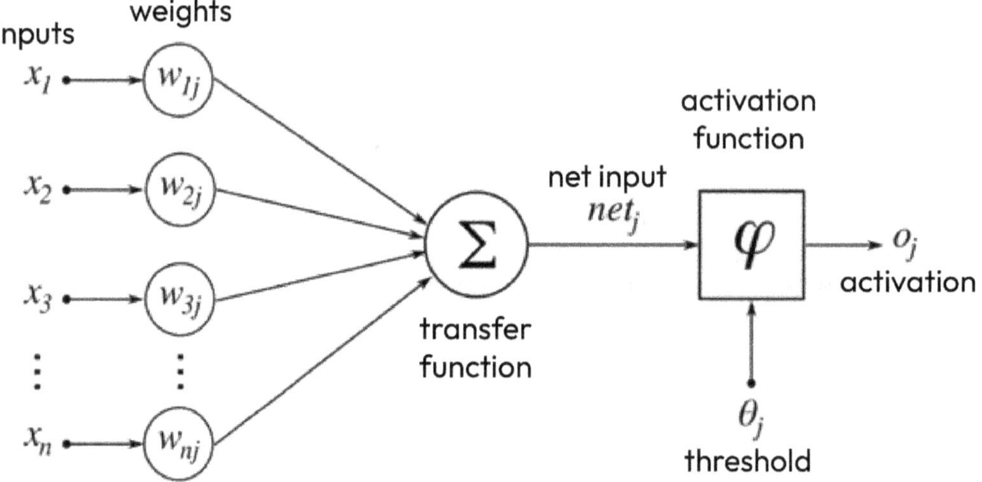

Figure 3.2 – Perceptron

- **Weights**: This will simulate the behavior of dendrites by multiplying the inputs or features with a set of weights (W in *Figure 3.2*, while j refers to any neuron in the model).
- **Sum** and **bias**: The combination of input signals done in the nucleus will be modeled as a sum with a bias (θ in *Figure 3.2*) and a processing function, called the **activation function**. (We will describe these functions in the following step.)
- **Output**: Either connections to other neurons, or the direct output of the whole model (o in *Figure 3.2*).

Comparing *Figure 3.1* and *Figure 3.2*, we can see the similarities between the simplified model of the biological neuron and the Perceptron. Furthermore, we can also see how all these parts are connected together, from processing the input to delivering the output.

Defining a regression model

Therefore, from a mathematical perspective using matrix multiplication, we can write the following equations for the model $y = f(W \cdot X + b)$, where W is the weight vector $[W_1, W_2, ... W_n]$, (n is the number of features), X is the feature vector $[X_1, X_2, ... X_n]$, b is the bias term, and $f()$ is the activation function. For the regression case we will be dealing with, we will work with a linear activation function, where the output is equal to the input.

Therefore, in our case, the activation function is the identity function (output equal to the input), we have $y = W \cdot X + b$.

We can easily achieve this with MXNet and its NDArray library:

```
# Perceptron Model
def perceptron(weights, bias, features):
return mx.nd.dot(features, weights) + bias
```

And that's it! That's our *y* neural network output in terms of the *X* inputs and the *W* and *b* parameters of the model.

> **Important Note**
>
> In the original paper by Rosenblatt, the expected output was either 0 or 1 (classification problem) and to fulfill this requirement, the activation function was defined as the step function (0 if the input is smaller than 0, or 1 if the input is larger than or equal to 0). This was one of the strongest limitations to Rosenblatt's neuron model, and different activation functions were proposed later that improve the behavior of the model.

In deep learning networks, we do not use a single neurons (Perceptrons) as our model. Typically, several layers of Perceptrons are stacked together, and the number of layers is also known as the **depth** of the network.

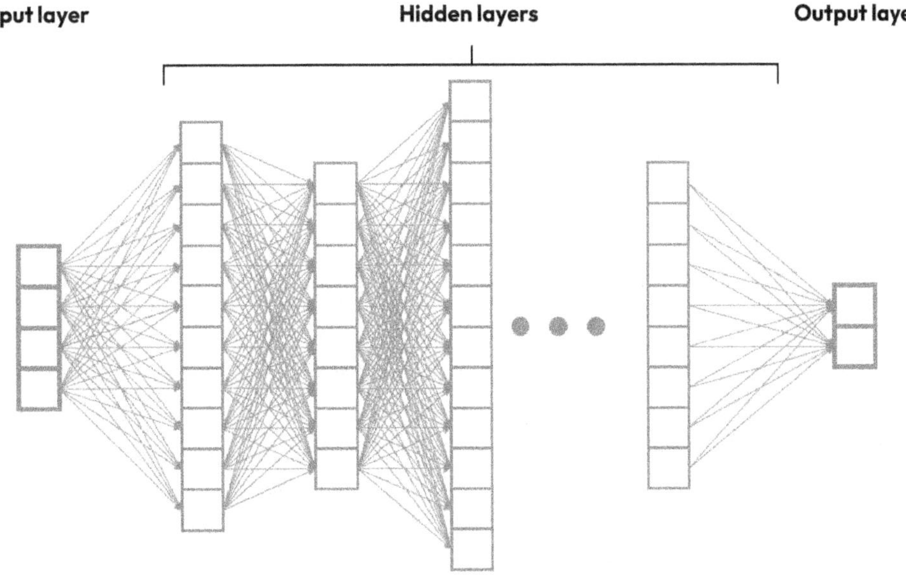

Figure 3.3 – Deep learning network

These networks are quite powerful and have been proven to match or surpass human-level performance in several fields, including image recognition in computer vision and sentiment analysis in natural language processing.

Describing basic activation functions

The most common activation functions for regression problems are the linear activation function and the **ReLU** activation function. Let's briefly describe them.

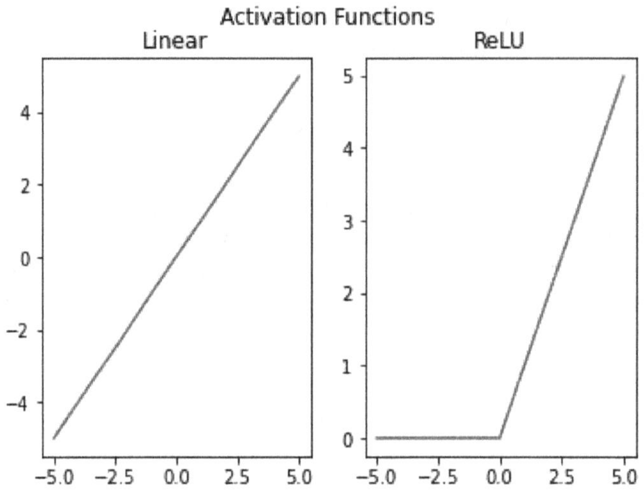

Figure 3.4 – Activation functions for regression

Linear Activation Function

In this function, the output is equal to the input. It is not bounded, and therefore it is suitable for unbounded numerical outputs, as is the case with the output for the regression problems.

ReLU

The **Rectified Linear Unit** activation function is very similar to the linear activation function: its output is equal to the input, but in this case, only when the input is larger than 0; the output is 0 otherwise. This function is suitable to only pass positive information to the next layers (sparse activation), and also provides better gradient propagation. Therefore its use is very common in intermediate layers on deep learning networks.

> **Important Note**
>
> As we will see in the next recipes, training involves computing iteratively new gradients and using these computations to update the model parameters. When using activation functions such as the sigmoid, the larger the number of layers, the smaller the gradient becomes. This problem is known as the vanishing gradient problem, and the ReLU activation function behaves better in those scenarios.

Defining the features

So far, we have defined our model and its behavior theoretically, but we have not used our problem framing or our dataset to define it. In this section, we will start working at a more practical level.

The next step in defining our model is to decide which features (inputs) we are going to work with. We will continue using the House Sales Dataset we met in *Recipe 1, Toy dataset for regression – load, manage, and visualize house sales dataset* in *Chapter 2, Working with MXNet and Visualizing Datasets: Gluon and DataLoader*. This dataset contained data for 21,613 houses, including prices and 19 input features. Although in our model we will work with all the input features, in the aforementioned recipe, we saw that the three non-correlated features that primarily contributed to the house price were as follows:

- Square feet of living space
- Grade
- Number of bathrooms

For an initial study, we will work with these three features. After selecting these features, if we show the first five houses, we will see the following:

	price	sqft_living	bathrooms	grade
0	221900.0	1180	1.00	7
1	538000.0	2570	2.25	7
2	180000.0	770	1.00	6
3	604000.0	1960	3.00	7
4	510000.0	1680	2.00	8

Figure 3.5 – Filtered features for house prices

One of the paths that we did not exploit when analyzing this dataset previously is that `grade` is not a continuous feature as the others are; its values come from a discrete set of values. This type of feature is called a categorical feature, and can be nominal or ordinal. A nominal feature is a class name – for example, we could have the architectural style of the house as a feature, and the values of that feature could be Victorian, Art Deco, Craftsman, and so on. Ordinal features, on the other hand, are class numbers. In our case, `grade` is an ordinal feature consisting of numerical values in an order (1 is worst, 13 is best).

In both cases, categorical features can be represented numerically in different ways that will help our models better learn the relationship between that particular feature and the output. There are several approaches to dealing with categorical features. In this example, we are going to work with one of the simplest approaches, a **one-hot encoding** scheme.

> **Tip**
> You will find more information on working with categorical data in the *There's more* section at the end of this recipe.

With a one-hot encoding scheme, each of the categories is decomposed into its own feature and will be assigned a binary value accordingly. In our case, `grade` contains integer values from 1 to 13, and therefore, we add 13 new features to our input vector. Each of these new features will have a value of 0 or 1. For example, for a house of grade 1, the feature vector looks as follows:

	price	sqft_living	bathrooms	1	3	4	5	6	7	8	9	10	11	12	13
0	221900.0	1180	1.00	0	0	0	0	0	1	0	0	0	0	0	0
1	538000.0	2570	2.25	0	0	0	0	0	1	0	0	0	0	0	0
2	180000.0	770	1.00	0	0	0	0	1	0	0	0	0	0	0	0
3	604000.0	1960	3.00	0	0	0	0	0	1	0	0	0	0	0	0
4	510000.0	1680	2.00	0	0	0	0	0	0	1	0	0	0	0	0

Figure 3.6 – One-hot encoding for grade feature

If we take a close look at *Figure 3.6*, we will see that there is no one-hot encoding for the grade column corresponding to the value 2. This is because no actual house in our dataset had that grade, and therefore, it has not been added as a feature.

Therefore, the final number of features is 14 and we have 1 output, the price of the house.

Initializing the model

Now that we have defined the input (the features) and output dimensions, we can initialize our model. We will explore this in more detail in the next recipes, but it is useful to show a glimpse of how this would look:

```
Weights:
  [[ 0.96975976]
  [-0.52853745]
  [-1.88909 ]
  [ 0.65479124]
  [-0.45481315]
  [ 0.32510808]
  [-1.3002341 ]
  [ 0.3679345 ]
```

```
 [ 1.4534262 ]
 [ 0.24154152]
 [ 0.47898006]
 [ 0.96885103]
 [-1.0218245 ]
 [-0.06812762]]
<NDArray 14x1 @cpu(0)>
Bias:
[-0.31868345]
<NDArray 1 @cpu(0)>
```

As expected, the `Weights` vector has 14 components (the number of features) and the `Bias` vector has 1 component.

Evaluating the model

Now that our model is initialized, we can use it to estimate the price of the first house, which can be seen in *Figure 3.6* to be around 2.2 million dollars. With our current model, the estimated house price is (in $):

```
2610.2383
```

As we have the expected price, we can compute some error metrics. In this case, I have chosen the absolute error and the error relative to the actual price. These quantities can be easily computed in Python:

```
error_abs = abs(expected_output - model_output)
error_perc = error_abs / expected_output * 100
print("Absolute Error:", error_abs)
print("Relative Error (%):", error_perc)
```

The values obtained for the errors are as follows:

```
Absolute Error: 219289.76171875
Relative Error (%): 98.82368711976115
```

As you can see, 2.6k dollars is a very small price for a house of 1,180 sqft, even though it only has 1 bathroom and is of an average grade (7). This means our error metrics have very large values suggesting a ~99% error rate. This means that either we are not evaluating our model properly (in this case, we just used 1 value, we might have been unlucky) or we only used the initialized parameters that did not give us an accurate estimation. We need to improve our model parameters using a process called **training** to improve our evaluation metrics. We will explore these topics in detail in the next recipes.

How it works...

Regression models can be as complex as the model designer wants. They can have as many layers as necessary to model adequately the relationships between the input features and the desired output values.

In this recipe, we described how a biological neuron works inside our brain and simplified it to derive a simple mathematical model that we could use for our regression problem. In our case, we only used one layer, typically called the input layer, and we defined the weights and bias as its parameters.

Moreover, we learned how to initialize our model, explored the effect initialization has on the weights and bias, and saw how we can use our data to evaluate the model. We will develop all these topics further in the next recipes.

There's more...

In this recipe, we briefly introduced several topics. We started by describing Rosenblatt's Perceptron. If you want to read the original paper, here is the link: `https://www.ling.upenn.edu/courses/cogs501/Rosenblatt1958.pdf`.

Although in this recipe and the following ones we are going to work with some equations, we will use libraries and code that will allow us to focus on the actual outputs and their relationships with the inputs. However, for the interested reader, here is a refresher: `https://machinelearningmastery.com/gentle-introduction-linear-algebra/`.

Moreover, we analyzed the input features in more detail, specifically working with `grade`, a categorical feature, using one-hot encoding. There are different ways to work with categorical data, which are explored at this link: `https://towardsdatascience.com/understanding-feature-engineering-part-2-categorical-data-f54324193e63`.

For more information regarding initialization and evaluation, please continue reading the recipes that follow in this chapter.

Defining loss functions and evaluation metrics for regression

In the previous recipe, we defined our input features, described our model, and initialized it. At that point, we passed the features vector of a house to predict the price, calculated the output, and compared it against the expected output.

At the end of the previous recipe, the comparison of the expected output and the actual output of the model intuitively provided us with an idea of how good our model was. This is what it means to "evaluate" our model: we assessed the model's performance. However, that evaluation is not complete for several reasons, as we did not correctly take into account several factors:

- We only evaluated the model on one house – what about the others? How can we take all houses into account in our evaluation?
- Is the difference between values an accurate measurement of model error? What other operations make sense?

In this recipe, we will cover how we can assess (that is, evaluate) the performance of our models, and study functions that are suitable for this matter.

Furthermore, we will introduce a concept that will be very important when we optimize (i.e., train) our models: loss functions.

Getting ready

Before defining some useful functions for evaluating regression models and computing their losses, let's specify two required and three desirable properties of the functions that we'll use to evaluate our model:

- [Required] **Continuous**: Obviously, we would like our evaluation functions to not be undefined on some potential error value, which will allow us to use these functions on a large set of pairs (expected output, actual model output).
- [Required] **Symmetric**: This is easily explained with an example. Let's say that the price of a house is 2.2 million dollars – we would like our evaluation function to evaluate the model in the same way whether the output was 2 million dollars or 2.4 million dollars, as both values are the same distance from the expected value, just in different directions.
- [Desirable] **Robust**: Again, this is easier to explain with an example. Taking the same example as in the previous point, imagine we have 2 outputs of 2.4 and 2.8 million dollars. The error is already going to be large when compared with the expected output of 2.2 million dollars, so we would not want to make the error even larger due to the loss/evaluation function. From a mathematical perspective, we don't want the error to grow exponentially, or it could make the computations diverge to **Not a Number (NaN)**. With robust functions, large errors do not make computations diverge.
- [Desirable] **Differentiable**: This is the least intuitive of all the properties. Typically, we would aim to achieve error rates of as close as possible to zero. However, that is a theoretical scenario that only happens when we have enough data to describe a problem perfectly, and when the model is large enough that it can represent the mapping from the data to the output values. In reality, we can never be sure that we are complying with neither of these previous assumptions, and therefore, the unrealistic expectation of zero error evolves to become the minimum error possible for our data and our model. We can only detect the minimum values of a function by calculating its differential function, hence this **differentiable** property. A small stroke of luck, though, is that differentiability implies continuity, therefore if our function can satisfy property #4, it automatically satisfies property #1.

- [Desirable] **Simple**: The simpler the function that satisfies all the properties, the better, because that way we can understand the results more intuitively and it will not be costly, computationally speaking.

> **Tip**
> Not only evaluation functions must satisfy these criteria. As we will see in the next recipe, they must also be satisfied by a very important function for training, the *loss function*.

How to do it...

Let's discuss some evaluation and loss functions and analyze their advantages and disadvantages. The functions we will describe are the following:

1. Mean absolute error
2. Mean squared error
3. Smooth L1 loss

Mean absolute error

The first function we are going to study is *almost* perfect according to the five properties described earlier. The intuitive idea of this function is to use the difference between values as an indicator of the distance or error between those values. We apply the `abs` function, meaning absolute value, to make it symmetrical:

$$MAE = \frac{1}{n}\sum_{j=1}^{n}\left|y_j - \hat{y}_j\right|$$

When plotted, the function produces the following graph:

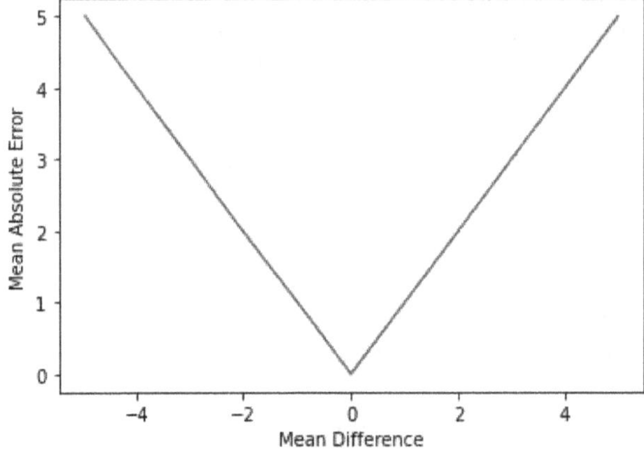

Figure 3.7 – MAE graph

If we analyze this function according to the properties defined previously, we see that all properties except #4 are fulfilled; unfortunately, the function is not differentiable at point 0. As we will see in the next recipe, this is particularly challenging when this function is used as a loss function. However, when evaluating our models, this is not required as the evaluation process does not need differentiation, and **Mean Absolute Error** (**MAE**) is considered a typical regression metric in evaluation.

> **Important Note**
>
> This function can be used for evaluation and can also be used as a loss function (by itself or as a regularization term). In this case, it is common to call it L1 loss or term explain how this relates to the rest of the sentence, as it computes the L1 distance of the vectors corresponding to the expected output and the actual output.

Another similar metric is the **Mean Absolute Percentage Error** (**MAPE**). In this metric, each output error is normalized by the expected output:

$$MAPE = \frac{100\%}{n} \sum_{i=1}^{n} \left| \frac{y_i - \hat{y}_i}{y_i} \right|$$

Mean squared error

To solve the differentiability issue, the **Mean Squared Error** (**MSE**) function is very similar to the MAE, but increases the order from 1 to 2 in the difference term. The intuitive idea is to use the simplest quadratic differentiable function (x^2):

$$MSE = \frac{1}{n} \sum_{i=1}^{n} \left(Y_i - \hat{Y} \right)^2$$

When plotted, the function produces the following graph:

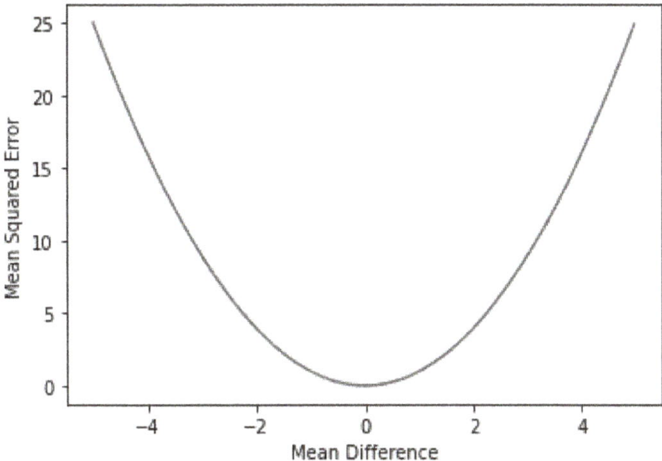

Figure 3.8 – MSE graph

If we analyze this function according to the properties defined, we see that all properties except #3 are fulfilled. Unfortunately, the function is not as robust as the MAE. Large errors grow exponentially, and therefore, this evaluation function is much more susceptible to outliers, as one single data point with a very large error can cause the squared error for that value to be quite large, and therefore it will make a large contribution to the MSE, leading to erroneous conclusions.

> **Important Note**
>
> This function can be used for loss functions (by itself or as a regularization term). In this case. It is common to call it L2 loss or term (ridge regression), as it computes the L2 distance of the vectors corresponding to the expected output and the actual output.

In order to have the same units as the output variable, the MSE can have the squared root applied. This evaluation metric is called **Root Mean Squared Error (RMSE)**:

$$RMSE = \sqrt{\sum_{i=1}^{n} \frac{(\hat{y}_i - y_i)^2}{n}}$$

Smooth mean absolute error/smooth L1 loss

Can't we have the best of both worlds? Of course we can!

By combining both functions – the MSE for small values of the error and the MAE for the large values of the error – we get the following:

$$smooth_{L_1}(x) = \begin{cases} 0.5x^2 & if\,|x| < 1 \\ |x| - 0.5 & otherwise \end{cases}$$

When plotted, the function produces the following graph:

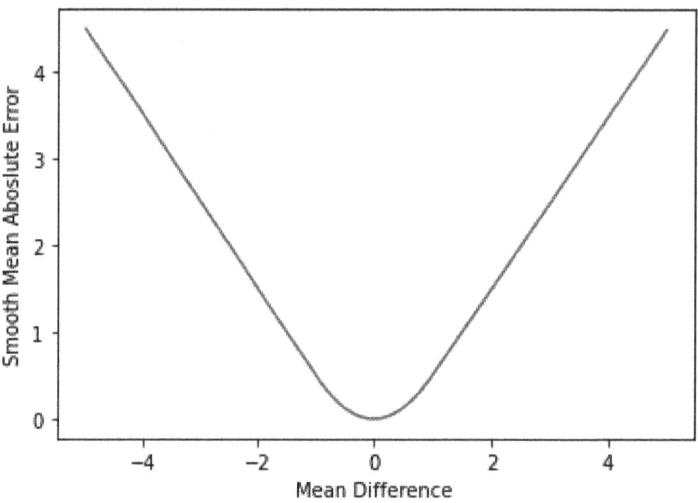

Figure 3.9 – Smooth mean absolute error graph

If we analyze this function according to the properties defined, we see that all properties are fulfilled.

> **Important Note**
> This function can be used for loss functions. In this case, it is common to call it smooth L1 loss.

How it works...

In the first recipe of this chapter, we designed our first regression model based on a simple biological neuron. We initialized its parameters randomly and performed our first naive evaluation. The result was not good, and we conjectured that this was due to two reasons: our evaluation mechanism was not robust enough and our model parameters had not been optimized.

In this recipe, we explored how to improve on the first reason: evaluation. We covered three of the most important evaluation metrics and mentioned their relationship with loss functions, which we will explore in detail in the next recipe.

Moreover, we discussed which evaluation metrics are better, exploring how the MAE is robust but unfortunately not differentiable, and the MSE is differentiable but allows outliers to steer the metric (which is not ideal). We concluded the recipe by combining the functions to get the best of both worlds.

There's more...

A very interesting set of evaluation functions that we did not explore in this recipe is the coefficient of determination and its extensions. However, this set is only used for linear regression modeling. More information can be found at the following links:

- **Coefficient of determination** `https://en.wikipedia.org/wiki/Coefficient_of_determination`
- **Statistics By Jim** `https://statisticsbyjim.com/regression/r-squared-invalid-nonlinear-regression/`

Furthermore, there are generally many different functions that can be used for evaluation and loss in regression problems; you can refer to this link for further details: `https://machine-learning-note.readthedocs.io/en/latest/basic/loss_functions.html`

Training regression models

In supervised learning, training is the process of optimizing the parameters of a model towards a specific objective. It is typically the most complex and the most time-consuming step in solving a deep learning problem statement.

In this recipe, we will visit the basic concepts involved in training a model. We will apply them to solve the regression model we previously defined in this chapter, combined with the usage of the functions we discussed.

We will predict house prices using the dataset seen in *Recipe 1, Toy dataset for regression – load, manage, and visualize house sales dataset* from *Chapter 2, Working with MXNet and Visualizing Datasets: Gluon and DataLoader.*

Getting ready

There are a number of concepts that we should get familiar with to understand this recipe. These concepts define how the training will proceed:

- **Loss function**: The training process is an iterative optimization process. As the training progresses, the model is expected to perform better against an operation that compares the expected output with the actual output of the model. That operation is the loss function, also known as the objective, target, or cost function, which is being optimized during the training process.
- **Optimizer**: With each iteration of the training process, each parameter of the model is updated by a quantity (calculated using the loss function). The optimizer is an algorithm that defines how that quantity is calculated. The most important hyperparameter of the optimizer is the **learning rate**, which is the multiplier applied to the calculated quantity to update the parameters.
- **Dataset split**: Stopping the training when the model can perform at its best in the real world is critical to achieving success in deep learning projects. One method to accomplish this is to split our dataset into training, validation, and test sets.
- **Epochs**: This is the number of iterations for which the training process will run.
- **Batch size**: The number of training samples analyzed at a time to produce an estimation of the gradient.

How to do it...

In this recipe, we will create our own training loop and evaluate how each hyperparameter influences the training. To achieve this, we will follow these steps:

1. Improving the model
2. Defining the loss function and optimizer
3. Splitting our dataset
4. Analyzing fairness and diversity
5. Defining the number of epochs and the batch size
6. Putting everything together for a training loop

Improving the model

To solve this problem, the architecture we explored in the previous recipes (Perceptron) will not be enough. We will stack several Perceptrons together and connect them through different layers. This architecture is known as a **Multi-Layer Perceptron** (**MLP**). We will define a network architecture with three hidden layers that are fully connected (dense) and use the **ReLU** activation function (introduced in the first recipe of the chapter) with 128, 1,024, and 128 neurons, respectively, and an output layer with the corresponding 1 single output. The last layer is left without an activation function, also called the linear activation function ($y = x$).

Furthermore, the house sales dataset is a very complex problem, and finding solutions that generalize well (i.e., work sufficiently well in the real world) isn't easy. For this purpose, we have included two new advanced features in the model:

- **Batch normalization**: With this step in the process, for each mini-batch, the input distribution is standardized. This helps with training convergence and generalization.
- **Dropout**: This method consists of disabling neurons of the network randomly (given a probability). This helps reduce overfitting (this concept will be explained in the next recipe) and improve generalization.

Our code is as follows:

```
def create_regression_network():
    # MultiLayer Perceptron Model (this time using Gluon)
    net = mx.gluon.nn.Sequential()
    net.add(mx.gluon.nn.Dense(128))
    net.add(mx.gluon.nn.BatchNorm(axis=1, center=True, scale=True))
    net.add(mx.gluon.nn.Activation('relu'))
    net.add(mx.gluon.nn.Dropout(.5))
    net.add(mx.gluon.nn.Dense(1024))
    net.add(mx.gluon.nn.BatchNorm(axis=1, center=True, scale=True))
    net.add(mx.gluon.nn.Activation('relu'))
    net.add(mx.gluon.nn.Dropout(.4))
    net.add(mx.gluon.nn.Dense(128))
    net.add(mx.gluon.nn.BatchNorm(axis=1, center=True, scale=True))
    net.add(mx.gluon.nn.Activation('relu'))
    net.add(mx.gluon.nn.Dropout(.3))
    net.add(mx.gluon.nn.Dense(1))
     return net
```

One important thing to note is that the input to our model has also been modified:

- The numerical inputs have been scaled to produce inputs with zero mean and unit variance. This improves the convergence of the training algorithm.
- The categorial input (grade) has been one-hot encoded. We introduced this concept in *Recipe 4, Toy dataset for text tasks – load, manage, and visualize enron emails dataset* from *Chapter 2, Working with MXNet and Visualizing Datasets: Gluon and DataLoader*

This increases the number of features to 30. As the dataset contains ~20k rows, this provides around 600k data points. Let's compare this to the number of parameters in our model:

```
[...]
Parameters in forward computation graph, duplicate included
    Total params: 272513
    Trainable params: 269953
    Non-trainable params: 2560
Shared params in forward computation graph: 0
Unique parameters in model: 272513
```

The number of trainable parameters in our model is ~270k. The number of data points is approximately two times the number of trainable parameters in our model. Typically, this is the minimum for a successful model and ideally, we would like to work with a dataset size of ~10 times the number of the parameters of the model.

> **Tip**
>
> Even though the comparison between the data points available and the number of parameters of the model is a very useful one, different architectures have different requirements in terms of data. As usual, experimentation (trial and error) is the key to finding the right balance.

The last important point about our model is the initialization method, as with multilayer networks, random initialization might not yield the best results. The most common methods nowadays are the following:

- **Xavier initialization**: Takes into account the number of inputs and the number of outputs when computing the variance
- **MSRA PReLU** or **Kaiming initialization**: The Xavier initialization method has some issues with ReLU activation functions, so this method is preferred

MXNet provides very simple access to these functionalities, in this case, for MSRA PReLU initialization:

```
net.collect_params().initialize(mx.init.MSRAPrelu(), ctx=ctx, force_
reinit=True)
```

> **Important Note**
>
> Initialization methods provide values for the weights and bias so that the model activation functions are not initialized on saturated (flat) regions. The intuition is to have these weights and biases with zero mean and unit variance. For the statistical analysis, a link is provided in *There's more....*

Defining the loss function and optimizer

As we saw in the previous recipe, the smooth L1 (also known as Huber) loss function will work quite well.

There are several optimizers that have been proven to work well for **supervised learning** problems:

- **Gradient descent**: The intuitive idea behind this method (and all optimizers) is that the gradient (the directional derivative toward the coordinate axes) provides an indicator of how to modify the model parameters to minimize the loss. To calculate it, this method computes all the data in the dataset, and therefore it is very robust, but also time-consuming. If you only take one random data point for the calculation, it is called **Stochastic Gradient Descent**, which is much faster, but also very noisy. To get the best of both worlds, we can use **Mini-Batch Gradient Descent**, where we define the size of a batch and calculate the gradients of that batch. This is connected to the `batch_size` parameter when we worked with **DataLoader** in *Chapter 2, Working with MXNet and Visualizing Datasets: Gluon and DataLoader.*
- **Momentum/Nesterov accelerated gradient**: Gradient descent can have a problem with stability and can start jumping and getting trapped in local minima. One method to avoid these issues is to consider the past steps that the algorithm has taken, which is achieved with these two optimizers.
- **Adagrad/Adadelta/RMSprop**: GD uses the same learning rate for all parameters, without taking into account the frequency with which they are updated. Adagrad and these other optimizers deal with this issue by adjusting the learning rate by parameter. However, Adagrad learning rates decrease over time and may yield values close to zero that do not perform any further updates. To solve this problem, Adadelta and RMSprop were developed.
- **Adam/AdaMax/Nadam**: These state-of-the-art optimizers combine both of the improvements from gradient descent: past-step calculations and adaptive learning rates. Adam uses the L2 norm for the exponentially weighted average of the gradients, whereas AdaMax uses the infinity norm (max operation). Nadam replaces the momentum component in Adam with the Nesterov momentum to accelerate convergence.

MXNet and Gluon provides very simple interfaces to define the loss function and the optimizer. With the following two lines of code, we are choosing the Huber loss function and the Adam optimizer:

```
# Define Loss Function
loss_fn = mx.gluon.loss.HuberLoss()
# Define Optimizer and Hyper Parameters
```

```
trainer = mx.gluon.Trainer(net.collect_params(), "adam", {"learning_
rate": 0.01})
```

Splitting our dataset

One of the most important things to consider in all data science projects is the performance of a trained model on data outside the dataset we are going to work with. In supervised learning, for training and evaluation, we work with data knowing the desired (expected) output, so how can we make sure that when using our model on new data, without a known output, it is going to perform as expected?

We deal with this issue by splitting our dataset into three parts:

1. **Training set**: The training set is used during training to compute the updates of the model parameters.
2. **Validation set**: The validation set is used during training to check every epoch how the model has improved (or not) with those updates previously calculated.
3. **Test set**: Finally, once the training has finished, we can compute its performance on *unseen data*, which is the test set, the only part of the dataset that was not used to improve the model during training.

Furthermore, in order to have stable training that will allow our model to work properly with data outside our dataset, the data split needs to be computed taking into account the following:

- **Size of the splits**: This depends on the amount of data available and the task. Typical percentage splits for training/validation/test data are 60/20/20 or 80/10/10.
- **Which data points to select for each of the splits**: The key here is to have a balanced dataset. For example, in our house prices dataset, we do not want to have only houses with two and three bedrooms on the training set, then four-bedroom houses in the validation set, and finally houses with five or more bedrooms in the test set. Ideally, each set should be an accurate representation of the dataset. This is very important for sensitive datasets where fairness and diversity must be considered.

We can achieve these splits easily, in this case, using a function from the well-known library **scikit-learn**:

```
# Dataset Split 80/10/10
from sklearn.model_selection import train_test_split
full_train_df, test_df = train_test_split(house_df, test_size=0.2,
random_state=42)
# To match correctly 10% size, we use previous size as reference
train_df, val_df = train_test_split(full_train_df, test_size=len(test_
df), random_state=42)
```

In the preceding code snippet, we do our three-way split for train, validation, and test data in two steps:

- We assign 20% of the dataset to the test set.
- The remaining 80% will be split equally between the validation and test sets.

Analyzing fairness and diversity

Imagine for a moment that we work for a real-estate website, and we manage the data science team. There is a compelling feature that will drive traffic to our website: when homeowners want to sell a property, they can fill in some data about their house and will be able to see a machine learning-optimized estimate of the price at which they should put their house up for selling, with data indicating that it will be sold within the next 3 months at that price. This feature sounds really cool as homeowners can fine-tune the asking price of the house they want to sell, and potential buyers will see reasonable prices according to market.

However, we suddenly realize we do not have enough data yet for houses with two or fewer bathrooms, and we know that feature is highly sensitive for our model. Deploying this model for real-life properties would mean that houses with two or fewer bathrooms could be valued closer to valuations of houses with more bathrooms by our model, simply because that is all the data our model has been able to see. This would mean that for the cheapest houses, the most affordable ones for low-income families, we will be increasing their prices *unfairly*, which would be a serious problem.

Our model cannot know better because we have not shown it better. In this scenario, what options do we have? These are the ones that might suit a real-world situation:

- Be confident about the robustness of our model and deploy it in production regardless.
- Convince business leaders not to deploy the model until we have all the data we need.
- Deploy the model in production, but only allow sellers to use it for houses with three bathrooms or more.

The first option is the least adequate of all, however, it is actually the most frequently applied one, due to the following:

- It is inconvenient to work on a project for several months and delay it right before it is expected to launch. Management typically does not expect nor want this kind of news and it could put some jobs at risk.
- However, the most common reason for this to happen is that the error in the data goes unnoticed. There is never enough time to verify that the data is accurate/fair/diverse enough and the focus shifts to delivering an optimized model as soon as possible.

The second option is difficult to argue for a similar reason as the first option, and the third option might look good on paper, but is actually very dangerous. If I found myself in that situation, I would not choose the third option, simply because we cannot be sure that the data is diverse and fair across all features, so a proper data quality assessment is required. If we found this kind of error this late in a project, it is because not enough focus was put on data quality. This typically happens with companies that have been recording or storing large amounts of data and now want to do some machine learning project with it, instead of designing data collection operations with a clear objective. This is one of the most common reasons why machine learning projects fail in such companies.

Let us take a look at how our dataset looks from the point of view of fairness and diversity:

First, as seen in *Recipe 1, Toy dataset for regression – load, manage and visualize house sales dataset* from *Chapter 2, Working with MXNet and Visualizing Datasets: Gluon and DataLoader*, we will start with the price distribution:

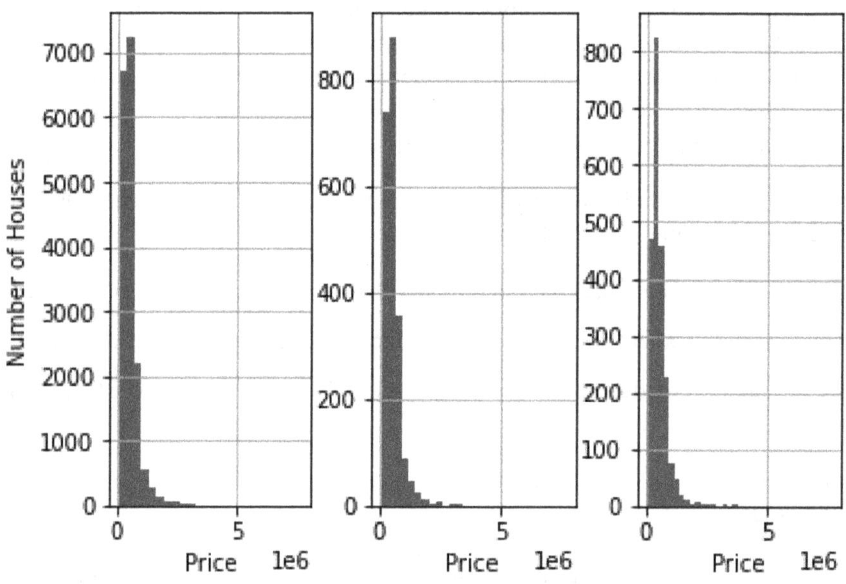

Figure 3.10 – Price distribution across the training, validation, and test sets

Although we can see a small dip in houses with prices lower than $500k, the price distribution is fairly represented across the three datasets and no manual modifications are necessary.

The living space in square foot looks as follows:

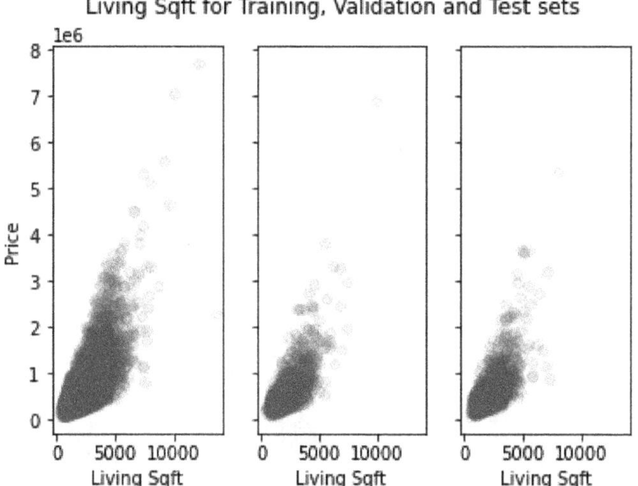

Figure 3.11 – Living Sqft plots for the training, validation, and test sets

The largest differences we can see here are due to a very small number of high-valued houses. We can even consider these outliers, and if the parameters of our training are well chosen, this should not harm our prediction capabilities.

The number of bathrooms looks as follows:

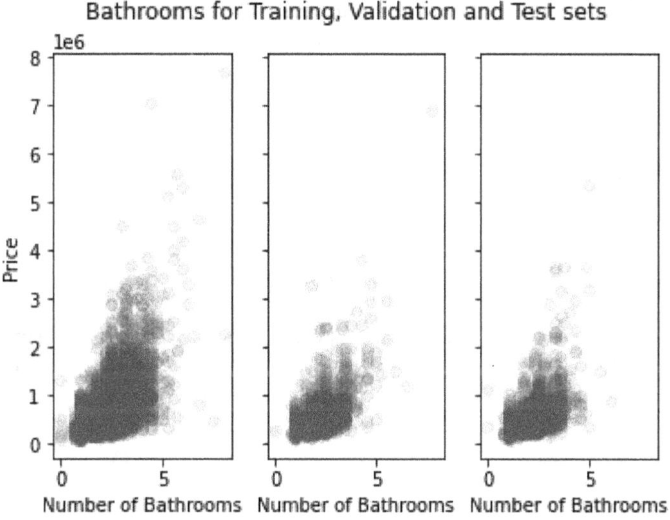

Figure 3.12 – Number of bathrooms across the training, validation, and test sets

This distribution is quite well represented in our validation and test sets.

Grade looks as follows:

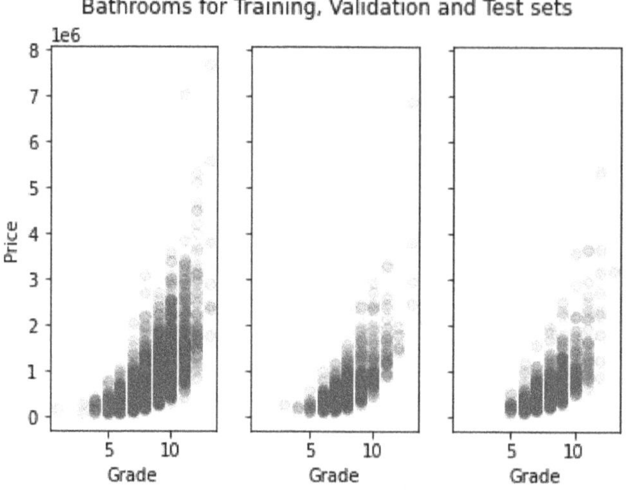

Figure 3.13 – Grade for training, validation, and test sets

This distribution is also well represented in our validation and test sets.

Grouping all the individual analyses, we can conclude that our training, validation, and test sets are fairly good representations of our full dataset. We must remember though that our dataset has its own limitations (although I have to say, for the selected features, it is quite well represented):

- Price: [75k$, 7.7M$]
- Living Sqft: [290, 13540]
- Number of Bathrooms [0, 8]
- Grade: [1, 13], but lacking 2 as we pointed out earlier

Defining the number of epochs and batch size

Number of epochs refers to the number of iterations the training algorithm will run. Depending on the complexity of the problem and the optimizer and hyperparameters chosen, this number can vary from very low (say 5-10) to very high (thousands of iterations).

Batch size refers to the number of training samples analyzed at the same time to estimate the error gradient. In *Recipe 3, Toy dataset for image tasks – load, manage and visualize iris dataset* in *Chapter 2, Working with MXNet and Visualizing Datasets: Gluon and DataLoader*, we introduced this concept as a means to optimize memory usage; the smaller the batch size, the less memory required. Furthermore, this speeds up computation of the gradient; the larger the batch size, the faster computations will run (if memory permits). Typical values range from 32 to 2,048 samples.

Putting everything together for a training loop

The training loop is the iterative process that runs the optimizer to calculate/estimate the gradient so that on each iteration, the error computed from the loss function (the objective of the optimizer) is reduced. As mentioned, each iteration is called an epoch. And for each iteration, the full training set is accessed in batches to compute the gradient.

Furthermore, as we will see, it is interesting to compute the loss function for the validation set. In our case, we will also compute the loss function for the test set, as it will provide us with specific details on how our model will perform.

In order to understand the difference in behavior when modifying the hyperparameters, we are going to run the training loop for our house price prediction dataset several times, modifying just one hyperparameter per table and keeping the rest of the variables constant (unless otherwise noted).

Optimizer and learning rate

As discussed earlier, the chosen optimizer for the training loop and the learning rate are intimately related, as for some optimizers (such as SGD) the learning rate is kept constant, whereas for others (such as Adam) it varies from a given starting point.

> Tip
>
> The best optimizer depends on several factors, and nothing trumps trial and error. I strongly suggest to try a few and see which one fits best. In my experience, SGD and Adam are typically the ones that work best, including this problem, the prediction of house prices.

Let's analyze how the training loss and validation loss vary for the SGD optimizer by varying the **learning rate** (**LR**) while keeping the other parameters constant: *Epochs = 100, Batch Size = 128, Loss fn = HuberLoss*:

Optimizer	LR	Training Loss/Validation Loss (last epoch)/ Validation Loss (Best Model)
SGD	10^{-5}	0.30/0.31/0.31
SGD	10^{-4}	0.19/0.20/0.19
SGD	10^{-3}	0.14/0.14/0.13
SGD	10^{-2}	0.10/0.12/0.11
SGD	10^{-1}	0.06/0.10/0.09
SGD	1.0	0.07/0.21/0.09
SGD	1.5	0.14/0.16/0.10
SGD	2.0	NaN/NaN/NaN (algorithm diverges)

Figure 3.14 – Loss for the SGD optimizer when varying the learning rate

From *Figure 3.14*, we can conclude that for the SGD optimizer, an LR value of between 10^{-1} and 1.0 is optimal. Furthermore, we can see that for very large values of LR (> 2.0) the algorithm diverges. That's why, when searching for optimal values for the LR, it's better to start small.

Let's analyze how the training loss and validation loss vary for the Adam optimizer by varying the **learning rate** (**LR**) and keeping the other parameters constant: *Epochs = 100, Batch Size = 128, Loss fn = HuberLoss*:

Optimizer	LR	Training Loss/Validation Loss (last epoch)/ Validation Loss (Best Model)
Adam	10^{-5}	0.11/0.13/0.13
Adam	10^{-4}	0.06/0.13/0.12
Adam	10^{-3}	0.05/0.10/0.08
Adam	10^{-2}	0.10/0.08/0.07
Adam	10^{-1}	0.30/0.35/0.33
Adam	1.0	0.30/0.35/0.33
Adam	1.5	0.30/0.34/0.33
Adam	2.0	0.30/0.34/0.33

Figure 3.15 – Loss for Adam optimizer when varying the learning rate

From *Figure 3.15*, we can conclude that for the Adam optimizer, an LR value of between 10^{-4} and 10^{-3} is optimal. As Adam calculates gradients differently, it is more difficult to make it diverge than SGD.

Adam requires smaller values for the learning rate because it *adapts* the value as the training process evolves.

Batch size

Let's analyze how the training loss and validation loss vary for the Adam optimizer by varying the batch size, keeping the other parameters constant: *Epochs = 100, LR = 10-2, Loss fn = HuberLoss*:

Batch Size	Training Loss/Validation Loss (last epoch)/ Validation Loss (Best Model)
16	0.20/0.18/0.14
32	0.17/0.15/0.10
64	0.12/0.08/0.07
128	0.10/0.09/0.07
256	0.10/0.12/0.08
512	0.08/0.11/0.08
1024	0.07/0.16/0.08
2048	0.07/0.15/0.11

Figure 3.16 – Loss for the Adam optimizer when varying batch size

From *Figure 3.16*, we can conclude that for the Adam optimizer, a batch size value of between 64 and 1,024 provides the best results.

Epochs

Another hyperparameter is the number of epochs, meaning the number of times the optimizer is going to process the full training set.

Let's analyze how the training loss and validation loss vary for the Adam optimizer varying the epochs, keeping the other parameters constant: *LR = 10-2, Batch Size = 128, Loss fn = HuberLoss*:

Epochs	Training Loss/Validation Loss (last epoch)/ Validation Loss (Best Model)
50	0.10/0.09/0.08
100	0.10/0.13/0.07
150	0.10/0.12/0.07
200	0.10/0.10/0.07
300	0.10/0.09/0.07
400	0.11/0.09/0.07
500	0.10/0.11/0.07
1000	0.11/0.11/0.07

Figure 3.17 – Loss for the Adam optimizer when varying the number of epochs

From *Figure 3.17*, we can conclude that around 100-200 epochs is good for our problem. With these values, it is very likely the best result will be achieved earlier than that.

How it works...

On our journey toward solving our regression problem, we learned in this recipe how to update our model hyperparameters optimally. We understood the role that each hyperparameter plays in the training loop and we performed some ablation studies for each individual hyperparameter. This helped us understand how our training and validation losses behaved when we modified each hyperparameter individually.

For our current problem and the chosen model, we verified that the best set of hyperparameters was as follows:

- Optimizer: Adam
- Learning Rate: 10^{-2}
- Batch Size: 128
- Number of epochs: 200

At the end of the training loop, these hyperparameters gave us a training loss of 0.10 and a validation loss of 0.10.

There's more...

With our model definition, we introduced three new concepts: **batch normalization**, **dropout**, and **scaling**. I find the following links useful to understand these advanced topics:

- **Introduction to batch normalization**: `https://machinelearningmastery.com/batch-normalization-for-training-of-deep-neural-networks/`
- **Batch normalization research paper**: `https://arxiv.org/abs/1502.03167`
- **Introduction to dropout**: `https://machinelearningmastery.com/dropout-for-regularizing-deep-neural-networks/`
- **Dropout research paper**: `https://jmlr.org/papers/v15/srivastava14a.html`
- **Scaling**: `https://machinelearningmastery.com/how-to-improve-neural-network-stability-and-modeling-performance-with-data-scaling/`

On the subject of initialization, this article explores the Xavier and Kaiming methods in detail (includes links to the research papers): `https://pouannes.github.io/blog/initialization/`.

In this recipe, we explored in depth how two optimizers, **SGD** and **Adam**, behave. These are two of the most important and best performant optimizers; however, there are many more, and some could work better for your specific problem.

One excellent resource to learn about which optimizers are implemented in MXNet and their characteristics is the official documentation: `https://mxnet.apache.org/versions/1.6/api/python/docs/tutorials/packages/optimizer/index.html`.

To compare the behavior and performance of each optimizer, I personally like the visualizations shown in this link (optimizers section): `https://towardsdatascience.com/on-optimization-of-deep-neural-networks-21de9e83e1`.

In this recipe, we worked with optimizers and their hyperparameters. Hyperparameter choice is a very complex problem, and it always requires a little bit of trial and error with each problem, verifying that the training loop works. A rule of thumb when selecting hyperparameters is to read research papers that tackle similar problems to yours and start with the hyperparameters proposed in those papers. You can then move from that starting point and see what works best for your particular case.

Apart from the training loss and the validation loss at the end of the training loop, we also provided a third loss value, *best validation loss*, we will explore what this value means and how it is calculated in the next recipe. This all maps to a question we have not answered properly yet: *when do I stop my training loop?* We will address this question in the next recipe.

Evaluating regression models

In the previous recipe, we learned how to choose our training hyperparameters to optimize our training. We also verified how those choices affected the training and validation losses. In this recipe, we are going to explore how those choices affect our actual evaluation in the real world. The observant reader will have noticed that we split the dataset into three different sets: training, validation, and test. However, during our training, we only used the training set and the validation set. In this recipe, we will emulate some real-world behavior of our model by running it on the unseen data, the test set.

Getting ready

When evaluating a model, we can perform qualitative evaluation and quantitative evaluation:

- **Qualitative evaluation** is the selection of one or more random (or not so random, depending on what we are looking for) samples and analyzing the result, verifying whether it matches our expectations.
- **Quantitative evaluation** deals with computing the outputs for a large number of inputs and calculating statistics about them (typically the mean), hence we will compute the MAE and MAPE.

Furthermore, we are going to take a look at how training can have a large influence on the evaluation.

How to do it...

Before jumping in to model evaluation, let's discuss how we can measure our model training performance. Therefore, the steps in this recipe are the following:

1. Measuring training performance – overfitting
2. Qualitative evaluation
3. Quantitative evaluation

Measuring training performance – overfitting

Deep learning networks are quite powerful, surpassing human-level performance on a variety of problems. However, when not kept in check, these networks can also provide incorrect and unexpected results. One of the most important and frequent errors happens when the network, using its full capability, memorizes the samples that are being shown (the training set), yielding excellent results for that data. However, in this scenario, the network has simply memorized the training samples, and when deployed in a real-world use case it is going to perform poorly. This type of error is called **overfitting**.

Thankfully, there is a very successful strategy to deal with overfitting, and we have already touched on it. It starts with splitting our full dataset into a training set and a validation set, which we did in the previous recipe.

From a theoretical point of view, training and validation losses typically have behaviors similar to that shown in the following graph:

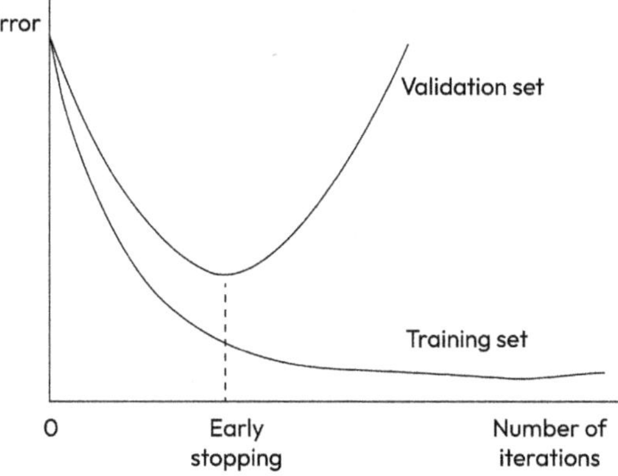

Figure 3.18 – Losses versus epochs – ideal

In *Figure 3.18*, we can see how the training and validation losses typically evolve (an idealized portrayal). The training loss continues decreasing as the training progresses, always optimizing (albeit more slowly as the number of epochs increases). The validation loss, however, reaches a point where it does not decrease further, but rather increases. At the lowest level of validation loss is where the model has reached its peak performance and where we should stop the learning process (early).

Let's examine how this kind of behavior looks in the real world. For our problem, the training loss and the validation loss evolve as the training progresses, as follows:

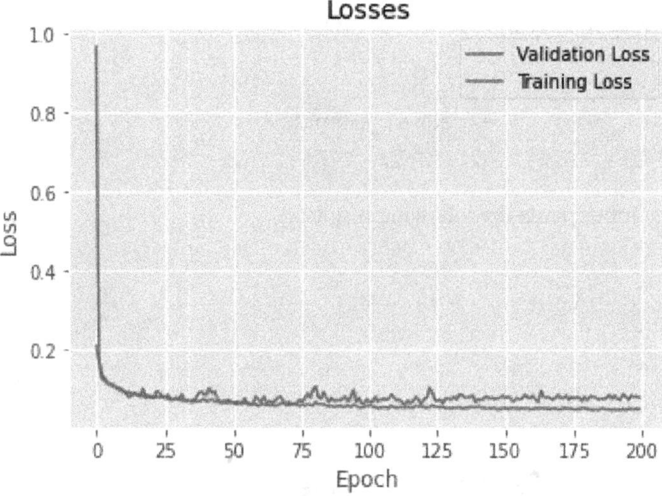

Figure 3.19 – Losses versus epochs – real

As we can see in *Figure 3.19*, the validation loss is much noisier than in the ideal world, and early stopping is much more difficult to achieve successfully. A very easy implementation is to save the model every time the validation loss decreases. This way, we are always certain that given a number of epochs the training is going to be run, the model with the best (lowest) validation loss will be saved. This was the method implemented in the previous recipe.

Qualitative evaluation

To verify our model is behaving similarly to what we expect (yielding a low error when predicting a house price), one easy approach is to run our model for a random input from the test set (unseen data). This can easily be done with the following code:

```
scaled_input = mx.nd.array([scaled_X_train_onehot_df.values[random_
index]])
# Unscaled Expected Output
expected_output = y_test[random_index]
 print("Unscaled Expected Output:", expected_output)
# Scaled Expected Output
scaled_expected_output = scaled_y_test[random_index]
 print("Scaled Expected Output:", scaled_expected_output)
# Model Output (scaled)
 output = net(scaled_input.as_in_context(ctx)).asnumpy()[0]
 print("Model Output (scaled):", output)
# Unscaled Output
unscaled_output = sc_y.inverse_transform(output)
 print("Unscaled Output:", unscaled_output)
# Absolute Error
abs_error = abs(expected_output - unscaled_output)
 print("Absolute error: ", abs_error)
# Percentage Error
perc_error = abs_error / expected_output * 100.0
print("Percentage Error: ", perc_error)
```

The preceding code snippet yields the following results:

```
Unscaled Expected Output: [380000.]
 Scaled Expected Output: [-0.4304741]
 Model Output (scaled): [-0.45450553]
 Unscaled Output: [370690.]
 Absolute error:  [9310.]
 Percentage Error:  [2.45]
```

As expected, the error rate is quite reasonable (just 2.45%!).

> **Important Note**
>
> Although I tried to keep the code as reproducible as possible, including setting the seeds for all random processes, there might be some sources of randomness. This means that your results might be different, but typically the order of magnitude of errors will be similar.

Quantitative evaluation – MAE

Let's calculate the **MAE** function, as described in the *Defining loss functions and evaluation metrics for regression* recipe earlier in this chapter:

```
Mean Absolute Error (MAE): [81103.97]
```

The MAE is $81k. Taking into account that prices varied from $75k to $7.7 million, this error seems reasonable. Do not forget that estimating house prices is a hard problem!

Quantitative evaluation – MAPE

The value provided by the MAE is good to get an idea of how small or large the errors are in our model's predictions. However, it does not provide a very meaningful merit figure, as the same MAE could have been achieved in different ways:

- *Small errors for all houses*: As houses increase in price, the absolute error number will be higher, and hence an $80k MAE might be quite good.
- *Very large errors for cheap houses*: In this case, an $80k MAE will mean that for the cheapest houses, the error might be 2-3 times, or even worse, the actual price of the house. This scenario would be very bad.

In general, we can add to the MAE another figure, similarly calculated to provide a **relative** error rate, instead of relying solely on an **absolute** value. For our model, we get the following:

```
Mean Absolute Percentage Error (MAPE): [16.008343]
```

Looks like our model is not behaving too badly, yielding a MAPE of 16%!

Quantitative evaluation – thresholds and percentage

Another question we could consider for evaluating our model is the following: *For how many houses (in %) did we accurately predict the price?*

Let's assume we consider that we accurately predicted the price of a house if the predicted price error is less than 25%. In our case, this gives us the following:

```
Houses with a predicted price error below 25.0 %: [81.23987971]
```

This calculation gives us an 81%, well done!

Furthermore, we could plot the percentage of houses we correctly predicted as a function of the error threshold:

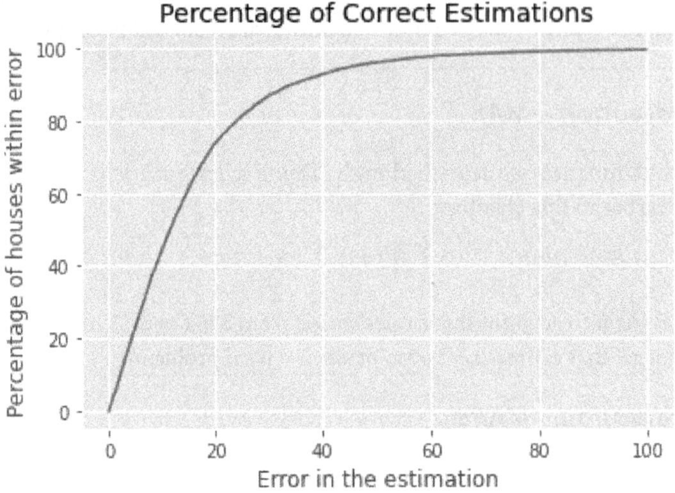

Figure 3.20 – Percentage of correct estimations

In *Figure 3.20*, we can see, as expected, that considering an error of 25% to deem a prediction accurate, our model yields 80%+ correct predictions.

How it works...

In this recipe, we explored how to evaluate our regression model. To properly do this, we revisited the decision made previously to split our full dataset into a training set, a validation set, and a test set.

During training, we used the training set to calculate the gradients to update our model parameters, and the validation set to confirm the real-world behavior. Afterward, to evaluate our model performance, we used the test set, which was the only remaining set of unseen data.

We discovered the value of describing our model behavior qualitatively by calculating the output of random samples, and of quantitatively evaluating our model performance by exploring calculations and graphs of MAE and MAPE.

We ended the recipe by defining what constitutes an accurate prediction by setting a threshold and plotting the behavior of the model by modifying the threshold.

There's more...

Deep learning has surpassed human-level performance on multiple tasks. However, evaluating models properly is paramount to verify how models will perform when deployed in production environments in the real world. I found interesting this small list of tasks where human-level performance has been reached by AI: `https://venturebeat.com/2017/12/08/6-areas-where-artificial-neural-networks-outperform-humans/`.

When evaluation is not performed properly, models may not behave as expected. Two of the most significant large-scale problems of this type (at Google in 2015 and Microsoft in 2016, respectively) are detailed in the following articles:

- **Google Mistakenly Tags Black People as 'Gorillas,' Showing Limits of Algorithms:** `https://www.wsj.com/articles/BL-DGB-42522`
- **Statistics By Jim:** `https://statisticsbyjim.com/regression/r-squared-invalid-nonlinear-regression/`

Unfortunately, although these issues are less and less frequent nowadays, they still exist. A database that contains these issues has been published and is updated whenever one of these issues is reported: `https://incidentdatabase.ai/`.

To prevent these issues, Google wrote a set of principles to develop Responsible AI. I strongly recommend all AI practitioners to abide by them: `https://ai.google/principles/`.

At this stage, we have completed our journey through a complete regression problem: we explored our regression dataset, decided on our evaluation metrics, and defined and initialized our model. We understood the best hyperparameter combination of optimizer, learning rate, batch size, and epochs, and trained it with early stopping. Lastly, we concluded by evaluating our model qualitatively and quantitatively.

4
Solving Classification Problems

In the previous chapters, we learned how to set up and run MXNet, how to work with Gluon and DataLoader, and how to visualize datasets for regression, classification, image, and text problems. We also discussed the different learning methodologies. In this chapter, we are going to focus on supervised learning with classification problems. We will learn why these problems are suitable for deep learning models with an overview of the equations that define these problems. We will learn how to create suitable models for them and how to train them, emphasizing the choice of hyperparameters. We will end each section by evaluating the models according to our data, as expected in supervised learning, and we will look at the different evaluation criteria for classification problems.

Specifically, we will cover the following recipes:

- Understanding math for classification models
- Defining loss functions and evaluation metrics for classification
- Training for classification models
- Evaluating classification models

Technical requirements

Apart from the technical requirements specified in the *Preface*, the following technical requirements apply:

- Ensure that you have completed the first recipe, *Installing MXNet, Gluon, GluonCV and GluonNLP*, from *Chapter 1, Up and Running with MXNet.*
- Ensure that you have completed the second recipe, *Toy dataset for classification – Loading, Managing, and Visualizing Iris Dataset*, from *Chapter 2, Working with MXNet and Visualizing Datasets: Gluon and DataLoader.*
- Most of the concepts for the model, the loss and evaluation functions, and the training were introduced in *Chapter 3, Solving Regression Problems*. Furthermore, as we will see in this chapter, classification can be seen as a special case of regression. Therefore, it is strongly recommended to complete *Chapter 3* first.

The code for this chapter can be found at the following GitHub URL: `https://github.com/PacktPublishing/Deep-Learning-with-MXNet-Cookbook/tree/main/ch04`.

Furthermore, you can access each recipe directly from Google Colab; for example, for the first recipe of this chapter, visit the following link: `https://colab.research.google.com/github/PacktPublishing/Deep-Learning-with-MXNet-Cookbook/blob/main/ch04/4_1_Understanding_Maths_for_Classification_Models.ipynb`.

Understanding math for classification models

As we saw in the previous chapter, **classification** problems are **supervised learning** problems whose output is a class from a set of classes (categorical assignments) – for example, the *iris* class of a flower.

As we will see throughout this recipe, classification models can be seen as individual cases of regression models. We will start by exploring a binary classification model. This is a model that will output one of two classes. We will label these classes `[0, 1]` for simplicity.

The simplest model we can use for such a binary classification problem is a **linear regression** model. This model will output a number; therefore, to modify the output to satisfy our new classification criteria, we will modify the activation function to a more suitable one.

As in the previous recipes, we will use a neural network as our model, and we will solve the iris dataset prediction problem we introduced in the second recipe, *Toy dataset for classification: Load, Manage and Visualize Iris Dataset*, in *Chapter 2, Working with MXNet and Visualizing Datasets: Gluon and DataLoader.*

Getting ready

We will be building upon the knowledge we gained in the previous recipes; therefore, it is highly recommended to read them. Moreover, as mentioned in the previous chapter, before jumping to understand our model, for the math part of this recipe, we will be using a little bit of matrix operations and linear algebra, but it will not be hard at all.

How to do it...

In this recipe, we will be looking at the following steps:

1. Defining a binary classification model
2. Defining a multi-label classification model
3. Defining the features
4. Initializing the model
5. Evaluating the model

Defining a binary classification model

This is the perceptron model we introduced in the previous recipes:

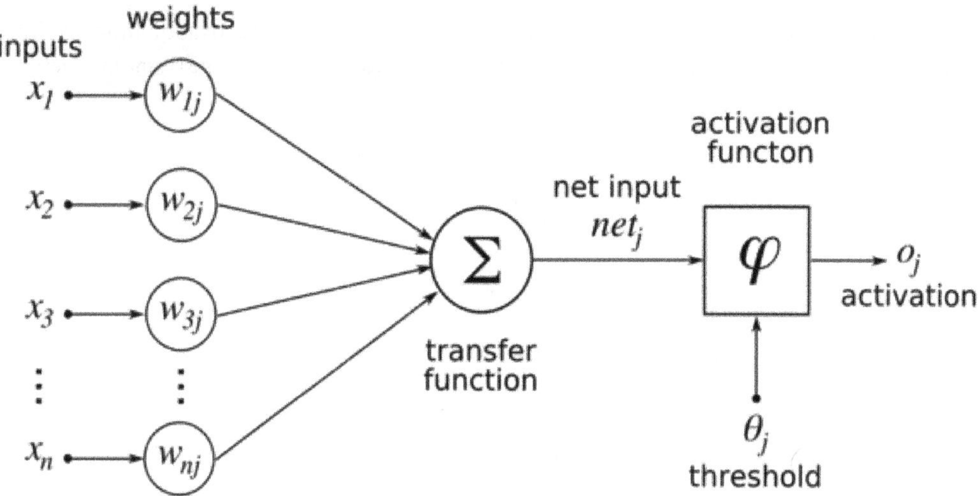

Figure 4.1 – Perceptron

This model could be mathematically described as $y = f(WX + b)$, where W is the weight vector $[W_1, W_2, ... W_n]$, (n is the number of features), X is the feature vector $[X_1, X_2, ... X_n]$, b is the bias term, and $f()$ is the activation function.

In the regression use case, we chose the activation function as the identity function, which provided an output equal to the input; therefore, we had $y = WX + b$.

For our binary classification use case, we want to have an output that will help us classify our input data points into two classes (`0` and `1`). In the original Perceptron paper in 1958, Rosenblatt, who studied a binary classification problem, chose the step function, which provided 0 and 1 as its only possible output.

If we recall from *Chapter 3, Solving Regression Problems*, in the third recipe, *Loss functions and evaluation metrics for regression*, we imposed certain properties on those functions. The fourth property, differentiability, was required due to the computations required by gradient descent. The same property applies to activation functions, and the step function Rosenblatt used does not comply with it.

Furthermore, if we could find a function that is also continuous between 0 and 1, we could assess that number as the probability or confidence of the output being `1`, as assessed by the model. This approach has advantages that we will explore later in the recipe.

Therefore, as the step function does not fulfill our properties, we need a new activation function. The most common activation function as the output of binary classification models is the **sigmoid** function:

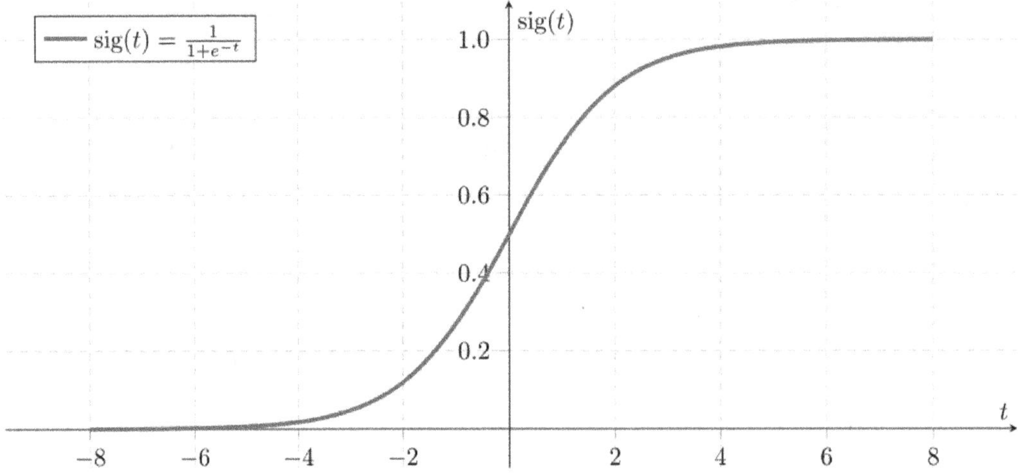

Figure 4.2 – Sigmoid activation function

The sigmoid function complies with all the required properties and the output quickly becomes `0` or `1`, which allows us to identify the output class suggested by the model.

Defining a multi-label classification model

What happens when we have multiple (let's call this number *k*) classes instead of having just two classes to classify as we just saw? In this case, we need a different network architecture for our model. On the one hand, one output will not suffice now, as we need to have *k* different outputs. On the other hand, although we could use the sigmoid function as the activation function for each of the outputs, it is very useful if each output can be assessed as probabilities of each class, as we saw for the binary case. Using sigmoid will not enforce the condition for probabilities that the sum of all of them must be 1 (which means that each of the inputs must correspond to one of the classes).

In this case, a function very similar to the sigmoid function that satisfies the conditions described is the softmax function:

$$\sigma(x_j) = \frac{e^{x_j}}{\sum_i e^{x_i}}$$

The class that will be selected is the one that will output the maximum value:

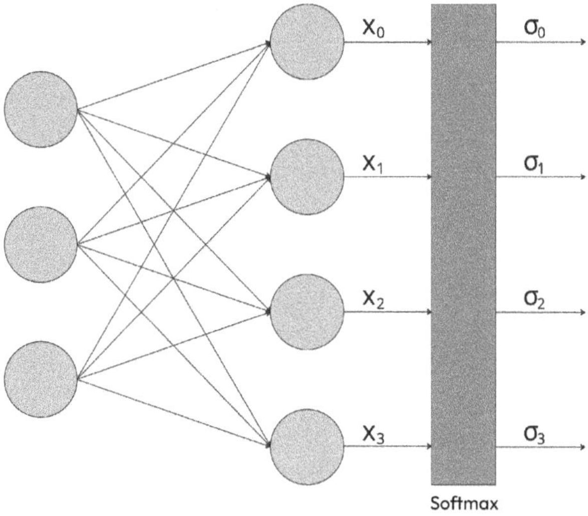

Figure 4.3 – Multi-label classification network

The full definition of our perceptron is $Y = f(WX + B)$, where W is now a weight matrix (with n x k shape: n being the number of features and k the number of outputs), X is the feature vector (n components), B is the bias vector (k components), and $f()$ is the softmax activation function. Y is now a vector of k outputs, where the class with the maximum value will be the class assigned to the input.

Defining the features

So far, we have defined our model and its behavior theoretically; we did not use our problem framing or our dataset to define it. In this section, we will start working at a more practical level.

The next step to continue defining our model is to decide on which features (inputs) we are going to work with. We will continue using the Iris dataset we already know from the second recipe, *Toy dataset for classification: loading, managing, and visualizing the House Sales dataset* in *Chapter 2, Working with MXNet and Visualizing Datasets: Gluon and DataLoader*. This dataset contained data for 150 flowers, including the class and 4 input features:

- Sepal length (in cm)
- Sepal width (in cm)
- Petal length (in cm)
- Petal width (in cm)

If we show the first five flowers, we will see the following:

	class	sepal length (cm)	sepal width (cm)	petal length (cm)	petal width (cm)
0	0	5.1	3.5	1.4	0.2
1	0	4.9	3.0	1.4	0.2
2	0	4.7	3.2	1.3	0.2
3	0	4.6	3.1	1.5	0.2
4	0	5.0	3.6	1.4	0.2

Figure 4.4 – Features of flowers (Iris dataset)

Furthermore, for the Iris dataset, we have different output classes:

- `Setosa (0)`
- `Versicolor (1)`
- `Virginica (2)`

Initializing the model

Now that we have defined the input dimension (number of features) and output dimensions, we can initialize our model using random initialization:

```
Weights:
[[-1.2347414  -1.771029    -0.45138445]
 [ 0.57938355 -1.856082    -1.9768796 ]
 [-0.20801921  0.2444218   -0.03716067]
 [-0.48774993 -0.02261727   0.57461417]]
 <NDArray 4x3 @cpu(0)>
 Bias:
 [1.4661262  0.6862904  0.35496104]
 <NDArray 3 @cpu(0)>
```

If we compare these values to the values we obtained in the first recipe, *Understanding math for regression models*, from *Chapter 3, Solving Regression Problems*, we can see how the weights are now represented as a matrix as we have several outputs, not just one (as it was in the regression case), and the bias is a vector instead of a number, for the same reason.

Evaluating the model

Now that our model is initialized, we can use it to estimate the class of the first flower, which can be seen in *Figure 3.24* to be *Setosa* (`0`). Here it is with our current model:

```
0
```

Nicely done! Unfortunately, this was pure chance as the model was randomly initialized.

In the next recipe, we will see how to properly evaluate our classification models.

How it works...

Like regression, classification models can have as many layers (depth) as needed, stacking as many of them as the problem's solution requires.

In this recipe, we described the modifications from the perceptron described in the first recipe, *Understanding maths for regression models*, in *Chapter 3, Solving Regression Problems*. There were two main modifications. The first one is that, as in this case, we want to categorize each input into a set of classes, we need one output per class. Moreover, in order to be able to understand the outputs of our model as probabilities, we needed a new activation function, softmax.

To finalize, we learned how to initialize our model, the effect initialization has on the weights and bias, and how we can use our data to evaluate it. We will develop all these topics further in the later recipes.

There's more...

At the beginning of the recipe, we walked through the activation function changes from Rosenblatt (step function) to regression (linear) to classification (sigmoid). One of the details we discussed was the non-differentiability of the step function. A deeper analysis can be found at the following link: `https://en.wikibooks.org/wiki/Signals_and_Systems/Engineering_Functions#Derivative`

Using multi-layer architectures and/or sigmoid (or other activation functions) provides neural networks with the capability to approximate any function, which is known as the **Universal Approximation Theorem**. More details can be found here: `https://en.wikipedia.org/wiki/Universal_approximation_theorem`

Defining loss functions and evaluation metrics for classification

In the previous recipe, we defined our input features, described our model, and initialized it. At that point, we passed a features vector of a flower to predict its iris species, calculated the output, and compared it against the expected class.

We also showed how those preliminary results did not represent a proper evaluation. In this recipe, we will explore the topic of evaluating our classification models.

Furthermore, we will also understand which loss functions fit best for the binary and multi-label classification problem.

Getting ready

Loss functions and evaluation functions need to satisfy the same properties that are described in *Chapter 3, Solving Regression Problems*, in the second recipe, *Defining Loss functions and evaluation metrics for regression*; therefore, I recommend reading that chapter first for a more thorough understanding.

We will start developing our topics by analyzing the binary classification approach (two output classes), and we will generalize afterward to the multiple-label classification approach.

How to do it...

Let's discuss some evaluation and loss functions and analyze their advantages and disadvantages. The functions we will describe are as follows:

- Cross-entropy loss function
- Evaluation – confusion matrix
- Evaluation – metrics

Cross-entropy loss function

As we discussed in our previous recipe, once the model has output a *probability* for each of our classes, we want to select the class that has the maximum probability as the output of our model.

When optimizing our model parameters, what we want is to find out which model parameters provide a maximum probability (`1`) for our desired class and a minimum probability for the rest (`0`). The derivation of the equation is out of the scope of the book, but you can find more information in *There's more...* section at the end of the recipe.

For the case of two output classes, we have the binary cross-entropy loss (for N samples):

$$BCE = -\frac{1}{N}\sum_{i=0}^{N} y_i . \log(\hat{y}_i) + (1 - y_i) . \log(1 - \hat{y}_i)$$

We can plot this function for one sample, and assume the expected output to be *1 (yi = 1)*:

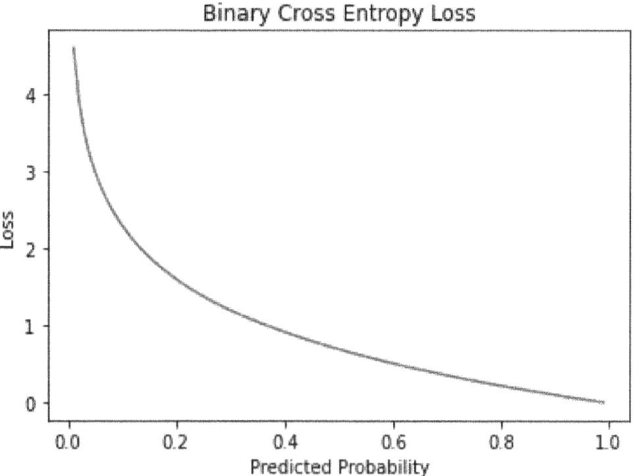

Figure 4.5 – Binary cross-entropy loss graph ($y_i = 1$)

For the general case of multiple labels, the multi-label or categorical cross-entropy loss for *M* classes becomes the following:

$$L\left(y,\hat{y}\right) = -\sum_{j=0}^{M}\sum_{i=0}^{M}\left(y_{ij} * log\left(\hat{y}_{ij}\right)\right)$$

This equation yields the same graph as *Figure 4.5* when comparing each pair of classes.

We will use this function in combination with an optimizer during the training loop in the next recipe.

Evaluation – confusion matrix

The confusion matrix helps us measure the performance of our model, comparing the results between the expected values (ground truth) and the actual values our model will provide. For a binary classification problem (where we defined the classes as `Positive` and `Negative`), we have the following:

		Predicted class	
		Positive	**Negative**
Actual class	**Positive**	TP	FN
	Negative	FP	TN

Figure 4.6 – Binary confusion matrix

In *Figure 4.6*, for each combination of predicted and actual class, we have the following terms (only valid in binary classification):

- **TP**: True positives.
- **FP**: False positives. It is also known as a *type I* error.
- **FN**: False negatives. It is also known as a *type II* error.
- **TN**: True negatives.

Ideally, we want TP and TN to be as close to 100% as possible, and FP and FN as close to 0% as possible.

When we have a multi-label classification problem, in the confusion matrix, we will have one row and one column per class. For *K* classes, we will have a *KxK* matrix where, in the matrix main diagonal, we are looking for 100% probabilities and 0% in the rest of the values. For example, using our Iris dataset (three output classes), we can compute the confusion matrix (*3x3*) for a model just randomly initialized, similar to the one we computed in the previous recipe. In this case, we obtain the following:

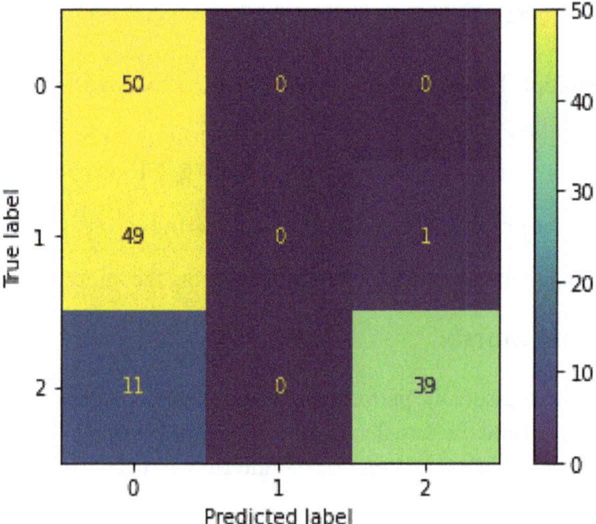

Figure 4.7 – Multi-label confusion matrix

As expected, the results shown are quite bad. Notably, not a single versicolor flower was correctly classified; however, this example helps us visualize a multi-label confusion matrix.

Evaluation – accuracy, precision, recall, specificity, and F1-score metrics

To characterize the performance of a model for a given binary classification problem, there are several interesting metrics:

$$Precision = \frac{TP}{TP + FP}$$

$$Recall = \frac{TP}{TP + FN}$$

$$F1 = \frac{2 \times Precision \times Recall}{Precision + Recall}$$

$$Accuracy = \frac{TP + TN}{TP + FN + TN + FP}$$

$$Specificity = \frac{TN}{TN + FP}$$

Each of these metrics serves the purpose of helping us understand the performance of our models:

- **Accuracy**: From all the values, which ones were correctly classified (for both classes)?
- **Precision**: This is the rate of positive predictions that were correctly classified. However, it does not provide any information regarding negative predictions.
- **Recall**: This is the rate of positive labels that were correctly classified by the model. This figure does not include any information regarding negative labels.
- **Specificity**: Similar to recall but for negative labels. This is the rate of negative labels that were correctly classified by the model. This figure does not include any information regarding positive labels. This metric is seldom used.
- **F1 score**: This is the harmonic mean of precision and recall. By combining both metrics, this metric provides a better assessment of the model considering positive and negative classes. To achieve a high F1 score, the model needs to have a high precision and a high recall. A low F1 score will indicate that either precision, recall, or both are low.

These metrics can be computed for the multi-label scenario. For example, for our randomly initialized model and the Iris datasets, these were the figures computed (except specificity, which does not have a metrics function in **scikit-learn**):

```
Accuracy  : 0.5933333333333334
Precision : 0.4765151515151515
Recall    : 0.5933333333333334
F1-score  : 0.49722222222222223
```

The values are very close to average results, as expected from a randomly initialized model.

Evaluation – Area Under the Curve (AUC)

For binary classification problems, the output we want to provide is a class (`Positive` or `Negative`); however, the output of our model is a number (the probability of a positive result). To transform this result into a class, we need to apply a threshold.

For example, if we define our threshold as `0.5`, every probability larger than 0.5 will get assigned the `Positive` class. By decreasing our threshold, more values will be considered positive, increasing the number of TPs (**True Positive Rate**, or **TPR**) and the number of FPs (**False Positive Rate**, or **FPR**). If the threshold is increased, the effect is the contrary: fewer values will be considered positive, hence a smaller TPR and FPR.

As we modify the value of the threshold, we will have different TPR and FPR values. If we plot those values in a graph, we obtain the following:

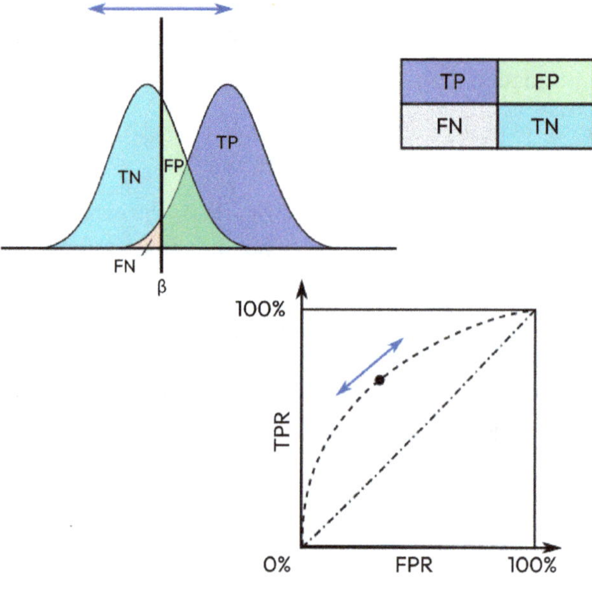

Figure 4.8 – AUC

If we calculate the area covered between the curve, the *x* axis, the $y = 0$ axis, and the $y = 1$ axis, we obtain a parameter that is not dependent on the threshold value; it defines the performance of our model for the given data.

TPR and FPR are only defined for binary classification cases. For the multi-label classification case, we can emulate binary classification cases. There are two possible approaches:

- One versus one
- One versus rest

If you are interested, you can find more information in the *There's more...* section of this recipe. These curves are also known as **receiver operating characteristic** curves.

How it works...

After understanding the differences between a regression model and a classification model, including the activation functions, in this recipe, we focused on the loss functions (useful for training) and the metrics (useful for evaluation). We explored both cases: binary classification and multi-label classification.

We computed the most common loss function for classification, the binary/categorical cross-entropy loss function, and we defined several evaluation metrics such as accuracy, precision, recall, and F1 score. Furthermore, we learned about the confusion matrix as an easy way to look at the per-class performance of our models.

We ended the recipe by taking a look at the AUC, which provides a visualization agnostic of thresholds.

There's more...

The mathematical formulas for cross-entropy loss were not derived. In these links, you can find more information:

- Binary Cross-Entropy Loss: `https://mxnet.apache.org/versions/1.7/api/python/docs/tutorials/packages/gluon/loss/loss.html#Cross-Entropy-Loss-with-Sigmoid`
- Categorical Cross-Entropy Loss: `https://mxnet.apache.org/versions/1.7/api/python/docs/tutorials/packages/gluon/loss/loss.html#Cross-Entropy-Loss-with-Softmax`

For a better understanding of how to compute multi-label classification metrics, I recommend the following link: `https://towardsdatascience.com/multi-class-metrics-made-simple-part-i-precision-and-recall-9250280bddc2`.

To conclude, reading this AUC explanation can provide further insight: `https://developers.google.com/machine-learning/crash-course/classification/roc-and-auc`.

For the multi-label case, these examples can help understand the one versus one/one versus rest approaches: `https://scikit-learn.org/stable/auto_examples/model_selection/plot_roc.html`.

Training for classification models

In this recipe, we will visit the basic concepts of training a model to solve a classification problem. We will apply them to optimize the classification model we previously defined in this chapter, combined with the usage of the loss functions and evaluation metrics we discussed.

We will predict the iris class of flowers using the dataset seen in the second recipe, *Toy dataset for classification – load, manage, and visualize Iris dataset*, from *Chapter 2, Working with MXNet and Visualizing Datasets: Gluon and DataLoader.*

Getting ready

In this recipe, we will follow a similar pattern as we did in *Chapter 3, Solving Regression Problems*, in the third recipe, *Training for regression models*, so it will be interesting to revisit the concepts of the loss function, optimizer, dataset split, epochs, and batch size.

How to do it...

In this recipe, we will create our own training loop and we will evaluate how each hyperparameter influences the training. To achieve this, we will follow these steps:

1. Improve the model.
2. Define the loss function and optimizer.
3. Split our dataset and analyze fairness and diversity.
4. Put everything together for a training loop.

Improving the model

To solve this problem, given the limited amount of data the dataset contains (150 samples), we will define a **Multi-Layer Perceptron** (**MLP**) network architecture, as we saw in *Chapter 3, Solving Regression Problems*, in the third recipe, *Training for regression models*. This will have 2 hidden layers fully connected (dense) and **ReLU** activation function with 10 neurons in each, and an output layer with the corresponding 3 outputs (1 per class). The last layer is left without an activation function, although softmax was expected. In the next section, we will understand why. For this network, the necessary code is as follows:

```
def create_classification_network(num_outputs = 3):
    # MLP with Gluon
    net = mx.gluon.nn.Sequential()
    net.add(mx.gluon.nn.Dense(10, activation="relu"))
    net.add(mx.gluon.nn.Dense(10, activation="relu"))
    net.add(mx.gluon.nn.Dense(num_outputs))
    # Note that the latest layer does not have an activation
```

```
    # function whereas Softmax was expected.
    # This is due to an optimization during training:
    # the loss function includes the softmax computation. return net
```

We have also applied scaling to our input features. The number of parameters for the model is as follows:

```
Parameters in forward computation graph, duplicate included
   Total params: 193
   Trainable params: 193
   Non-trainable params: 0
Shared params in forward computation graph: 0
Unique parameters in model: 193
```

The number of trainable parameters is ~200. For our (small) dataset, we have 4 features for each of the 150 rows; therefore, our dataset is ~3 times the number of parameters of the model. Typically, this is the minimum for a successful model and, ideally, we would like to work with a dataset size of ~10 times the number of the parameters of the model.

> **Important Note**
> Even though the comparison between the data points available and the number of parameters of the model is a very useful one, different architectures have different requirements in terms of data. As usual, experimentation (trial and error) is the key to finding the right balance.

Defining the loss function and optimizer

As discussed in the previous recipe, we will compute the **Categorical Cross-Entropy** (**CCE**) loss function. However, there is an optimization detail when computing the CCE loss with the softmax activation function; therefore, the computation for the softmax function during training is included in the loss function. For **inference**, we need to add it externally. Similar to what we did for regression problems, we will focus our analysis on the **Stochastic Gradient Descent** (**SGD**) and Adam optimizers.

Splitting our dataset

One of the strongest disadvantages of the Iris dataset is its size; with 150 samples, it is a small dataset. For this reason, we will apply a 50/40/10 split for the training, validation, and test sets.

If we analyze the splits to verify fairness and diversity, we obtain the following:

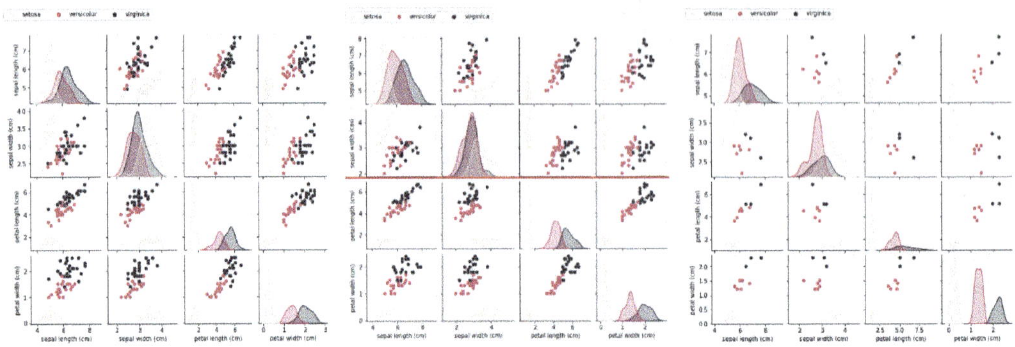

Figure 4.9 – Distribution for a training set (left), a validation set (middle), and a test set (right)

We can see that each of the classes is well represented across all features, the small size being the only reason for the discrepancies.

Putting it all together for a training loop

The training loop for the classification case is very similar to the regression case, and we will follow a similar analysis as done earlier in this chapter: we will compare each of the hyperparameters, keeping the others constant (unless otherwise noted).

Optimizer and learning rate

As discussed earlier, the chosen optimizer for the training loop and the learning rate are related because, for some optimizers (such as SGD), the learning rate is kept constant, whereas for others (such as Adam), it varies from a starting point.

> **Tip**
>
> The best optimizer depends on several factors, and nothing trumps trial and error; I strongly suggest trying a few to see which one fits best. In my experience, typically, SGD and Adam are the ones that work best, including in this problem.

Let's analyze how the training loss and validation loss vary for the SGD optimizer when we change the **learning rate** (**LR**), keeping the other parameters constant: epochs = 100, batch size = 64, and loss function = softmax cross-entropy:

Optimizer	LR	Training Loss/Validation Loss (last epoch)/ Validation Loss (Best Model)
SGD	10^{-5}	2.05/1.41/1.41
SGD	10^{-4}	1.90/1.33/1.33
SGD	10^{-3}	2.00/1.20/1.20
SGD	10^{-2}	0.80/0.64/0.64
SGD	10^{-1}	0.17/0.15/0.15
SGD	1.0	0.20/0.14 0.07
SGD	1.5	0.09/0.11/0.07
SGD	2.0	0.08/0.12/0.07
SGD	3.0	0.37/0.06/0.06
SGD	5.0	0.84/0.54/0.24

Figure 4.10 – Loss for SGD optimizer with different LRs

From *Figure 4.10*, we can conclude that for the SGD optimizer, an LR value between 1.0 and 3.0 is optimal. Furthermore, we can see that for very large values of LR (> 2.0), the algorithm still converges, whereas for the regression case, SGD and very large LRs made the model diverge.

Let's analyze how the training loss and validation loss vary for the Adam optimizer when we change the LR, keeping the other parameters constant: epochs = 100, batch size = 64, and loss function = softmax cross-entropy:

Optimizer	LR	Training Loss/Validation Loss (last epoch)/ Validation Loss (Best Model)
Adam	10^{-5}	1.99/1.33/1.33
Adam	10^{-4}	1.60/1.13/1.13
Adam	10^{-3}	0.33/0.29/0.29
Adam	10^{-2}	0.12/0.08/0.08
Adam	10^{-1}	0.06/0.25/0.08
Adam	1.0	0.79/0.53/0.53
Adam	1.5	1.10/10.99/10.25
Adam	2.0	1.88/1.20/0.71

Figure 4.11 – Loss for the Adam optimizer with different LRs

From *Figure 4.11*, we can conclude that for the Adam optimizer, an LR value between 10^{-2} and 10^{-1} is optimal.

Although, in this case, SGD with an LR of 3.0 is yielding the best results (smallest loss), the evolution of the optimization process is much noisier than for Adam, possibly due to the limited amount of data available (the batch size did not influence this). A smooth optimization process is also an indication of how well a model can generalize; hence, we will choose Adam as our optimizer for the rest of our tests.

Batch size

Let's analyze how the training loss and validation loss vary for the Adam optimizer by changing the batch size, keeping the other parameters constant: epochs = 100, LR = 10^{-2}, and loss function = softmax cross-entropy:

Batch Size	Training Loss/Validation Loss (last epoch)/ Validation Loss (Best Model)
16	0.02/0.13/0.08
32	0.02/0.08/0.06
64	0.06/0.09/0.08
128	0.08/0.20/0.17
256	0.15/0.56/0.56

Figure 4.12 – Loss for the Adam optimizer by varying the batch size

From *Figure 4.12*, we can conclude that for the Adam optimizer, a batch size value between 32 and 64 provides the best results.

Epochs

Let's analyze how the training loss and validation loss vary for the Adam optimizer by varying the epochs, keeping the other parameters constant: LR = 10^{-2}, batch size = 32, and loss function = softmax cross-entropy:

Epochs	Training Loss/Validation Loss (last epoch)/ Validation Loss (Best Model)
50	0.21/0.23/0.23
100	0.07/0.10/0.09
150	0.03/0.10/0.09
200	0.02/0.10/0.09
300	0.01/0.13/0.09
400	>0.00/0.14/0.09

Figure 4.13 – Loss for the Adam optimizer by varying the number of epochs

From *Figure 4.13*, we can conclude that a range of 200–300 epochs is good for our problem. With these values, it is very likely the best result will be achieved earlier than that.

How it works...

On our path to solving our classification problem, in this recipe, we learned how to update our model hyperparameters optimally. We revisited the role that each hyperparameter plays in the training loop and we performed some ablation studies for each individual hyperparameter. This helped us understand how our training and validation losses behaved when we modified each hyperparameter individually.

For our current problem and the chosen model, we verified that the best set of hyperparameters was as follows:

- Optimizer: Adam
- LR: 10^{-2}
- Batch size: 32
- Number of epochs: 300

At the end of the training loop, these hyperparameters gave us a training loss of `0.01` and a validation loss of `0.1`.

There's more...

In this recipe, we mostly put together concepts we have been learning about in the previous recipes and chapters.

We did pass through how, in our model definition, we did not explicitly use the softmax activation function. This is due to how cross-entropy loss and the softmax activation function work together during training (its joint derivative). A good reference to understand this point is the following:

`https://peterroelants.github.io/posts/cross-entropy-softmax/`

Evaluating classification models

In the previous recipe, we learned how to choose our training hyperparameters to optimize our training. We also verified how those choices affected the training and validation losses. In this recipe, we are going to explore how those choices affect our actual evaluation in the real world. You will have noticed that we split the dataset into three different sets: training, validation, and test sets. However, during our training, we only used the training set and the validation set. In this recipe, we will emulate real-world behavior by using the unseen data from our model, the test set.

Getting ready

When evaluating a model, we can perform qualitative evaluation and quantitative evaluation.

Qualitative evaluation is the selection of one or more random (or not so random, depending on what we are looking for) samples and analyzing the result, verifying whether it matches our expectations.

In this recipe, we will compute the evaluation metrics we defined in the second recipe, *Defining loss functions and evaluation metrics for classification models*, in this chapter.

Furthermore, we are going to take a look at how training can have a large influence on the evaluation.

How to do it...

Before jumping into model evaluation, we will discuss how we can measure our model training performance. Therefore, the steps of this recipe are as follows:

1. Measuring training performance – losses and accuracy
2. Qualitative evaluation
3. Quantitative evaluation

Measuring training performance – losses and accuracy

As we saw for regression, a good way to prevent overfitting was early stopping. When training our classification model in the previous recipe, we stored the training loss, validation loss, and validation accuracy. Let's see how the training loss, validation loss, and validation accuracy evolve as the training progresses:

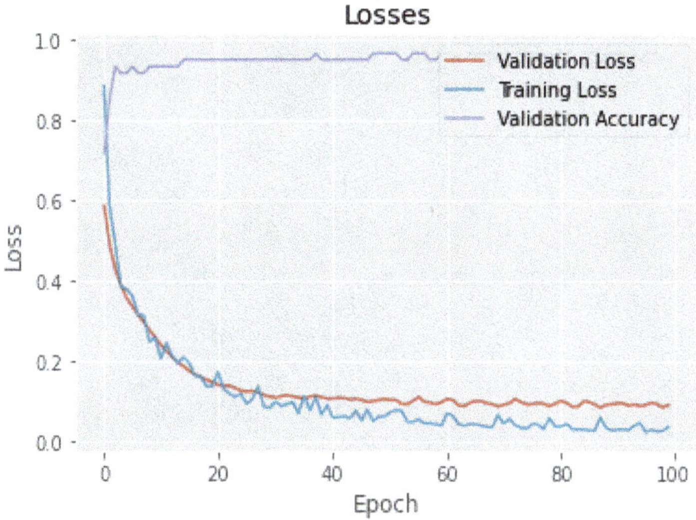

Figure 4.14 – Losses and accuracy versus epochs (Adam)

As we can see in *Figure 4.14*, at around epoch 50, the validation loss starts increasing although the training loss continues to decrease. Moreover, the validation accuracy also seems to plateau (close to 1.0/100%) around that epoch. We saved the model for the best accuracy and those are the values reported during training in the third recipe, *Training for classification models*, in this chapter.

> **Important Note**
>
> As a reminder, if you want to use early stopping as is, MXNet provides a callback for it: `https://mxnet.apache.org/versions/1.6/api/r/docs/api/mx.callback.early.stop.html`.

An important thing to mention is that, in the previous recipe, we mentioned that SGD did not provide very smooth training. If we plot the values, we obtain the following:

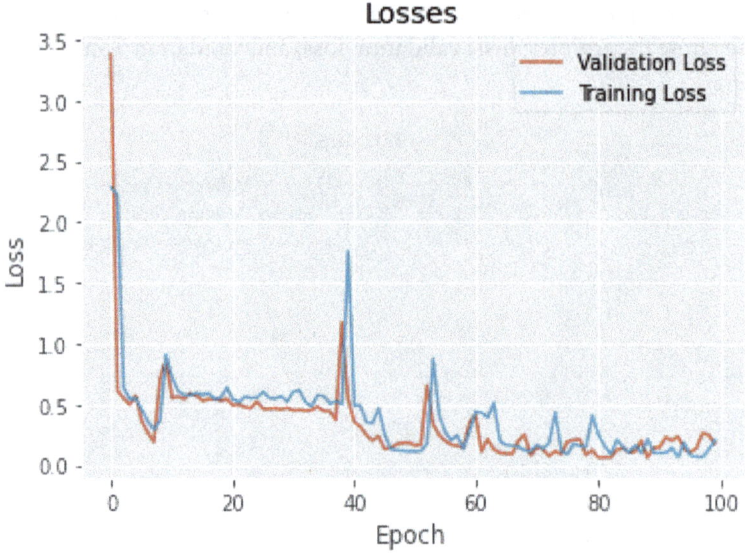

Figure 4.15 – Losses versus epochs (SGD)

As we can see, the training is not stable at all.

Qualitative evaluation

To verify that our model is behaving similarly to what we expect (yielding a high accuracy when predicting the iris species of a flower), a simple qualitative approach is to run our model for a random input from the test set (unseen data). In our case, this is as follows:

```
Expected Output: 0
Model Output: [[9.999949e-01 4.748235e-06 4.116847e-07]]
 Class Output: 0
Accuracy     : 1.0
```

For this example, as this was just one random input, accuracy can be either 100% or 0% (accurate or not), and we got the right class.

> **Important Note**
>
> Although I tried to keep the code as reproducible as possible (including setting the seeds for all random processes), there might be some sources of randomness. This means that your results might be different, but typically, the order of magnitude of errors will be similar.

Quantitative evaluation – confusion matrix

For the stored results, the confusion matrix obtained is as follows:

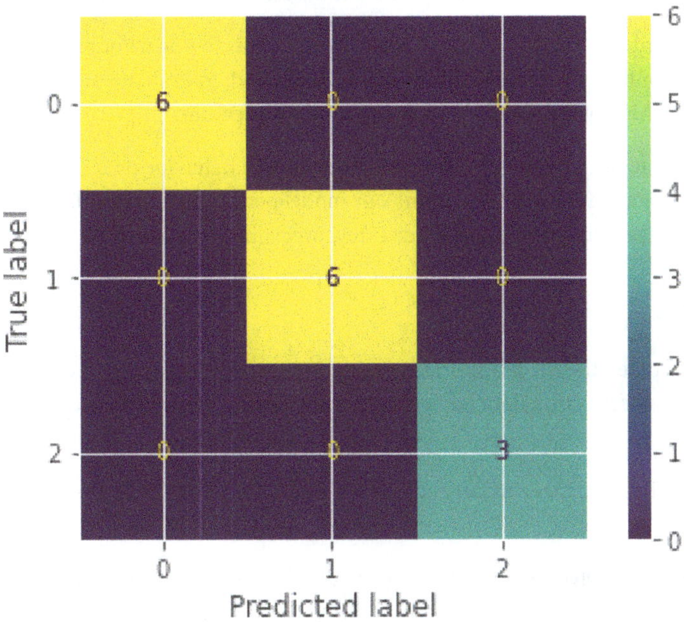

Figure 4.16 – Confusion matrix

These results are excellent, as there are only values different from zero in the main diagonal. This means our model yielded perfect results for the test set.

Quantitative evaluation – accuracy, precision, recall, and F1 score

During our previous recipes, we defined these metrics and worked with them to optimize the training. These evaluations were done for the training set and the validation set. For the test set, we obtained the following values:

- Accuracy: 1.0
- Precision: 1.0
- Recall: 1.0
- F1 score: 1.0

How it works...

In this recipe, we explored how to evaluate our classification model. To properly do this, we revisited the right decision of splitting our full dataset into a training set, a validation set, and a test set.

During training, we used the training set to calculate the gradients to update our model parameters, and the validation set to confirm the real-world behavior. Afterward, to evaluate our model performance, we used the test set, which was the only remaining set of unseen data.

We discovered the value of qualitatively describing our model behavior by calculating the output of random samples, and of quantitatively describing our model performance by exploring several numbers and graphs including the confusion matrix, accuracy, precision, recall, and F1 score.

There's more...

In this recipe, we computed the most important evaluation metrics for balanced classification datasets. However, when the dataset is imbalanced, we need to be careful. This is a good tutorial I like about this topic:

`https://machinelearningmastery.com/probability-metrics-for-imbalanced-classification/`.

At this stage, we have completed our path through a complete classification problem: we explored our classification dataset, decided on our evaluation metrics, and defined and initialized our model. We understood the best hyperparameter combination of optimizer, learning rate, batch size, and epochs, and trained it with early stopping. We ended by evaluating it qualitatively and quantitatively.

5
Analyzing Images with Computer Vision

Computer vision is one of the fields in which deep learning has progressed enormously, surpassing human-level performance in several tasks such as image classification and object recognition. Furthermore, the field has moved from academia to real-world applications, and the industry is recognizing its practitioners as adding high value to businesses.

In this chapter, we will learn how to use GluonCV, a MXNet Gluon library specific to computer vision, how to build our own networks, and how to use GluonCV's model zoo to use pretrained models for several applications.

Specifically, we will cover the following topics:

- Understanding convolutional neural networks
- Classifying images with AlexNet and ResNet
- Detecting objects with Faster R-CNN and YOLO
- Segmenting objects in images with PSPNet and DeepLab-v3

Technical requirements

Apart from the technical requirements specified in the *Preface*, the following technical requirements apply in this chapter:

- Ensure that you have completed *Installing MXNet, Gluon, GluonCV and GluonNLP*, the first recipe from *Chapter 1, Up and Running with MXNet*
- Ensure that you have completed *A toy dataset for regression – load, manage, and visualize a house sales dataset*, the first recipe from *Chapter 2, Working with MXNet and Visualizing Datasets: Gluon and DataLoader*

The code for this chapter can be found at the following GitHub URL: `https://github.com/PacktPublishing/Deep-Learning-with-MXNet-Cookbook/tree/main/ch05`.

Furthermore, you can access each recipe directly from Google Colab – for example, for the first recipe of this chapter: `https://colab.research.google.com/github/PacktPublishing/Deep-Learning-with-MXNet-Cookbook/blob/main/ch05/5_1_Understanding_Convolutional_Neural_Networks.ipynb`.

Understanding convolutional neural networks

In the previous chapters, we have used *fully connected* **Multi-Layer Perceptron** (**MLP**) networks to solve our regression and classification problem. However, as we will see, these networks are not optimal for solving image-related problems.

Images are highly dimensional entities – for example, each pixel in a color image has three features (red, green, and blue values), and a 1,024x1,024 image has more than 1 million pixels (a 1 megapixel image) and, therefore, more than 3 million features ($3 * 10^6$). If we connect all these points in the input layer, to a second layer of 100 neurons for a *fully connected* network, we will require more than 10^8 parameters, and that would be only for the first layer. Processing images is, therefore, a time-intensive operation.

Furthermore, imagine that we are trying to detect eyes in faces; if a pixel belongs to an eye, the likelihood of nearby pixels belonging to the eye is very high (think of the pixels that make up the iris, for example). When inputting all our pixels directly into our network, all the information connected to pixel location is lost.

An architecture called a **Convolutional Neural Network** (**CNN**) was developed to tackle these problems, and we will analyze the most important features of CNNs and how to implement them in this recipe to solve image-related problems.

Getting ready

As with the previous chapters, in this recipe, we will use a few matrix operations and linear algebra, but it will not be too difficult.

How to do it...

In this recipe, we will take the following steps:

1. Introduce convolutional layer equations.
2. Understand the convolution parameters and receptive field.
3. Run a convolutional layer example with MXNet.
4. Introduce pooling layer equations.

5. Run a pooling layer example with MXNet.
6. Summarize CNNs.

Introducing convolutional layer equations

The location problems described in the recipe introduction are formally known as **translation invariance** and **locality**. In CNNs, these problems is solved by using the convolution/cross-correlation operation in the so-called convolutional layers.

In the convolution, we have an input image (or **feature map** – see the following *Important note* relating to convolutional layer equations) that is combined with a **kernel**, which are the learnable parameters of this layer. The simplest way to see how this operation works is with an example – if we had a 3x3 input and we wanted to combine it with a 2x2 kernel, it would look like a sliding window, as shown in the following diagram:

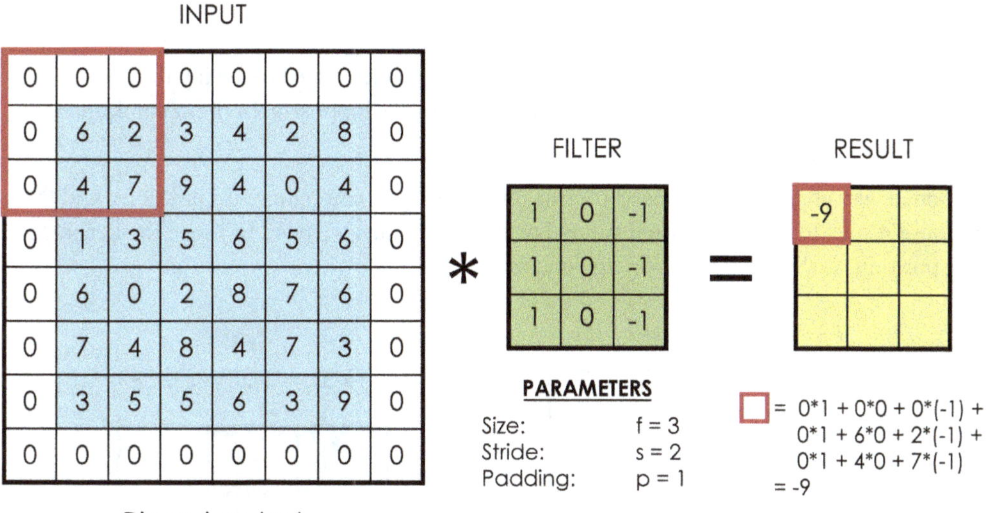

Figure 5.1 – The convolutional layer

As shown in *Figure 5.1*, to compute 1 pixel of the output, we can intuitively place the kernel over the input and compute multiplications, and then add all these values to obtain the final result. This helps a network learn features from the image.

Furthermore, this operation is less computing intensive than the *fully connected* layers, addressing the computing problem we identified.

Important note

When using a convolution layer as the first step (typical in CNNs), the input is the full image. Moreover, the output can be understood as another image of a lower dimension with certain properties, given by the kernel. As kernels are learned to highlight certain features of the image, these outputs are known as *feature maps*. In the layer close to the input, each pixel of these feature maps is a combination of a small number of pixels of the image (for example, horizontal or vertical lines). As data travels through convolutional layers, these feature maps represent higher levels of abstraction (for example, eyes on a face).

Understanding the convolution parameters and receptive field

The example shown in *Figure 5.1* is quite simple, for illustrative purposes; the input size is 6x6 and the kernel size is 3x3. These sizes are variable and depend on the network architecture. However, there are three parameters that are very important to calculate the output size; these are padding, stride, and dilation.

Padding is the number of zero-valued pixels (rows/columns) that are added to the input. The larger the padding, the larger the output, and effectively, this increases the input size. In the example, padding is `1` (represented as *p*).

As mentioned, we can intuitively place the kernel over the input, compute multiplications, and then add all these results to obtain the final value. For the next value, we need to move the kernel to a different position, as shown in the following diagram.

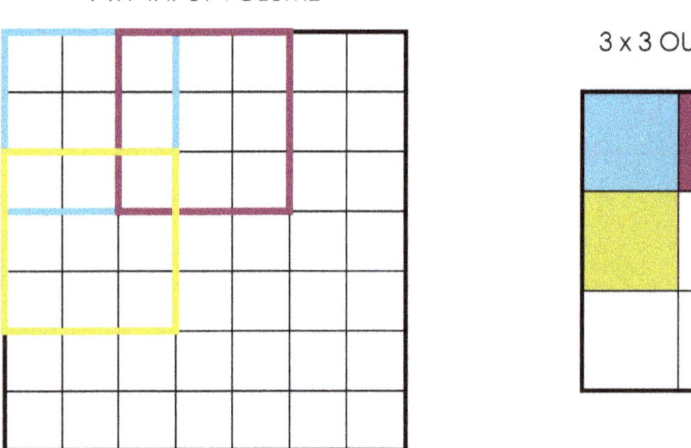

Figure 5.2 – The stride parameter

The number of spaces the kernel moves is defined by the stride. In *Figure 5.2*, we can see an example of the stride being 2, with each 3x3 kernel separated by 2 values.

Dilation is defined by how separated each input value that is used in the convolution with the kernel is.

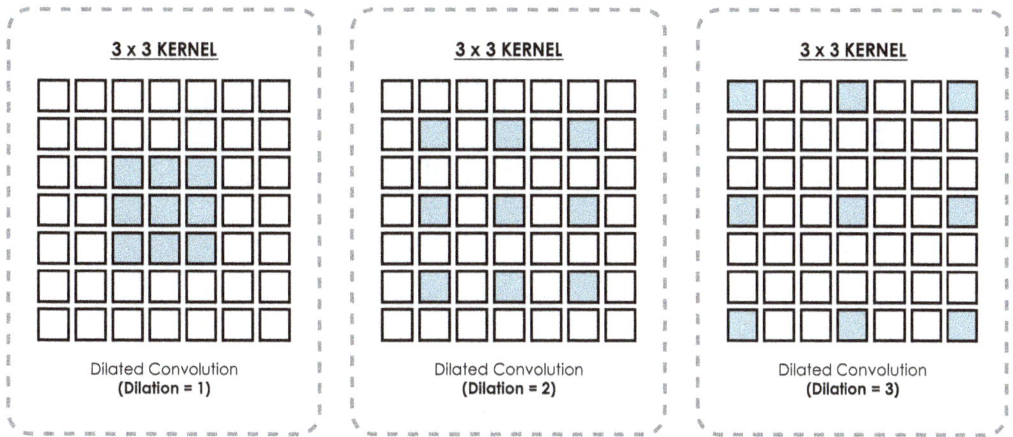

Figure 5.3 – The dilation parameter

As we can see in *Figure 5.3*, different dilation parameters combine elements of the input.

These parameters define what is known as the **receptive field**, which is the size of the region of the input that produces an activation. It is an important parameter because only a feature present in the receptive field of our model will be represented in the output.

In the example in *Figure 5.1*, a 6x6 input, combined with a 3x3 kernel, with a stride of 2, padding of 1, and dilation of 1, yields a 3x3 output and has a receptive field of the full input (all pixels in the input are at least used once).

Another very interesting property of these parameters is that given the right combinations, you can have an output that is the same size as the input. The equation that gives the dimensions of the output is as follows (the equation needs to be applied to the height and the width, respectively):

$$o = [i + 2*p - k - (k - 1)*(d - 1)]/s + 1$$

In the preceding equation, o is the output dimension (height or width if working with 2D images), i is the input dimension (height/width), p is padding, k is the kernel size, d is dilation, and s is stride. There are different combinations that will give the input and the output the same size, one of which is $p = 1, k = 3, d = 1, s = 1$.

Running a convolutional layer example with MXNet

We can implement the following example using MXNet capabilities (note that padding is 0, stride is 1, and dilation is 1):

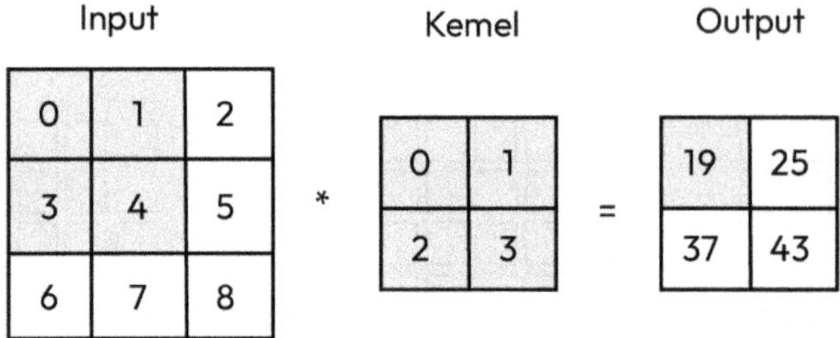

Figure 5.4 – A convolution example

If we want to follow the example depicted in *Figure 5.4*, applying the convolution operation to a 3x3 matrix with a 2x2 kernel, we can use the following code:

```
X = mx.np.array([[0.0, 1.0, 2.0], [3.0, 4.0, 5.0], [6.0, 7.0, 8.0]])
 K = mx.np.array([[0.0, 1.0], [2.0, 3.0]]) convolution(X, K)
```

These operations give the following as a result:

```
array([[19., 25.],
       [37., 43.]])
```

This is the expected result for the defined convolution step (taking into account the given padding, stride, and dilation parameters).

Introducing the pooling layer equations

As described earlier, a desirable property of neural network models when working with images is that as we traverse a network, we can increasingly process higher-level features, or equivalently, each pixel in the deep feature maps has a higher receptive field from the input.

The operation performed in these layers is similar to the convolutional layers, in the sense that we take a kernel of constant dimensions and apply a sliding window. However, in this case, the kernel parameters are constant and, therefore, not learned during the training of the network. This kernel is seen as an operation (a function) and is typically either the max function (the **max pooling layer**) or the average function (the **average pooling layer**).

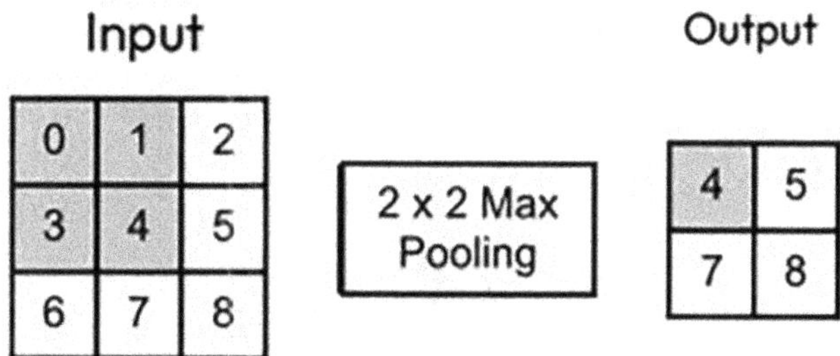

Figure 5.5 – The max pooling layer

Furthermore, by combining pixels that are nearby, we also achieve local invariance, another desirable property to process images.

Running a pooling layer example with MXNet

We can implement the example shown in *Figure 5.5* using MXNet's capabilities, as shown here:

```
X = mx.np.array([[0.0, 1.0, 2.0], [3.0, 4.0, 5.0], [6.0, 7.0, 8.0]])
maxpool(X, (2, 2))
```

These operations give the following as a result:

```
array([[4., 5.],
       [7., 8.]])
```

This is the expected result for the defined 2x2 max pooling step.

Summarizing CNNs

A typical CNN for **image classification** has, therefore, two different parts:

- **Feature extraction**: This is also known as the network backbone. It is built with combinations of the convolutional and pooling layers seen in this recipe. The input to each layer is feature maps (an image and all its channels in the input layer), and the output is feature maps of reduced dimensions but with a larger number of channels. Layers are typically stacked – a convolutional layer, an activation function (typically, **Rectified Linear Unit** (**ReLU**)), and a max pooling layer.
- **Classifier**: The last feature map is reshaped into a vector that can then be applied to *fully connected* layers, as discussed in *Chapter 4, Solving Classification Problems*. The number of neurons in the output layer is equivalent to the number of classes that an image needs to be classified as, using the `softmax` function as the activation. The number of layers in the classifier depends on the specific problem.

Therefore, a CNN architecture can be the following:

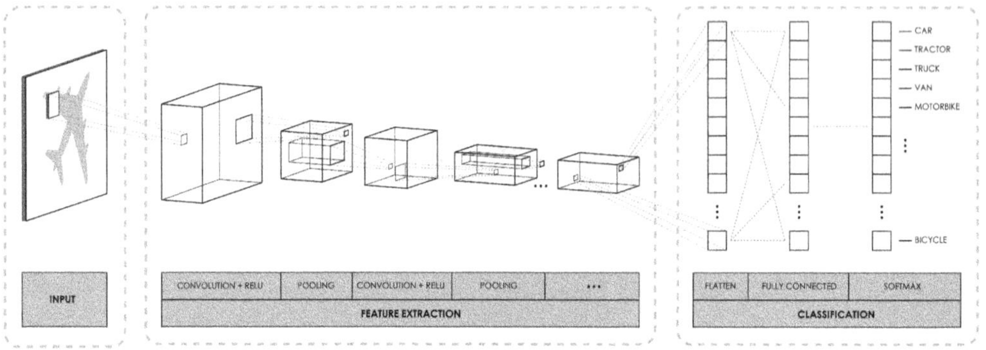

Figure 5.6 – CNN architecture

In *Figure 5.6*, we can see a CNN architecture for *image classification*, composed of two stages for feature extraction, each stage combining a convolution layer (with the ReLU activation function) and a max pooling layer. Then, for the classifier, the remaining feature map is flattened into a vector and passed through a *fully connected* layer to finally provide the output with the *softmax* activation function.

How it works...

In this recipe, we have introduced CNNs. This architecture has been developed since early the 2000s and is responsible for the revolution in **computer vision** applications, and making **deep learning** become the spotlight in most data-oriented tasks including **natural language processing**, **speech recognition**, **image generation**, and so on, reaching state-of-the-art in all of them.

We have understood how CNNs work internally, exploring the concepts of *feature maps*, *receptive field*, and the mathematical concepts behind the main layers of these architectures, the *convolutional* and *max pooling* layers, and how they are combined to build a complete CNN model.

There's more...

CNNs have evolved rapidly; in 1998, one of the first CNNs was published, solving a practical problem: `http://yann.lecun.com/exdb/publis/pdf/lecun-01a.pdf`

After that, it was not until 2012, with AlexNet, that CNNs gained worldwide attention, and since then, progress quickly developed, until they surpassed human-level performance. For more information on the history of CNNs, refer to this article: `https://towardsdatascience.com/from-lenet-to-efficientnet-the-evolution-of-cnns-3a57eb34672f`.

We briefly touched on the topics of translation invariance and locality. For more information on these topics, visit `https://d2l.ai/chapter_convolutional-neural-networks/why-conv.html`.

The relationship between convolution and cross-correlation is discussed here: `https://towardsdatascience.com/convolution-vs-correlation-af868b6b4fb5`.

For a better understanding of matrix dimensions, padding, stride, dilation, and receptive field, a good explanation is provided here: `https://theaisummer.com/receptive-field/`.

CNNs have been state-of-the-art for image classification until quite recently; in October 2020, **Transformers** were applied to computer vision tasks by Google Brain with **ViT**: `https://ai.googleblog.com/2020/12/transformers-for-image-recognition-at.html`.

In a nutshell, instead of forcing a network with the locality principle, Transformer architectures allow the same model to decide which features matter most at any layer, local or global. This behavior is called **self-attention**. Transformers are state-of-the-art for image classification at the time of writing.

In this chapter, we are going to analyze in detail the following tasks – **image classification**, **object detection**, and **image segmentation**. However, *MXNet GluonCV Model Zoo* contains lots of models pre-trained for a large number of tasks. You are encouraged to explore the different examples provided at `https://cv.gluon.ai/model_zoo/index.html`.

Classifying images with MXNet – GluonCV Model Zoo, AlexNet, and ResNet

MXNet provides a variety of tools to compose custom deep learning models. In this recipe, we will see how to use MXNet to build a model from scratch, train it, and use it to classify images from a dataset. We will also see that although this approach works fine, it is time-consuming.

Another option, and one of the highest value features that MXNet and GluonCV provide, is their **Model Zoo**. GluonCV Model Zoo is a set of pre-trained, ready-to-go models, for use with your own applications. We will see how to use Model Zoo with two very important models for image classification – **AlexNet** and **ResNet**.

In this recipe, we will analyze and compare these approaches to classify images on a reduced version of the *Dogs vs. Cats* dataset.

Getting ready

As with previous chapters, in this recipe, we will use a few matrix operations and linear algebra, but it will not be too difficult.

Furthermore, we will be classifying image datasets; therefore, we will revisit some concepts that we've already seen:

- *Understanding image datasets- load, manage, and visualize the Fashion MNIST dataset*, the third recipe from *Chapter 2, Working with MXNet and Visualizing Datasets: Gluon and DataLoader*
- *Chapter 4, Solving Classification Problems*

How to do it...

In this recipe, we will take the following steps:

1. Explore a reduced version of the *Dogs vs. Cats* dataset.
2. Create an AlexNet custom model from scratch.
3. Train an *AlexNet* custom model.
4. Evaluate an *AlexNet* custom model.
5. Introduce Model Zoo.
6. Introduce ImageNet pre-trained models.
7. Load an *AlexNet* pre-trained model from *Model Zoo.*
8. Evaluate an *AlexNet* pre-trained model from *Model Zoo.*
9. Load a ResNet pre-trained model from *Model Zoo.*
10. Evaluate a *ResNet* pre-trained model from *Model Zoo.*

Exploring the reduced version of the dataset Dogs vs. Cats

For our image classification experiments, we will work with a new dataset, *Dogs vs. Cats.* This is a Kaggle dataset (`https://www.kaggle.com/c/dogs-vs-cats`) that can be downloaded manually. In this recipe, we will work with a reduced version of this dataset that can be downloaded from **Zenodo** (`https://zenodo.org/records/5226945`)

From the set of images in the dataset (either a cat or a dog is depicted), our model will need to correctly classify these images. In the first step, as we saw in previous chapters, we are going to do some **Exploratory Data Analysis** (**EDA**).

Figure 5.7 – The Dogs vs. Cats dataset

As can be seen in *Figure 5.7*, each image in the dataset is in color, and they are resized to 224 px by 224 px (width and height). There are 1,000 images in the training and validation set and 400 images in the test set.

As we did in *Understanding image datasets – load, manage, and visualize the Fashion MNIST dataset*, the third recipe from *Chapter 2, Working with MXNet and Visualizing Datasets: Gluon and DataLoader*, we can compute some visualizations using the dimensionality reduction techniques – **Principal Component Analysis** (**PCA**), **t-distributed stochastic neighbor embedding** (**t-SNE**), and **Uniform Manifold Approximation and Projection** (**UMAP**):

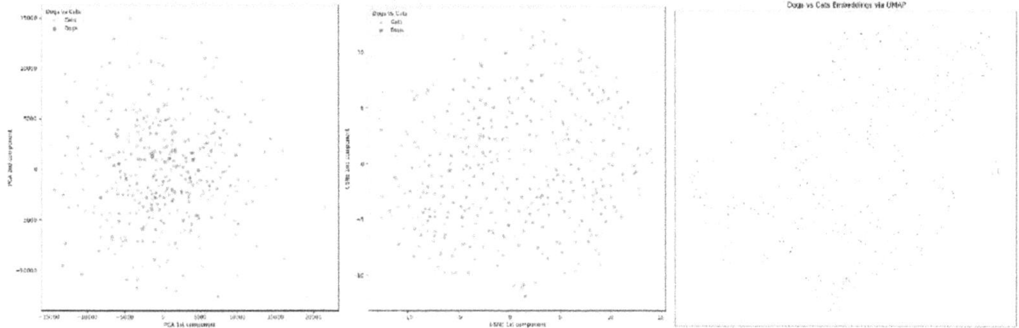

Figure 5.8 – Dogs versus cats visualizations – PCA, t-SNE, and UMAP

In *Figure 5.8*, there is no clear boundary region to separate dogs versus cats. However, as we will see in the following sections, an architecture introduced in the previous chapter, CNNs, will achieve very good results on the task.

Creating an AlexNet custom model from scratch

AlexNet was a deep neural network that was developed by Alex Krizhevsky, Ilya Sutskever, and Geoffrey Hinton in 2012. It was designed to compete in the **ImageNet Large Scale Visual Recognition Challenge** (**ILSVRC**) in 2012 and was the first CNN-based model to win this competition.

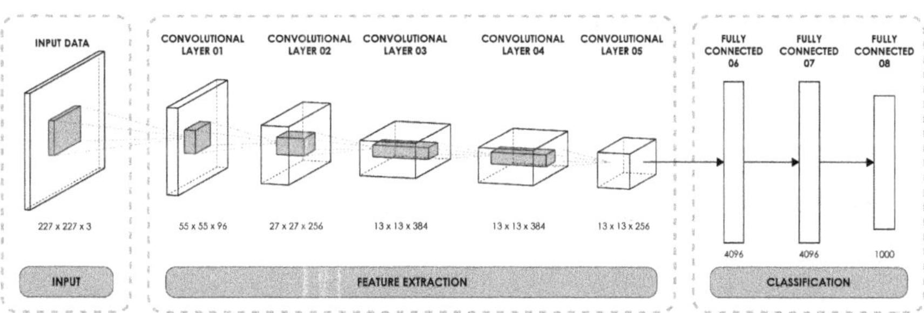

Figure 5.9 – AlexNet

The network uses five convolutional layers and three *fully connected* layers. The activation function used is the ReLU, and it contains about 63 million trainable parameters.

To generate this network from scratch with MXNet, we can use the following code:

```
def create_alexnet_network(num_classes=2):
    # Returns AlexNet architecture, as defined in MXNet source code
    net = nn.Sequential()
    net.add(
        nn.Conv2D(64, kernel_size=11, strides=4, activation='relu'),
        nn.MaxPool2D(pool_size=3, strides=2),
        nn.Conv2D(256, kernel_size=5, padding=2, activation='relu'),
        nn.MaxPool2D(pool_size=3, strides=2),
        nn.Conv2D(384, kernel_size=3, padding=1, activation='relu'),
        nn.Conv2D(384, kernel_size=3, padding=1, activation='relu'),
        nn.Conv2D(256, kernel_size=3, padding=1, activation='relu'),
        nn.MaxPool2D(pool_size=3, strides=2),          nn.Flatten(),
        # Last 3 layers is classifier
        # Adding dropout for regularization
        nn.Dense(4096, activation='relu'),
        nn.Dropout(0.5),
        nn.Dense(4096, activation='relu'),
        nn.Dropout(0.5),
        nn.Dense(num_classes)
    )
  return net
```

This code uses MXNet functions to add the corresponding 2D *convolutional*, *max-pooling*, and *fully connected* layers with their corresponding activation functions, generating an AlexNet architecture.

Training an AlexNet custom model

The task we are dealing with is an image classification task, which is a classification problem where input data is images, and therefore, we can use the training loop we saw in *Chapter 4, Solving Classification Problems* – slightly modified for this task.

The parameters chosen are as follows:

- **Number of epochs**: 20
- **Batch size**: 16 samples
- **Optimizer**: Adam
- **Learning rate**: 0.0001

With these parameters, we obtain the following results (the best model was achieved on epoch 11):

- **Training loss**: 0.36
- **Training accuracy**: 0.83
- **Validation loss**: 0.55
- **Validation accuracy**: 0.785

Evaluating an AlexNet custom model

The accuracy obtained with the best model (in this case, the one corresponding to the last training iteration) on the test set is as follows:

```
('accuracy', 0.7275)
```

That's quite a decent number for just five epochs of training.

Moreover, the confusion matrix computed is shown in the following figure (class **0** corresponds to cats and **1** to dogs):

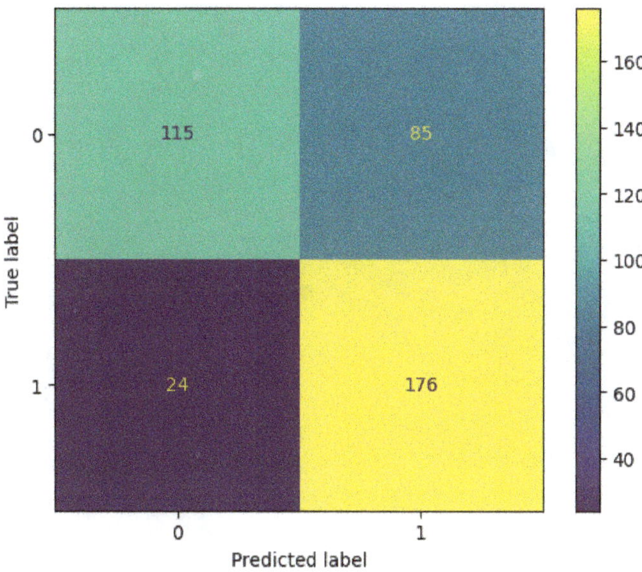

Figure 5.10 – A trained custom AlexNet confusion matrix

As shown in *Figure 5.10*, the model mostly predicts accurately the expected classes, with the following per-class errors:

- **Cats detected as dogs**: **85**/200 (43% of cat images were misclassified)
- **Dogs detected as cats**: **24**/200 (12% of dog images were misclassified)

Let's move on to the next heading.

Introducing Model Zoo

One of the best features that MXNet and GluonCV provide is their large pool of pre-trained models, readily available for its users to use and deploy in their own applications. This model library is called **Model Zoo**.

Moreover, depending on the task at hand, MXNet has some very interesting charts that compare the different pre-trained models optimized for tasks. For image classification (based on ImageNet), we have the following:

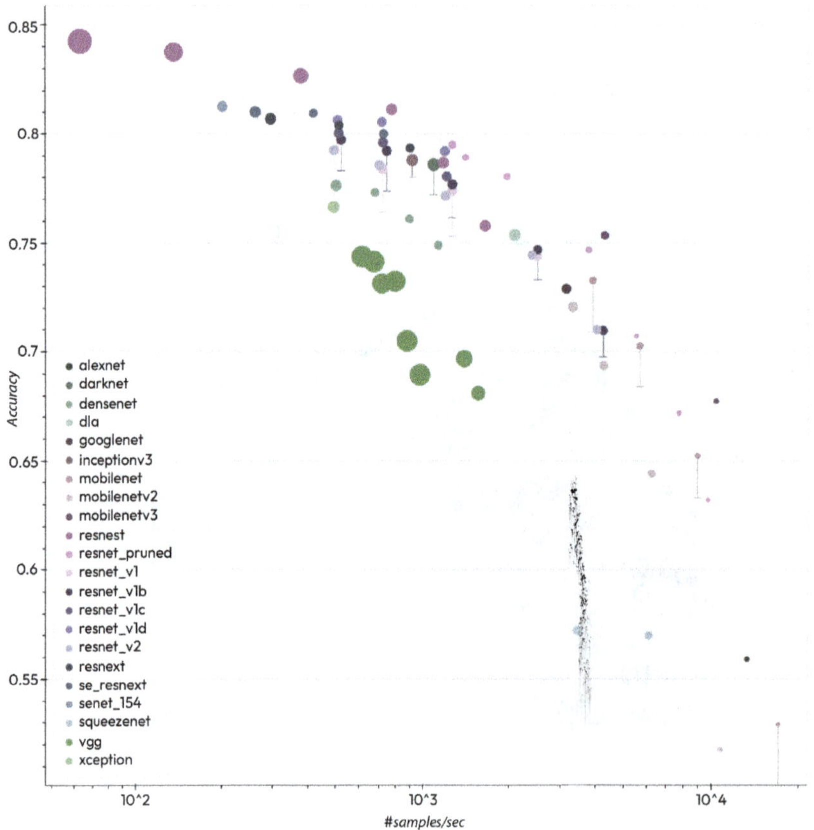

Figure 5.11 – Model Zoo for image classification (ImageNet)

> **Note**
>
> Source: `https://cv.gluon.ai/model_zoo/classification.html`

Figure 5.11 displays the most important pre-trained models in **GluonCV Model Zoo**, according to accuracy (the vertical axis) and inference performance (the samples per second and horizontal axis). There are no models (yet) in the top-right quadrant, meaning that, currently, we need to balance these characteristics.

Using models from GluonCV Model Zoo can be done with just a couple of lines of code, and we will explore this path to solve our reduced *Dogs vs. Cats* dataset in the following steps.

ImageNet pre-trained models

Models from the **GluonCV Model Zoo** for **image classification** tasks have been pre-trained in the *ImageNet* dataset. This dataset is one of the most well-known datasets in computer vision. It was the first large-scale image dataset and was part of the deep learning revolution when, in 2012, AlexNet won the ILSVRC.

This dataset has two variants:

- **Full dataset**: More than 20,000 classes in about 14 million images
- **ImageNet-1k**: 1,000 classes in about 1 million images

Due to the size and large number of classes, the full dataset is rarely used on benchmarks, with *ImageNet1k* the de facto ImageNet dataset (unless otherwise noted in the research papers, articles, and so on). Images in the dataset are in color and have a size of 224px by 224px (width and height).

All image classification pre-trained models in GluonCV Model Zoo have been pre-trained with ImageNet-1k, and therefore, they have 1,000 outputs. The outputs will be post-processed so that all ImageNet classes corresponding to cats point to class 0, all ImageNet classes corresponding to dogs point to class 1, and all other outputs point to class 2, which we will consider as unknown.

Loading an AlexNet pre-trained model from Model Zoo

To compare the advantages and disadvantages of using a custom-trained model and a pre-trained model from Model Zoo, in the following sections, we will work with a version of the AlexNet architecture that has been pre-trained on the *ImageNet* dataset, acquired from GluonCV Model Zoo.

Loading a pre-trained model is very easy and can be done with a single line of code:

```
alexnet = gcv.model_zoo.get_model("alexnet", pretrained=True, ctx=ctx)
```

The `get_model` GluonCV function receives three parameters:

- **The name of the model**: In this case, `alexnet`
- **pretrained**: A Boolean indicating whether we want to load the pre-trained weights and biases (if set to `False`, only the uninitialized architecture will be loaded)
- **ctx**: The context – typically, `mx.cpu()` or `mx.gpu()`, if available

This call will download the chosen model and, if required, its pre-trained weights and biases.

Evaluating an AlexNet pre-trained model from Model Zoo

With the loaded model from the previous section, we can now evaluate and compare our previous results, such as for `accuracy`:

```
('accuracy', 0.725)
```

As we can see, this number is slightly lower than the accuracy we achieved with our custom-trained AlexNet model.

After computing the confusion matrix, we obtain the following values:

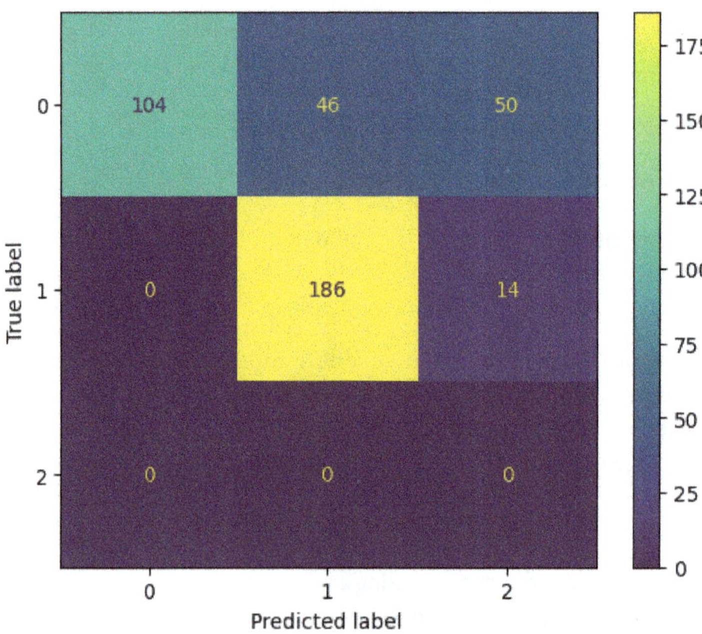

Figure 5.12 – A pre-trained AlexNet Confusion Matrix

When we analyze *Figure 5.12*, the most significant difference is that our previous confusion matrix was a 2x2 matrix (with the options of **0** or **1** for true labels and **0** or **1** for predicted labels). However, with our pre-trained model, we have obtained a 3x3 confusion matrix. This is because, as mentioned previously, pre-trained models have been trained in ImageNet, and they output 1,000 classes (instead of the two required for our dataset). These outputs have been post-processed, so all *ImageNet* classes corresponding to cats point to class **0**, all ImageNet classes corresponding to dogs point to class **1**, and all other outputs point to class **2**, which we will consider as unknown. When taking into account this unknown class **2**, the 3x3 matrix is computed. Please note how there are no images that produce a true label of **2**; the last row is all zeros.

The model mostly behaves accurately with the expected classes, with the following per-class errors (we need to add numbers from the two wrong columns):

- **Cats not detected as cats**: **96**/200 (48% of cats images were misclassified)
- **Dogs not detected as dogs**: **14**/200 (7% of dogs images were misclassified)

There is a significant difference in the per-class results, and this is due to the pre-trained dataset used, *ImageNet*, because it has a large number of classes associated with dog breeds and, therefore, has been trained more extensively on dog images.

Loading a ResNet pre-trained model from Model Zoo

Looking at AlexNet and later models with a higher depth, such as VGGNet, it became clear that deeper layers could help when classifying images. However, when training these deep networks, by using *backpropagation* and the *chain rule*, the training algorithm starts computing smaller and smaller values for the gradients because of the large number of multiplications of small numbers (the activation function outputs are in the [0, 1] range), and therefore, when the gradients for the early layers are computed, the updated weights are seldom modified. This is known as the **vanishing gradient** problem, and different **ResNet** architectures were developed to avoid it. Specifically, ResNet models use residual blocks that add direct lines to connect layers, providing shortcuts that can be leveraged during training to avoid vanishing gradients.

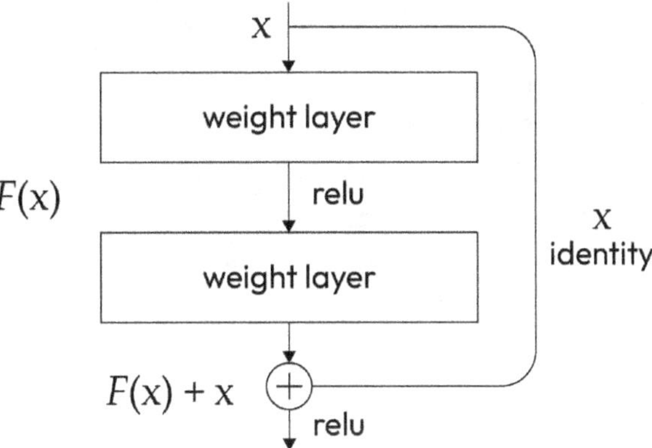

Figure 5.13 – A ResNet residual block

The architecture shown in *Figure 5.13* allows you to stack layers in a more scalable way, with known architectures with 18, 50, 101, and 152 layers. This approach proved very successful, and ResNet152 won the ILSVRC in 2015.

In this case, we will load the `v1d` version of `resNet50`, with a single line of code:

```
resnet50 = gcv.model_zoo.get_model("resnet50_v1d", pretrained=True,
ctx=ctx)
```

The model then downloads successfully.

Evaluating a ResNet pre-trained model from Model Zoo

With the loaded model from the previous section, we can now evaluate and compare with our previous results, such as for `accuracy`:

```
('accuracy', 0.925)
```

As we can see, this number is significantly higher than previous models.

After computing the confusion matrix, we obtain the following values:

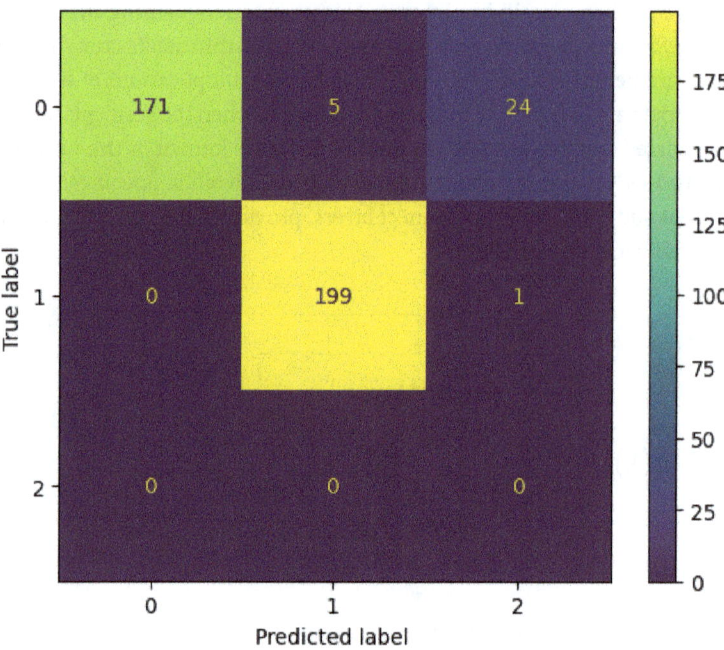

Figure 5.14 – A pre-trained ResNet confusion matrix

According to *Figure 5.14*, the per-class error rate is as follows:

- **Cats not detected as cats**: **29**/200 (14.5% of cat images were misclassified)
- **Dogs not detected as dogs**: **1**/200 (0.5% of dog images were misclassified)

There is a significant difference in the per-class results, and this is again due to the pre-trained dataset used, *ImageNet*, because it has a large number of classes associated with dog breeds and, therefore, has been trained more extensively on dog images.

How it works...

In this recipe, we compared two approaches to using computer vision models for image classification:

- Training a custom model from scratch
- Using pre-trained models from GluonCV Model Zoo

We applied both approaches with AlexNet architecture and compared the results with the **ResNet-101 Model Zoo** version.

Both approaches have advantages and disadvantages. Training from scratch provides us direct control on the number of output classes, and we can fully handle the training process and the evolution of loss and accuracy for both the training and validation datasets.

However, in order to train a model, we need sufficient data, and that might not be always available. Furthermore, adjusting the training hyperparameters (epochs, batch size, optimizer, and learning rate) and the training itself are time-consuming processes that, if not done properly, can yield suboptimal accuracy values (or other metrics).

In our examples, we used a reduced version of the *Dogs vs. Cats* dataset from Kaggle, and we used pre-trained models on *ImageNet*. The Kaggle dataset contains 25,000 images (more than 10 times more) and the reader is encouraged to try the proposed solution on that dataset (all helper functions have also been tested with the full dataset).

Furthermore, the choice of the *ImageNet* dataset was not casual; *ImageNet* has dog classes and cat classes, and therefore, the expectation was that these pre-trained models would perform well, as they had already seen images from the `dataset` class. However, when this is not possible, and we apply pre-trained models from a dataset to another dataset, the data probability distribution will typically be very different; hence, the accuracy obtained can be very low. This is known as the **domain gap** or **domain adaptation problem** between the source dataset (the data the model has been pre-trained on) and the target dataset (the data the model is evaluated on).

One way to tackle these issues for supervised learning problems is fine-tuning. This approach is explored in detail in *Chapter 7*.

We finalized the recipe evaluating our two pre-trained models, *AlexNet* and *ResNet*, and saw how CNN models have evolved through the years, increasing the accuracy obtained.

There's more...

In this recipe, we used *ImageNet* pre-trained models; for more information about *ImageNet*, and the ILSVRC, I suggest this article: `https://machinelearningmastery.com/introduction-to-the-imagenet-large-scale-visual-recognition-challenge-ilsvrc/`.

Although primarily about the ILSVRC, this previous link also includes some history regarding CNNs, including *AlexNet*, VGGNet, and *ResNet*.

However, computer vision datasets have been under strict scrutiny recently regarding data quality, and ImageNet is no exception, as this article describes: `https://venturebeat.com/2021/03/28/mit-study-finds-systematic-labeling-errors-in-popular-ai-benchmark-datasets/`.

Figure 5.11 shows a static image corresponding to the accuracy versus samples per second graph for the Model Zoo for image classification (on *ImageNet*). A snapshot of a dynamic version is available at this link and is worth taking a look at: `https://cv.gluon.ai/model_zoo/classification.html`.

At this link, results from different models available in GluonCV Model Zoo are included, and it is suggested that you reproduce these results, as it is an interesting exercise.

Apart from ImageNet, GluonCV Model Zoo provides models pre-trained on **CIFAR10**. A list of these models can be found at `https://cv.gluon.ai/model_zoo/classification.html#cifar10`.

For a deeper explanation of the vanishing gradient problem, Wikipedia provides a good start: `https://en.wikipedia.org/wiki/Vanishing_gradient_problem`.

Lastly, regarding ResNet and its current research relevance, in a recently published paper, it was shown that ResNet can still achieve **State-of-the-Art** (**SOTA**) results when the latest researched training techniques are applied to it, highlighting the importance of datasets and a training algorithm (versus optimization only for model architecture): `https://gdude.de/blog/2021-03-15/Revisiting-Resnets`.

Detecting objects with MXNet – Faster R-CNN and YOLO

In this recipe, we will see how to use MXNet and GluonCV on a pre-trained model to detect objects from a dataset. We will see how to use GluonCV Model Zoo with two very important models for **object detection** – **Faster R-CNN** and **YOLOv3**.

In this recipe, we will compare the performance of these two pre-trained models to detect objects on the *Penn-Fudan Pedestrians* dataset.

Getting ready

As for previous chapters, in this recipe, we will be using a few matrix operations and linear algebra, but it will not be too difficult.

As we will unpack in this recipe, object detection combines classification and regression, and therefore, chapters and recipes where we explored the foundations of these topics are recommended to revisit. Furthermore, we will be detecting objects on image datasets. This recipe will combine what we learned in the following chapters:

- *Understanding image datasets: load, manage, and visualize the Fashion MNIST dataset*, the third recipe from *Chapter 2, Working with MXNet and Visualizing Datasets: Gluon and DataLoader*
- *Chapter 3, Solving Regression Problems*
- *Chapter 4, Solving Classification Problems*

How to do it...

In this recipe, we will take the following steps:

1. Introduce object detection.
2. Evaluate object detectors.
3. Compare *Single-stage* and *Two-stage* object detectors.
4. Explore the *Penn-Fudan Pedestrians* dataset.
5. Introduce the Object Detection Model Zoo.
6. Worke with *MS COCO* pre-trained models.
7. Load a Faster R-CNN pre-trained model from *Model Zoo*.
8. Evaluate a *Faster R-CNN* pre-trained model from *Model Zoo*.
9. Load a YOLOv3 pre-trained model from *Model Zoo*.
10. Evaluate a *YOLOv3* pre-trained model from *Model Zoo*.
11. Conclude what we have learned.

Introducing object detection

In some of the previous chapters and recipes, we analyzed image classification problems, where the task of our models was to take an image and define the class most likely associated with it. In **object detection**, however, there can be multiple objects per image, corresponding to different classes, and in different locations of the image, and therefore, the output is now two lists, one providing the most likely class of each detected object and another indicating the estimated location of the object. The class output can be modeled as a classification problem, and the bounding box output can be modeled as a regression problem. Typically, locations are represented with what is called a bounding box. An example of bounding boxes is shown as follows:

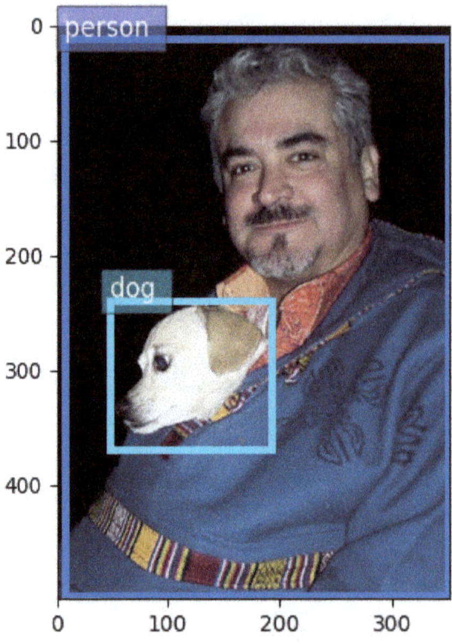

Figure 5.15 – Bounding box examples

In *Figure 5.15*, we can see two examples of bounding boxes for two different classes – `person` and `dog`.

Evaluating object detectors

In image classification, we defined a correct classification if the class identified in one image was the right one. However, in object detection, there are two parameters – the class and the bounding box. Intuitively, we can define a correct classification if, for each object that should be detected, there is a *similar enough* bounding box that has been classified properly. To define what *similar enough* means, we compute **Intersection over Union** (**IoU**), the ratio of the intersection of the area of the bounding boxes over the area of the union of the bounding boxes.

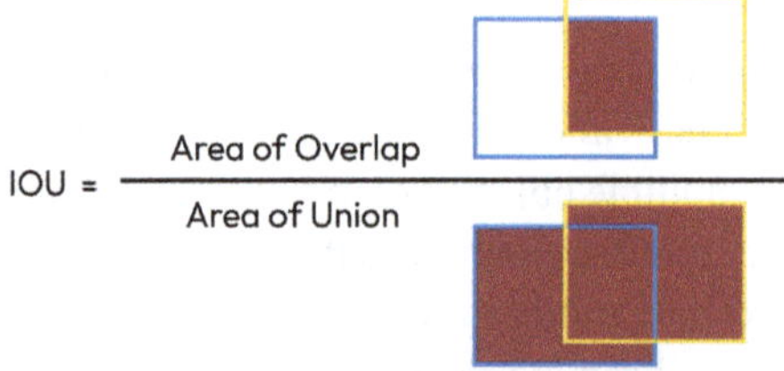

Figure 5.16 – IoU

A graphical interpretation of IoU can be seen in *Figure 5.16*. When the IoU is above a determined threshold, the bounding boxes are said to match. By using IoU and its threshold (typically 0.5), a detection can be classified as correct (given the object was also correctly classified), and metrics such as accuracy, precision, and recall can be computed (per class).

Furthermore, in *Chapter 4, Solving Classification Problems*, we discussed several options to evaluate *classification* problems. We introduced **Area Under the Curve** (**AUC**), and we saw how changing the threshold had an influence on **precision** and **recall**. When plotting precision and recall together (the **PR curve**), we can see the effect of the threshold, in the same way as we did for AUC. If we calculate the area covered between the curve (the *x* axis, *y = 0 axis*, and *y = 1 axis*), we obtain a parameter that is not dependent on the threshold value; it defines the performance of our model with the given data:

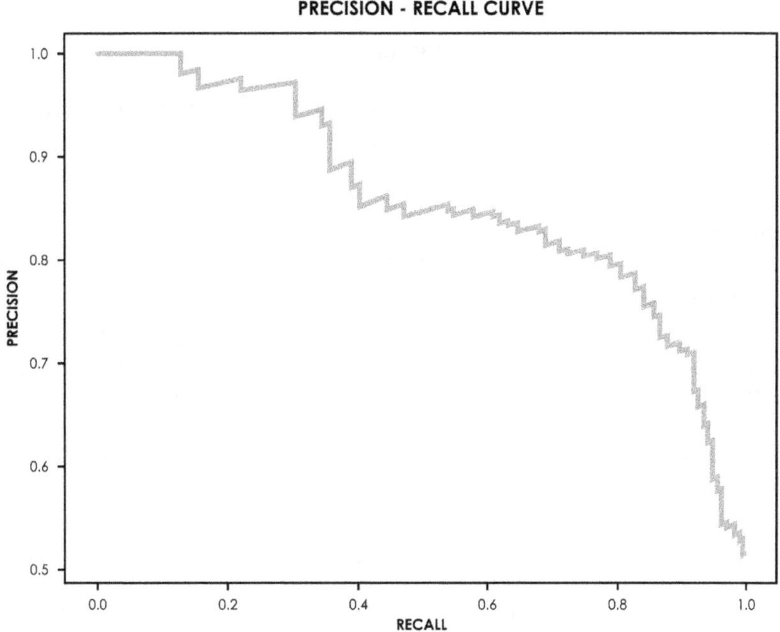

Figure 5.17 – The PR curve

One of the characteristics of this curve that we can see clearly in *Figure 5.17* is its zig-zag pattern. As we decrease the threshold, the curve goes down with false positives and goes up again with true positives.

However, in order to be able to compare different models easily, instead of comparing the PR curves for each class, a single number metric was developed, the **mean Average Precision** (**mAP**). In short, it is the mean of all the areas under the PR curves for a model:

$$mAP = \frac{1}{N}\sum_{i=1}^{N} AP_i$$

To compute *mAP*, the first step is to calculate the average precision for each class, which is the area under the PR curve and then compute its arithmetic mean. This value provides a single number where object detection models evaluated on the same dataset can be compared.

Comparing single-stage and two-stage object detectors

We can think of object detectors as cropping a specific part of an image and passing that cropped image through an image classifier, similar to what we did with full images in the previous recipe. Using this approach, our object detector will have two steps:

1. **Region proposal network**: This is the module that will indicate the regions where an object could be located.
2. **Object classifier**: The regions will be classified by the model, with the regions previously cropped and resized to match the model input constraints.

This approach is known as **two-stage object detection**, and its most important characteristic is its accuracy, although it is slow due to its complex architecture, and fully accurate (non-approximate) training cannot be done end to end.

Faster **Region-based Convolutional Neural Network** (**R-CNN**) is one of the models that follow this approach. One of the most important differences from previous versions of the model (**R-CNN** and **Fast R-CNN**) is that in order to provide faster computations, it uses pre-computed bounding boxes called *anchor boxes*, where the scale and the aspect ratio of the bounding box are pre-defined. This approach allows the networks to be modeled to compute the *offset* related to an anchor box, instead of the full bounding box coordinates, which simplifies the regression problem.

Another algorithm to improve computation times is **Non-Maximum Suppression** (**NMS**). Typically, thousands of regions will be proposed for the next step of the object detection pipeline. Many of these regions overlap with each other, and NMS is the algorithm that takes into account the confidence of the prediction, removing all overlapping regions over an IoU threshold.

Another approach for object detectors is to design architectures that make predictions of bounding boxes and class probabilities together, allowing end-to-end training in one step. Architectures following this approach are known as **single-stage object detectors**. These architectures also make use of anchor boxes and NMS to improve the regression task. The two most famous architectures using this approach are as follows:

- **You Only Look Once** (**YOLO**): The image is processed just once using a custom CNN architecture (a combination of convolutional and max-pooling layers), ending with two fully connected layers. These architectures have been developed continuously, with *YOLOv3* being one of the most popular.
- **Single Shot Detector** (**SSD**): The image is processed using a CNN backbone architecture (such as VGG-16) to compute feature maps, and the generated multi-scale feature maps are then classified. SSD512 (using *VGG-16* as the backbone) is one of the models that follow this architecture.

YOLOv3 model is the fastest and yields reasonable accuracy metrics, the SSD512 model is a good trade-off between speed and accuracy, and the Faster R-CNN model has the highest accuracy and is the slowest of the three.

Exploring the Penn-Fudan Pedestrians dataset

For our object detection experiments, we will work with a new dataset - *Penn-Fudan Pedestrians*. This is a publicly available dataset (`https://www.cis.upenn.edu/~jshi/ped_html/`) and is a collaboration between the universities of Pennsylvania and Fudan, and it must be downloaded manually.

The dataset has 423 pedestrians annotated from 170 images; 345 pedestrians were annotated for the release of the dataset (2007) and 78 pedestrians were added later, as the previous ones were either small or occluded.

From the set of images of the datasets, our models will need to correctly detect the pedestrians present in the images and localize them, using bounding boxes. To understand the problem better, as we saw in previous chapters, we are going to do some EDA.

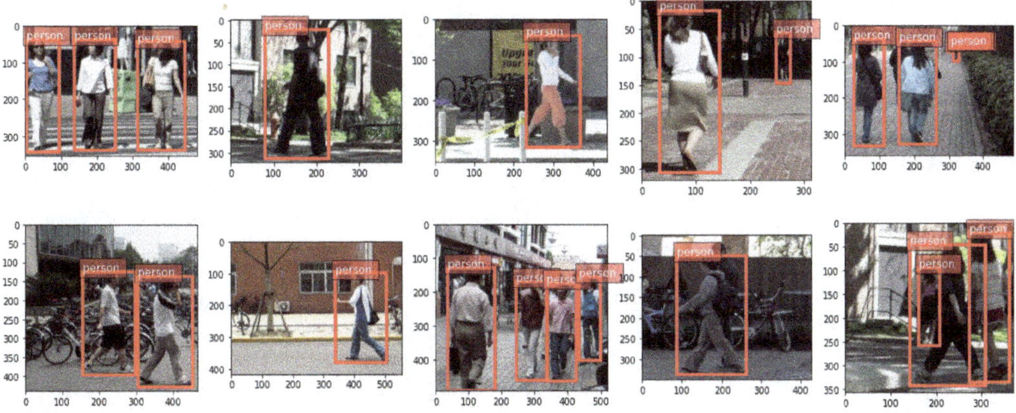

Figure 5.18 – The Penn-Fudan Pedestrians dataset

As we can see in *Figure 5.18*, each image can have one or more pedestrians, with their corresponding bounding boxes. Each image in the dataset is in color, and they have a variable width and height, which are later resized depending on the model requirements. This figure was computed using the GluonCV visualization utils package (the `plot_bbox` function):

```
gcv.utils.viz.plot_bbox(image, gt_bboxes, class_names=["person"],
ax=axes)
```

As for this datase,t there are no different classes to be classified, and no further visualizations are computed.

Introducing object detection Model Zoo

GluonCV also provides pre-trained models for object detection in its Model Zoo. For the *MS COCO* dataset, this is the accuracy (mAP) versus performance (samples per second) chart:

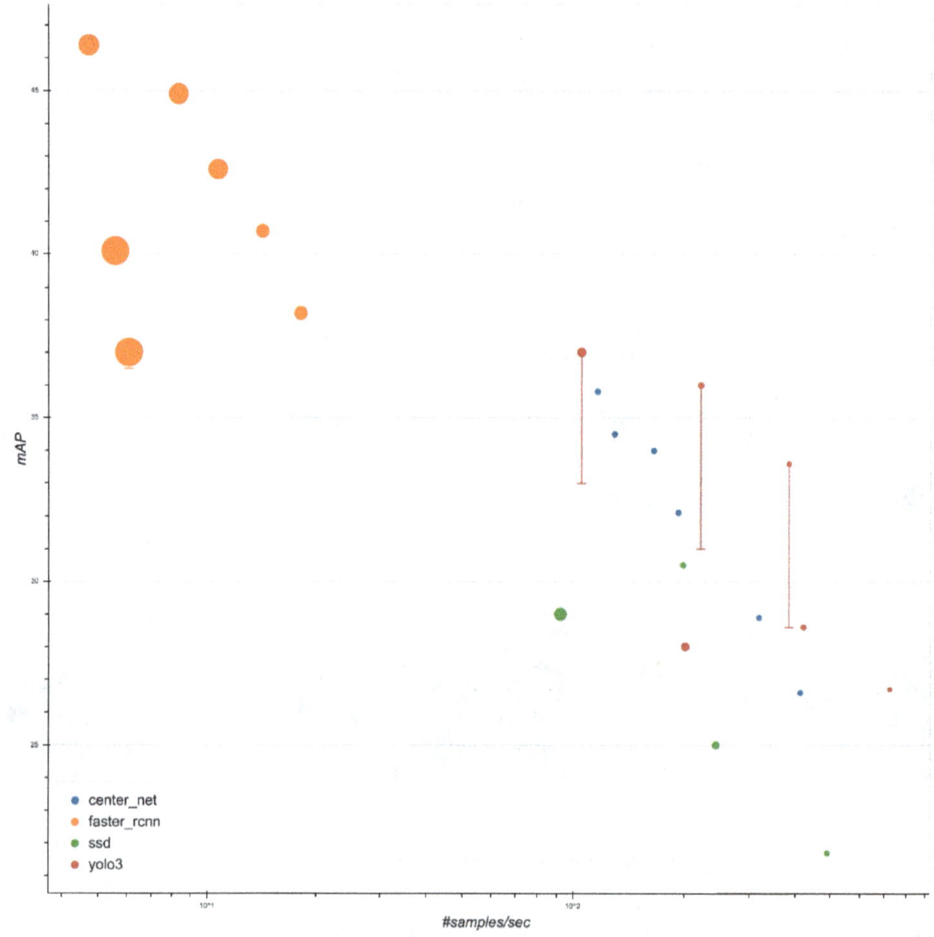

Figure 5.19 – Model Zoo for object detection (MS COCO)

> **Note**
>
> Image adapted from the following source: `https://cv.gluon.ai/model_zoo/detection.html`

Figure 5.19 displays the most important pre-trained models in GluonCV Model Zoo, comparing accuracy (mAP on the vertical axis) and inference performance (samples per second on the horizontal axis). There are no models (yet) in the top-right quadrant, meaning that, currently, we need to balance between both characteristics.

Using models from GluonCV Model Zoo can be done with just a couple of lines of code, and we will explore this path to solve our *Penn-Fudan Pedestrians* dataset in the following steps.

Working with MS COCO pre-trained models

Models from the GluonCV Model Zoo for object detection tasks have been pre-trained in the *MS COCO* dataset. This dataset is one of the most popular object detection datasets in computer vision. It was developed by Microsoft in 2015 and was updated until 2017. In its most recent update, it contains 80 classes (plus background) and is composed of the following:

- **Training/validation sets**: 118,000/5,000 images
- **A test set**: 41,000 images

Several object detection pre-trained models in GluonCV Model Zoo have been pre-trained with *MS COCO*, and therefore, each object detected will be classified among 80 classes. As there can be several objects in each image, in MXNet GluonCV implementations, the outputs of an object detection model are structured as follows:

- **An array of indices**: For each object detected, this array gives the index of the class of the detected object. The shape of this array is *BxNx1*, where *B* is the batch size and *N* is the number of objects detected per image (depending on the model).
- **An array of probabilities**: For each object detected, this array gives the probability associated with the detected object of being the detected class in the **array of indices**. The shape of this array is BxNx1, where B is the batch size and N is the number of objects detected per image (depending on the model).
- **An array of bounding boxes**: For each object detected, this array gives the bounding box coordinates associated with the detected object. The shape of this array is *BxNx4*, where *B* is the batch size, N is the number of objects detected per image (depending on the model), and 4 is the coordinates in the format *[x-min, y-min, x-max, y-max]*.

Looking at *Figure 5.19*, two separated groups can be seen – the Faster R-CNN family, which has the highest accuracy but is slow, and the YOLO family, which is very fast but has a lower accuracy. For our experiments, we have selected two of the most popular models, each of them corresponding to a different Faster R-CNN (the backbone of ResNet-101 with the FPN version) and *YOLOv3* (the backbone of Darknet53, a 53 CNN).

Furthermore, MS COCO contains the `person` class, and therefore, models pre-trained in MS COCO are well suited for the *Penn-Fudan Pedestrian* dataset.

Loading a Faster R-CNN pre-trained model from Model Zoo

Faster R-CNN is a two-stage object detection architecture, meaning that, first, it provides regions where objects could be located, and second, by analyzing those regions, it provides the classes and locations of the detected objects. It was developed by Ren et al. (Microsoft Research) in 2014.

It is the third iteration of a series of architectures that have evolved from each other – R-CNN, Fast R-CNN, and Faster R-CNN.

From the research papers (There's more section), we can see an R-CNN architecture:

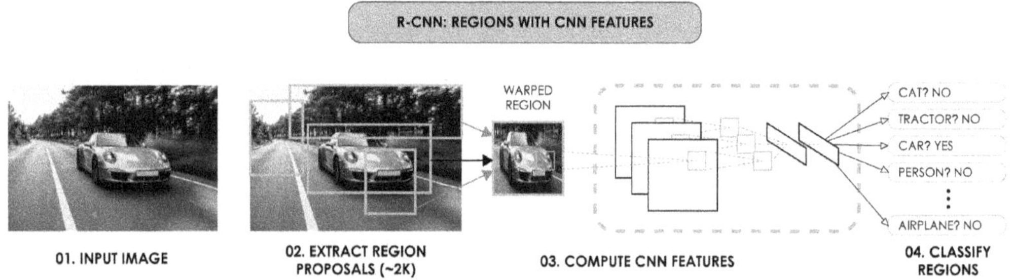

Figure 5.20 – The high-level architecture of R-CNN

> **Note**
>
> Source: The car image has the following source: *azerbaijan_stockers* on Freepik: `https://www.freepik.com/free-photo/mini-coupe-high-speed-drive-road-with-front-lights_6159501.htm#query=CAR&position=48&from_view=search&track=sph&uuid=e82c3ce9-2fe8-40ef-9d39-e8d27781fdf2`

Secondly, we can see the Fast R-CNN architecture:

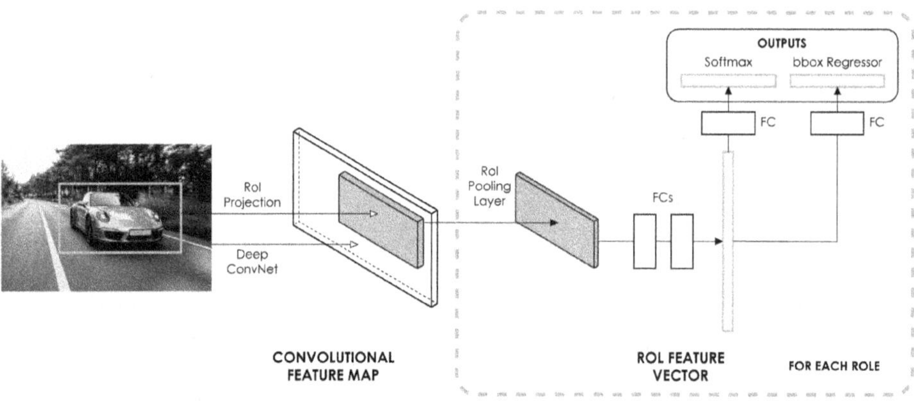

Figure 5.21 – The high-level architecture of Fast R-CNN

> **Note**
>
> Source: The car image has the following source: *azerbaijan_stockers* on Freepik: `https://www.freepik.com/free-photo/mini-coupe-high-speed-drive-road-with-front-lights_6159501.htm#query=CAR&position=48&from_view=search&track=sph&uuid=e82c3ce9-2fe8-40ef-9d39-e8d27781fdf2`

Finally, we can see the Faster R-CNN architecture:

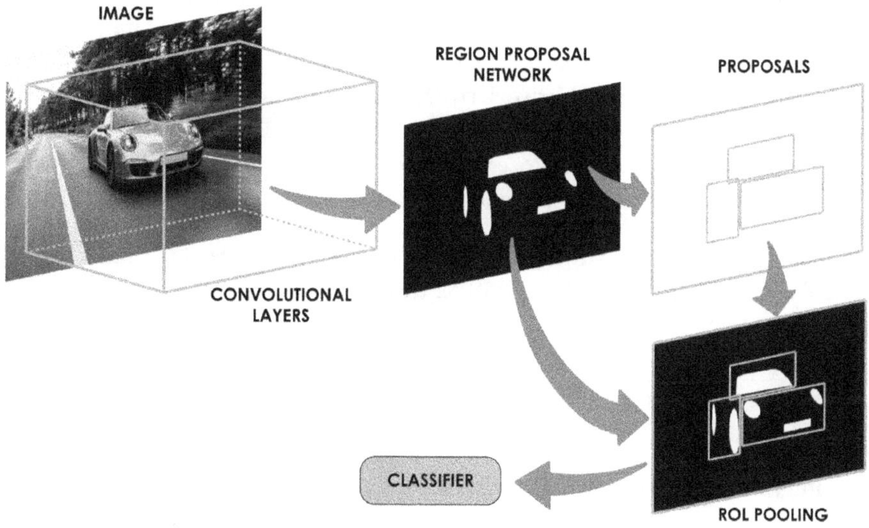

Figure 5.22 – The high-level architecture of Faster R-CNN

> **Note**
>
> Source: The car image has the following source: *azerbaijan_stockers* on Freepik: `https://www.freepik.com/free-photo/mini-coupe-high-speed-drive-road-with-front-lights_6159501.htm#query=CAR&position=48&from_view=search&track=sph&uuid=e82c3ce9-2fe8-40ef-9d39-e8d27781fdf2`

The different architectures have similarities and differences, namely the following:

- **R-CNN**: This uses the selective search algorithm to provide 2,000 region proposals. Each of these regions is then fed into a CNN, which generates a feature vector for each object of 4,096 features and the four-coordinate's bounding box. These feature vectors are the input to the **Support Vector Machine** (**SVM**) classifier.
- **Fast R-CNN**: In this iteration, instead of feeding each region to a CNN, the whole image is fed once, and a single feature map of the full image is computed, accelerating the process.

Then, **Regions Of Interest** (**ROIs**) are computed over this feature map (instead of the image), obtained similarly with the *selective search* algorithm. These are then passed through an *ROI pooling* layer, where each object in a proposed region is assigned a feature map of the same shape. This is an efficient method in which the output can now be fed into the two networks for the regressor and classifier, which provide the location and class of each object respectively. These two networks are based on fully connected layers. The regressor computes offsets from the ROIs, and for the classifier, the output activation function is *softmax*.

- **Faster R-CNN**: In this last iteration, three changes are introduced. Firstly, a backbone CNN is also used to compute the feature map of the image; however, instead of using the selective search algorithm to propose regions, some *fully convolutional* layers are added on top of the backbone CNN called a **Region Proposal Network** (**RPN**), yielding a much smaller computation time. Secondly, these region proposals are computed as offsets associated with anchor boxes. Lastly, to reduce the number of regions to process, **Non-Maximum Suppression** (**NMS**) is used. These three changes provide faster inference and higher accuracy.

For our experiments, we will use as the backbone the `v1d` version of weights from a ResNet-101 network, with a **feature pyramid network** as the RPN. We can load the model with a single line of code:

```
faster_rcnn = gcv.model_zoo.get_model("faster_rcnn_fpn_resnet101_v1d_
coco", pretrained=True, ctx=ctx)
```

The model then downloads successfully.

The GluonCV implementation of this model is capable of detecting 80,000 distinct objects.

Evaluating a Faster R-CNN pre-trained model from Model Zoo

Using the *Penn-Fudan Pedestrian* dataset, we can now perform qualitative and quantitative evaluation of the loaded model from the previous section.

Qualitatively, we can choose an image from the dataset and compare the output of the model with the ground-truth output from the dataset:

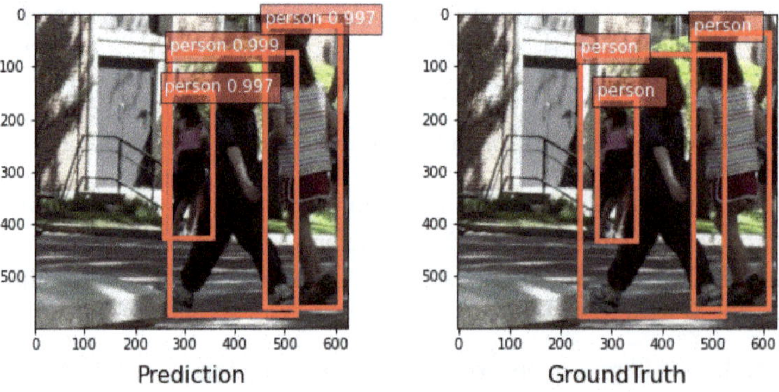

Figure 5.23 – Comparing predictions and ground-truth for Faster R-CNN

From *Figure 5.23*, we can see a very strong correlation between the predicted segmentation masks and the expected results from the ground-truth, as well as strong confidence (+99%) and perfect class accuracy from the model. This figure was computed using the GluonCV visualization `utils` package (the `plot_bbox` function).

Quantitatively, we can perform a mAP evaluation and the runtime spent computing this metric:

```
('VOCMeanAP', 0.6716161702078043)
 Elapsed Time:  249.30912852287292 secs
```

The computed mAP for this model for *MS COCO* (see the object detection model zoo) is 40.7; therefore, given the value of 0.67 for our model, we can conclude our model performs the task accurately. However, it did take some time to complete (~250 seconds), which is expected for Faster R-CNN architectures.

Loading a YOLOv3 pre-trained model from Model Zoo

You Only Look Once Version 3 (**YOLOv3**) is a single-stage object detection architecture, meaning it uses an end-to-end approach that makes predictions of bounding boxes and class probabilities in a single step. It was developed by Redmon et al. (at the University of Washington) from 2016 (YOLO) to 2018 (YOLOv3).

It is the third iteration of a series of architectures that have evolved from each other – YOLO, YOLOv2, and YOLOv3.

From the research papers, we can see the YOLO architecture:

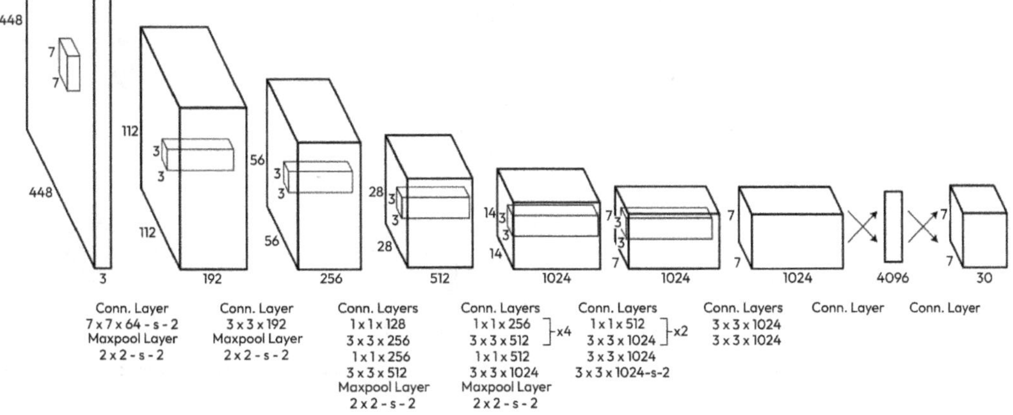

Figure 5.24 – The architecture of YOLOv1

Secondly, we can see the YOLOv2 architecture:

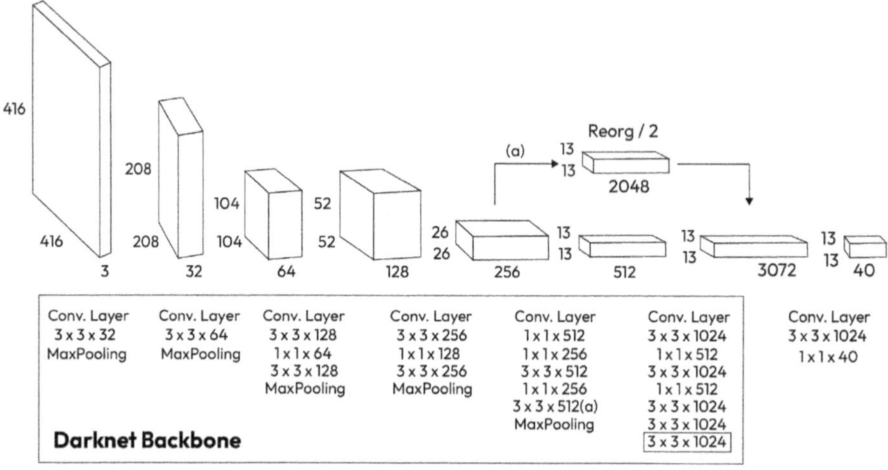

Figure 5.25 – The architecture of YOLOv2

And finally, we can see the YOLOv3 architecture:

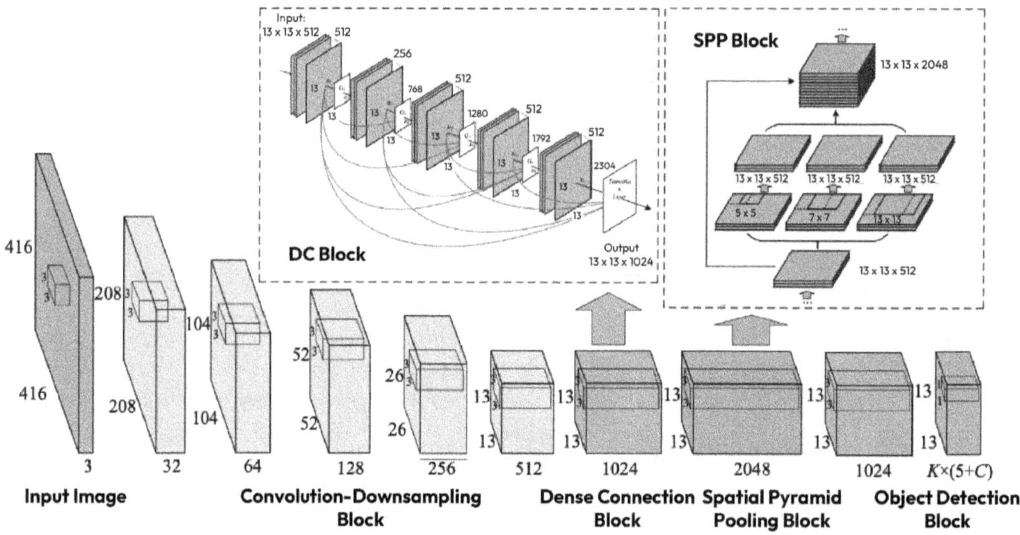

Figure 5.26 – The architecture of YOLOv3

The different architectures have similarities and differences, namely the following:

- **YOLO**: The initial model decomposes each image into grid cells of equal size. Each cell is responsible for detecting an object if the location of the center of the object is within the cell. Each cell can predict two bounding boxes, their class, and their confidence score, but only one object (with a different size and location). All the predictions are made simultaneously, using a CNN composed of 24 convolutional layers and 2 fully connected layers. There is a large number of overlapping bounding boxes, and NMS is used to reach the final output.
- **YOLOv2**: YOLO architecture struggles to detect small objects in groups. To solve this issue, on this iteration, a number of changes are introduced – batch normalization to improve training accuracy, anchor boxes for regression, increased detection capabilities to five bounding boxes per cell, and a new backbone network, DarkNet-19, with 19 convolutional layers and 5 max pooling layers, with 11 more layers for detection.
- **YOLOv3**: In this last iteration, three changes are introduced. Firstly, to increase further the detection accuracy of small objects, the backbone network is updated to DarkNet-53, with 53 convolutional layers and 53 layers for the detection head, allowing for predictions on three different scales. Secondly, the number of bounding boxes per cell is reduced from five to three; however, taking into account the three different scales, this provides nine anchor boxes.

For our experiments, we will use DarkNet-53 as a backbone. We can load the model with a single line of code:

```
yolo = gcv.model_zoo.get_model("yolo3_darknet53_coco",
pretrained=True, ctx=ctx)
```

The model then downloads successfully.

The GluonCV implementation of this model is capable of detecting 100 distinct objects.

Evaluating a YOLOv3 pre-trained model from Model Zoo

Using the *Penn-Fudan Pedestrian* dataset, we can now perform qualitative and quantitative evaluation of the loaded model from the previous section.

Qualitatively, we can choose an image from the dataset and compare the output of the model with the ground-truth output from the dataset:

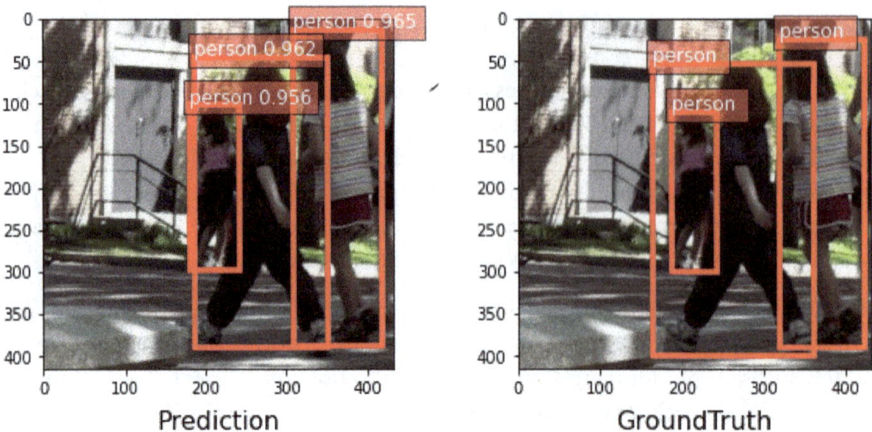

Figure 5.27 – Comparing predictions and ground-truth for YOLOv3

From *Figure 5.27*, we can see a very strong correlation between the bounding boxes of the expected results from the ground-truth and the actual outputs of the model, as well as strong confidence (+95%) and perfect class accuracy from the model. This figure was computed using the GluonCV visualization utils package (the `plot_bbox` function).

Quantitatively, we can perform a mAP evaluation and the runtime spent computing this metric:

```
('VOCMeanAP', 0.5339115787945962)
 Elapsed Time:  113.4049768447876 secs
```

The computed mAP for this model for *MS COCO* (see the object detection model zoo) is 36.0; therefore, given the value of `0.53` for our model, we can conclude that it is performing the task accurately. It took `113` seconds to complete.

How it works...

In this recipe, we tackled the object detection problem. We analyzed the differences between image classification problem and object detection problem for evaluation, network architectures, and datasets.

In our examples, we used a publicly available dataset, the *Penn-Fudan Pedestrians* dataset, and we used pre-trained models on MS COCO. This choice was not casual; *MS COCO* has a `person` class, and therefore, the expectation was that these pre-trained models would perform well, as they had already seen images from the dataset class. However, as mentioned in the previous recipe, when this is not possible, and we apply pre-trained models from a dataset to another dataset, the data probability distribution will typically be very different; hence, the accuracy obtained can be very low. This is known as the domain gap or domain adaptation problem between the source dataset (the images that a model has been pre-trained on) and the target dataset (the images that the model is evaluated on).

One way to tackle these issues for supervised learning problems is fine-tuning. This approach is explored in detail in *Chapter 7*.

We ended the recipe by evaluating our two pre-trained models, Faster R-CNN and YOLOv3, and we were able to confirm that our Faster R-CNN model was very accurate but slow, while YOLOv3 was much faster (2x) with a slightly lower accuracy (~20% decrease).

There's more...

Figure 5.19 showed a static image corresponding to the mAP versus samples per second graph for the Model Zoo for object detection (on *MS COCO*). There is also a dynamic version at this link that is worth taking a look at: `https://cv.gluon.ai/model_zoo/detection.html`. On this page, results from different models available in *GluonCV Model Zoo* are included; I suggest that you reproduce these result,s as it is an interesting exercise.

To understand more about the *MS COCO* dataset, the original paper is available at `https://arxiv.org/pdf/1405.0312.pdf`.

Furthermore, it is very interesting to read the original research papers and see how object detectors have evolved from an academic point of view:

- **R-CNN**: `https://www.cv-foundation.org/openaccess/content_cvpr_2014/papers/Girshick_Rich_Feature_Hierarchies_2014_CVPR_paper.pdf`
- **Fast R-CNN**: `https://arxiv.org/pdf/1504.08083.pdf`
- **Faster R-CNN**: `https://proceedings.neurips.cc/paper/2015/file/14bfa6bb14875e45bba028a21ed38046-Paper.pdf`
- **YOLO**: `https://arxiv.org/pdf/1506.02640.pdf`
- **YOLOv2**: `https://arxiv.org/pdf/1612.08242.pdf`
- **YOLOv3**: `https://arxiv.org/pdf/1804.02767.pdf`

In this recipe, we have seen how Faster R-CNN was more accurate but slower, and YOLOv3 was faster but less accurate. To balance the trade-off between accuracy and inference time for object detection problems, there are different possibilities.

One option is to estimate the difficulty of an image and apply a different object detector. If it is a challenging image, use the Faster R-CNN family; if it is simpler, use YOLOv3. This approach is explored in detail in this paper: `https://arxiv.org/pdf/1803.08707.pdf`; however, *fine-tuning* a fast model such as YOLOv3 is recommended as an initial approach to this issue.

Segmenting objects in images with MXNet – PSPNet and DeepLab-v3

In this recipe, we will see how to use MXNet and GluonCV on a pre-trained model, segmenting objects in images from a dataset. This means that we will be able to split objects into different classes, such as `person`, `cat`, and `dog`. When framing the problem as segmentation, the expected output is an image of the same size as the input image, with each pixel value being the classified label (we will analyze how this works in the following sections). We will see how to use GluonCV Model Zoo with two very important models for **semantic segmentation** – **PSPNet** and **DeepLab-v3**.

In this recipe, we will compare the performance of these two pre-trained models to segment objects semantically on the dataset introduced in the previous chapter, *Penn-Fudan Pedestrians*, as its ground-truth also includes segmentation masks.

Getting ready

As with previous chapters, in this recipe, we will use a few matrix operations and linear algebra, but it will not be too difficult.

As we will unpack in this recipe, semantic segmentation is similar to classification and object detection problems, and therefore, chapters and recipes where we explored the foundations of these topics are recommended to revisit. Furthermore, we will be working on image datasets. This recipe will combine what we learned in the following chapters:

- *Understanding image datasets: load, manage, and visualize the Fashion MNIST dataset*, the third recipe from *Chapter 2, Working with MXNet and Visualizing Datasets: Gluon and DataLoader*
- *Chapter 4, Solving Classification Problems*

How to do it...

In this recipe, we will take the following steps:

1. Introduce semantic segmentation.
2. Evaluate segmentation models.
3. Compare network architectures for semantic segmentation.
4. Explore the *Penn-Fudan Pedestrians* dataset with segmentation ground-truth.
5. Introduce Semantic segmentation Model Zoo.
6. Load a PSPNet pre-trained model from Model Zoo.
7. Evaluate a PSPNet pre-trained model from Model Zoo.
8. Load a DeepLab-v3 pre-trained model from Model Zoo.

9. Evaluate a DeepLab-v3 pre-trained model from Model Zoo.

Introducing semantic segmentation

In some of the previous chapters and recipes, we analyzed image classification problems where the task of our models was to take an image and define the class most likely associated with it. In semantic segmentation, however, there can be multiple objects per image, corresponding to different classes, and in different locations of the image. In object detection, the generated output to solve this problem was two lists, one providing the most likely class of each detected object and another indicating the estimated location of the object. For semantic segmentation, the output for each image is a set of binary images, one per class expected to be detected (dataset classes), where each pixel can have a value of 1 (active) if that pixel has been classified with that label, or 0 (inactive) otherwise. Each of these images is a **binary segmentation mask**:

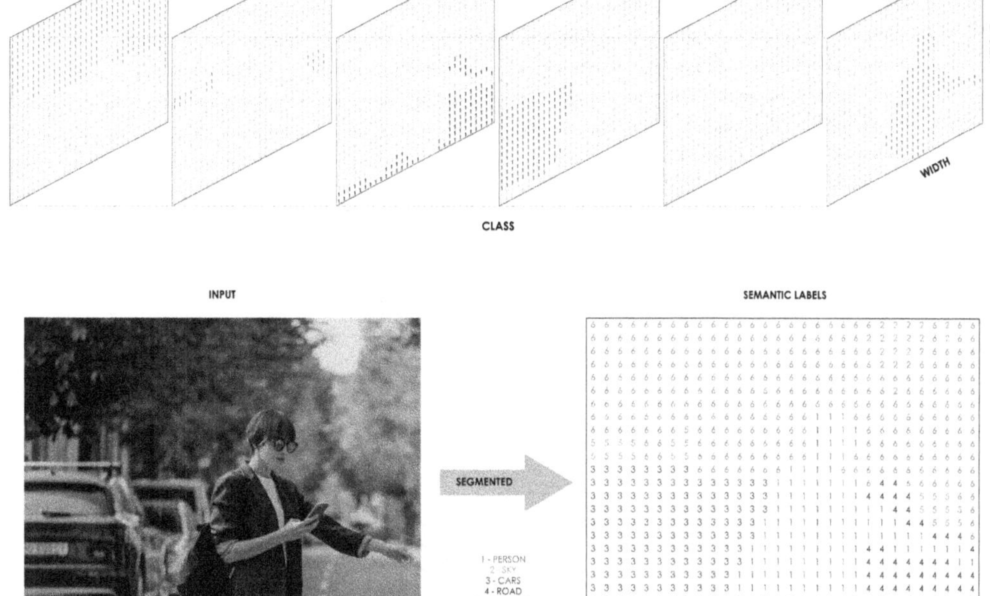

Figure 5.28 – Binary segmentation masks

> **Note**
>
> The person's image has been taken as an example here from the following source: *azerbaijan_stockers* on Freepik: `https://www.freepik.com/free-photo/young-woman-crossing-road-using-phone_10705234.htm#query=person%20in%20the%20street&position=13`

&from_view=search&track=ais&uuid=c3458125-63b6-4899-96e5-df07c307fb46The *masks* shown in *Figure 5.28* can be seen as one-hot embeddings of the classes, and they can be combined by associating each class with a different number (its class index, for example) to form a new image:

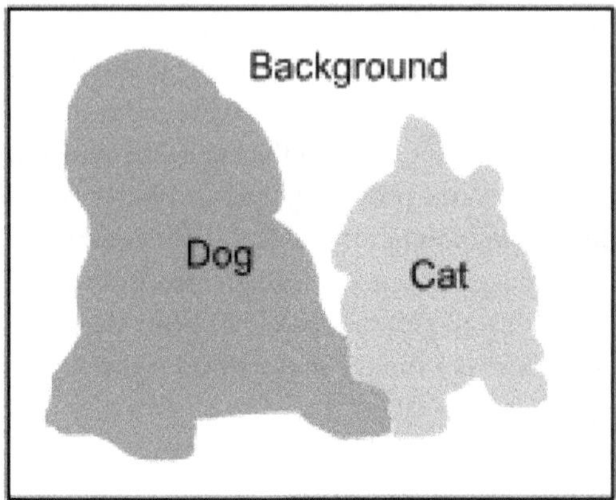

Figure 5.29 – Semantic segmentation

The output of a semantic segmentation model is, therefore, the different binary segmentation masks (see an example in *Figure 5.29*), and the number depends on the number of classes that the model has been trained on. Therefore, for each image input to the model with the shape *[H, W]* (*H* being the height and *W* being the *width*), the output array will have a shape of *[N, H, W]* (with *N* being the number of classes).

Evaluating segmentation models

An intuitive approach to evaluate models for a semantic segmentation task is to report the percentage of pixels that have been correctly classified. This metric is commonly used and is known as *pixel accuracy*.

$$Pixel\ Accuracy = \frac{\#TP + \#TN}{\#TP + \#TN + \#FP + \#FN}$$

However, pixel accuracy has a problem; when objects to be segmented are small in comparison to the image, this metric emphasizes the large number of pixels that have been correctly classified as not being the object (the inactive detection). For example, in a 1,000x1,000 image, we have a 100x100 object, and our model classifies the image as the background for all the pixels, with a pixel accuracy of 99%.

To fix these issues, we can use another metric to evaluate semantic segmentation models, **mean Intersection over Union** (**mIoU**).

Figure 5.30 – IoU for segmentation masks

The computation of this metric is similar to the IoU we saw in the previous recipe for object detection, in *Figure 5.16*. However, for object detection, the analysis was based on bounding boxes, whereas for semantic segmentation, as shown in *Figure 5.30*, it evaluates the number of pixels common (the intersection) between the target and the predicted masks, divided by the total number of pixels present across both masks (the union), and then its arithmetic mean is computed for all classes in the dataset.

Comparing network architectures for semantic segmentation

Semantic segmentation models' output is the different segmentation masks, which are the same size as the image input. To reach this objective, several architectures have been proposed, with the main difference with CNNs being that there are no fully connected layers. Appropriately, this network architecture is named **Fully Connected Networks** (**FCNs**). Several models evolved from this initial architecture and were state-of-the-art when they were first proposed:

- **Encoder–decoder**: The computation of feature maps by the convolutional and max pooling layers of the CNNs can be seen as an encoder, as image information is encoded in a multidimensional entity (feature maps). In this architecture, after the CNN feature maps, a series of upsampling layers (the decoder) are cascaded until images of the same size are computed. **U-Net** is an example of this architecture.
- **Spatial pyramid pooling**: One problem with FCNs is that the encoder does not provide enough global scene cues to the downstream layers, resulting in objects being misclassified due to a lack of global context (for example, boats labeled as cars in water-based images, where boats are expected and not cars). In this type of architecture, the feature map is aggregated to the output, along with feature maps at different grid scales computed by different modules. **PSPNet** is an example of this architecture.
- **Context modules**: Another option to capture multi-scale information is to add extra modules cascaded on top of the original network. **DeepLab-v3** can be seen as a combination of this type of network and spatial pyramid pooling networks.

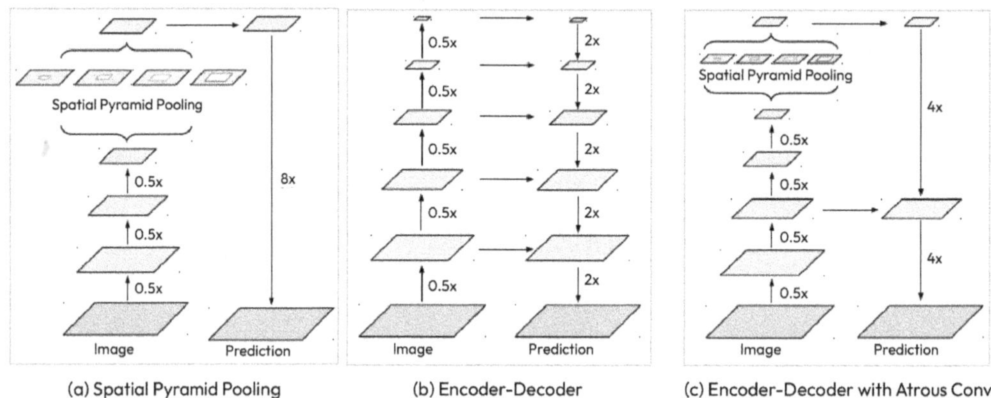

Figure 5.31 – Network architecture for semantic segmentation

For our experiments in this recipe, we will use pre-trained versions of PSPNet and DeepLab-v3.

Exploring the Penn-Fudan Pedestrians dataset with segmentation ground-truth

This dataset is the one that we worked with in the previous recipe. However, the ground-truth we will use in this recipe is not the bounding boxes required for object detection but, instead, the masks required for semantic segmentation.

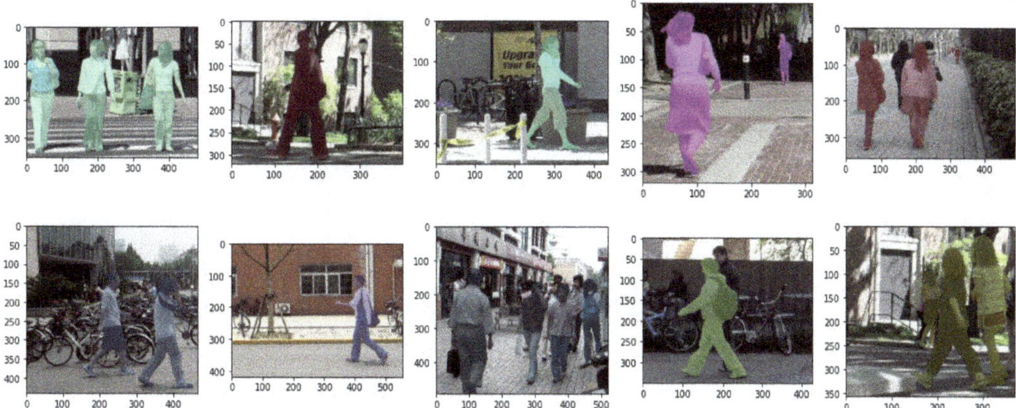

Figure 5.32 – The Penn-Fudan Pedestrians dataset with mask ground-truth

As we can see in *Figure 5.32*, each image can have one or more pedestrians, with their corresponding masks. This figure was computed using the GluonCV visualization utils package (the `plot_mask` function).

In this dataset, there are no different classes to be classified, and no further visualizations are computed.

Introducing Semantic Segmentation Model Zoo

GluonCV also provides pre-trained models for semantic segmentation in its Model Zoo. For the *MS COCO* dataset, this is the accuracy (mIoU) versus performance (samples per second) chart:

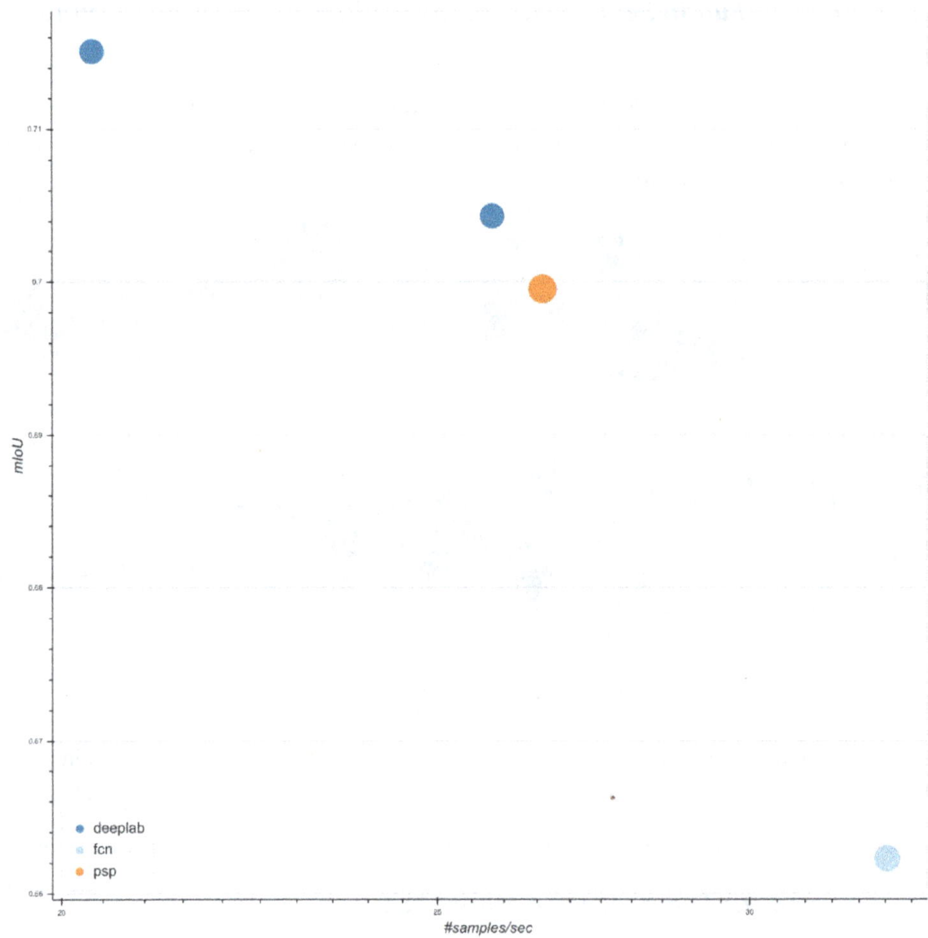

Figure 5.33 – Model Zoo for semantic segmentation (MS COCO)

> **Note**
>
> Source: `https://cv.gluon.ai/model_zoo/detection.html`

Figure 5.33 displays the most important pre-trained models in GluonCV Model Zoo, comparing accuracy (mIoU on the vertical axis) and inference performance (samples per second on the horizontal axis). There are no models (yet) in the top-right quadrant, meaning that, currently, we need to balance between both characteristics.

Using models from GluonCV Model Zoo can be done with just a couple of lines of code, and we will explore this path to solve our *Penn-Fudan Pedestrians* dataset for the segmentation task in the following steps.

Loading PSPNet pre-trained model from Model Zoo

Pyramid Scene Parsing Network (**PSPNet**) is a spatial pyramid pooling semantic segmentation architecture, which means that a pyramid pooling module is added to provide global cues. It was developed by Zhao et al. (the Chinese University of Hong Kong) in 2016. It achieved first place in the 2016 **ILSVRC Scene Parsing Challenge**.

Apart from the global pooling module, it differentiates from FCNs by using dilated convolutions.

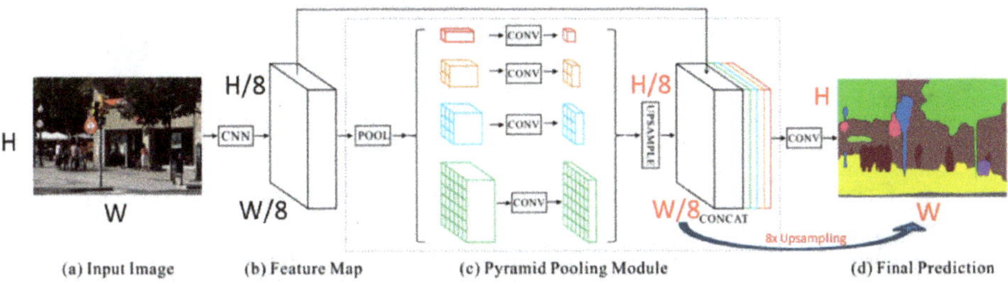

Figure 5.34 – The PSPNet architecture

> **Note**
> Source: `https://github.com/hszhao/PSPNet/issues/101`

In *Figure 5.34*, we can see the overall architecture of PSPNet:

- **Feature Map**: A ResNet-based CNN architecture, with dilated convolutions, is used to compute a feature map with 1/8 the size of the original image.
- **Pyramid Pooling Module**: Using a four-level pyramid (whole, half, quarter, and eighth), global context is computed. It is concatenated with the original feature map.
- **Final Prediction**: A final computation using a convolutional layer is done to generate the final predictions.

For our experiments, we will use a ResNet-101 network as a backbone. We can load the model with a single line of code:

```
pspnet = gcv.model_zoo.get_model('psp_resnet101_coco',
pretrained=True, ctx=ctx)
```

The model then downloads successfully.

Evaluating a PSPNet pre-trained model from Model Zoo

By using the *Penn-Fudan Pedestrian* dataset for a segmentation task, we can now perform qualitative and quantitative evaluation of the loaded model from the previous section.

Qualitatively, we can choose an image from the dataset and compare the output of the model with the ground-truth output from the dataset:

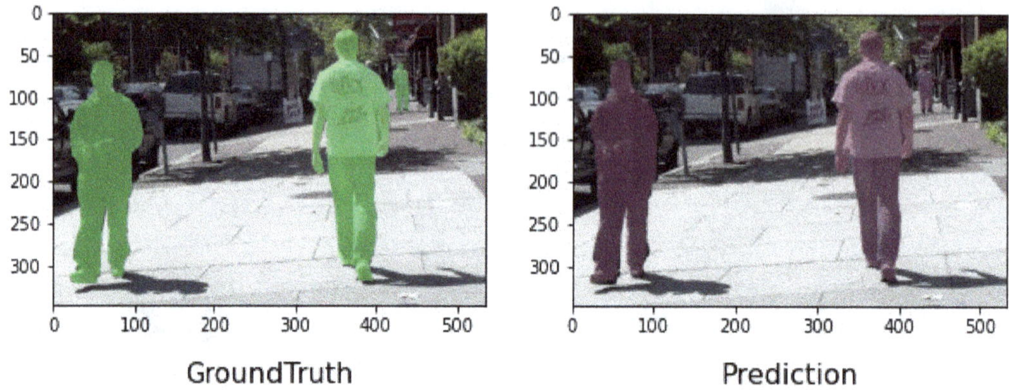

Figure 5.35 – Comparing predicted masks and ground-truth for PSPNet

From *Figure 5.35*, we can see a very strong correlation between the predicted segmentation masks and the expected results from the ground-truth. This figure was computed using the GluonCV visualization utils package (the `plot_mask` function).

Quantitatively, we can perform an mIoU evaluation and the runtime spent computing this metric:

```
PixAcc:  0.4650485574278924
mIoU  :  0.5612896701751177
Elapsed Time:  341.7681269645691 secs
```

The computed mIoU for this model for *MS COCO* (see *Semantic Segmentation Model Zoo*) is `0.70`; therefore, given the value of `0.56` for our model, we can conclude that our model performs the task accurately. However, it did take some time to complete it (~340 seconds).

Loading a DeepLab-v3 pre-trained model from Model Zoo

DeepLab-v3 is a semantic segmentation architecture that combines both *encoder-decoder* and **spatial pyramid pooling** architectures. It was developed by Chen et al. (Google) from 2015 (**DeepLab-v1**) to 2017 (**DeepLab-v3**).

It is the third iteration of a series of architectures that have evolved from each other – DeepLab-v1, DeepLab-v2, and DeepLab-v3.

From the research papers, also included in the There's more section we can see the DeepLab-v1 architecture:

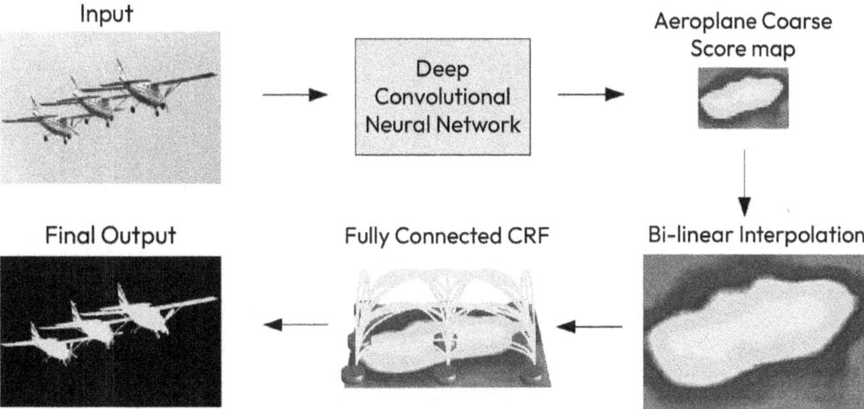

Figure 5.36 – The architecture of DeepLab-v1

Secondly, we can see the DeepLab- v2 architecture:

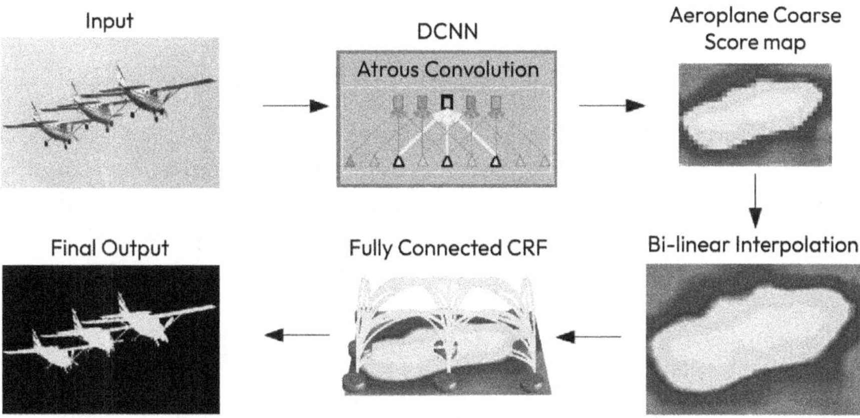

Figure 5.37 – The architecture of DeepLab-v2

And finally, we can see the DeepLab-v3+ architecture:

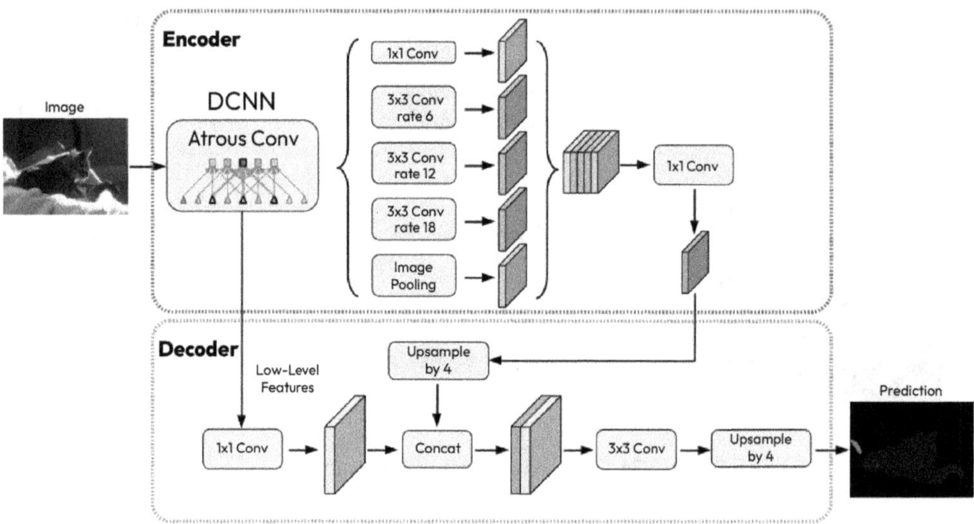

Figure 5.38 – The architecture of DeepLab-v3+

The different architectures have similarities and differences, namely the following:

- **DeepLab-v1**: This architecture evolves from the original FCN and uses a VGG-16 backbone network as well. The most important innovation is the usage of atrous or diluted convolutions instead of standard convolutions. We discussed this convolution parameter when we first introduced *convolutional layers* in *Chapter 3, Solving Regression Problems*, showing how it increased the receptive field. This architecture also uses fully connected **Conditional Random Fields (CRFs)** for post-processing to polish the final segmentation masks, although it is very slow and cannot be trained end to end.
- **DeepLab-v2**: The most important innovation for this model is a new module called **Atrous Spatial Pyramid Pooling (ASPP)**. With this module, on one hand, the network can encode multi-scale features into a fixed-size feature map (which is flexible to different input sizes), and on the ther hand, by using atrous convolutions, it increases the receptive field, optimizing the computation cost. In the original implementation, 4 to 6 scales were used. Furthermore, it uses ResNet as the backbone network.
- **DeepLab-v3**: In this last iteration, several changes are introduced. Firstly, the network is modified to use batch normalization and dropout. Secondly, the ASPP module is modified to add a new scale in a separate channel for global image pooling, to use fine-grained details. Lastly, the multi-scale part of the training is removed and is only applied to inference. A by-product of these small improvements is that the CRF step is no longer needed, providing much faster results.

For our experiments, we will use a ResNet-152 network as a backbone. We can load the model with a single line of code:

```
deeplab = gcv.model_zoo.get_model('deeplab_resnet152_coco',
pretrained=True, ctx=ctx)
```

The model then downloads successfully.

Evaluating a DeepLab-v3 pre-trained model from Model Zoo

Using the *Penn-Fudan Pedestrian* dataset for the segmentation task, we can now perform qualitative and quantitative evaluation of the loaded model from the previous section.

Qualitatively, we can choose an image from the dataset and compare the output of the model with the ground-truth output from the dataset:

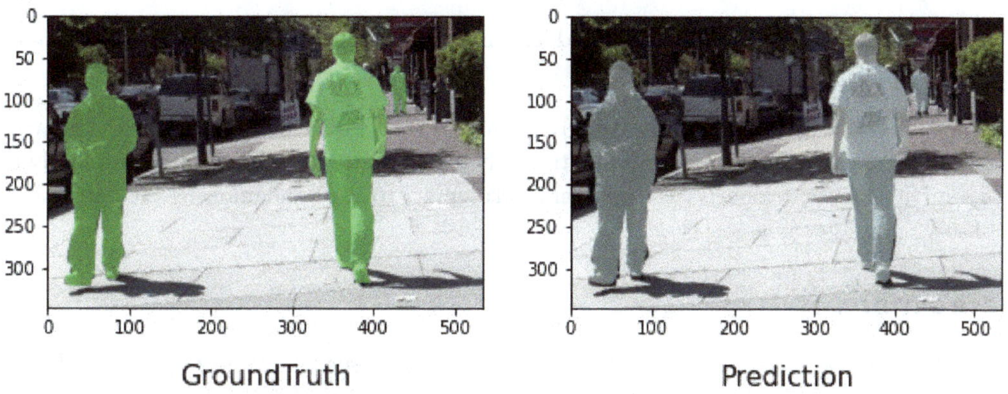

Figure 5.39 – Comparing predicted masks and ground-truth for DeepLab-v3

From *Figure 5.39*, we can see a very strong correlation between the predicted segmentation masks and the expected results from the ground-truth. This figure was computed using the GluonCV visualization utils package (the `plot_mask` function).

Quantitatively, we can perform mIoU evaluation and the runtime spent computing this metric:

```
PixAcc:  0.4653841191754281
mIoU  :  0.5616023247999165
Elapsed Time:  74.66736197471619 secs
```

The computed mIoU for this model for *MS COCO* (see *Semantic Segmentation Model Zoo*) is `0.715`; therefore, given the value of `0.56` for our model, we can conclude our model performs the task accurately, with a similar value to PSPNet. However, it is much faster (~70 secs).

How it works...

In this recipe, we tackled the semantic segmentation problem. We analyzed the differences between image classification and object detection for evaluation, network architectures, and datasets.

In our examples, we used a publicly available dataset, the *Penn-Fudan Pedestrians* dataset, and we used pre-trained models on *MS COCO*. This choice was not casual; *MS COCO* has a `person` class, and therefore, the expectation was that these pre-trained models would perform well, as they had already seen images from the dataset class.

However, as mentioned in the previous recipe, when this is not possible, and we apply pre-trained models from a dataset to another dataset, the data probability distribution will typically be very different; hence, the accuracy obtained can be very low. This is known as the **domain gap** or **domain adaptation problem** between the source dataset (the images that the model has been pre-trained on) and the target dataset (the images that the model is evaluated on).

One way to tackle these issues for supervised learning problems is fine-tuning. This approach is explored in detail in a later chapter.

We ended the recipe by evaluating our two pre-trained models, PSPNet and DeepLab-v3, and we were able to compare qualitatively and quantitatively their results for accuracy and computation speed, verifying how both models yield a similar pixel accuracy (`0.46`) and mIoU (`0.56`), although DeepLab-v3 was faster (`~4.5x`).

There's more...

Figure 5.33 showed a static image corresponding to the mIoU versus samples per second graph for the Model Zoo for Semantic Segmentation (on MS COCO); there is a dynamic version at this link that is worth taking a look at: `https://cv.gluon.ai/model_zoo/segmentation.html`. On this page, results from different models available in GluonCV Model Zoo are included; I suggest that you reproduce these results, as it is an interesting exercise.

During our discussion of the evolution of network architectures, most notably DeepLab, we mentioned how dilated/atrous convolutions help with multi-scale context aggregation. This research paper explores this topic in depth: `https://arxiv.org/pdf/1511.07122.pdf`.

Furthermore, it is very interesting to read the original research papers and see how the segmentation task has evolved from an academic point of view:

- **FCN**: `https://arxiv.org/pdf/1605.06211v1.pdf`
- **U-Net**: `https://arxiv.org/pdf/1505.04597.pdf`
- **PSPNet**: `https://arxiv.org/pdf/1612.01105.pdf`
- **DeepLab-v1**: `https://arxiv.org/pdf/1412.7062.pdf`

- **DeepLab-v2**: `https://arxiv.org/pdf/1606.00915v2.pdf`
- **DeepLab-v3**: `https://arxiv.org/pdf/1706.05587.pdf`

Semantic segmentation is an active area of research and, as such, is constantly evolving, with new networks appearing and redefining the state of the art, such as the following:

- **DeepLab-v3+**: The next step from DeepLab-v3, developed by Chen et al. (Google, 2018): `https://arxiv.org/pdf/1802.02611.pdf`
- **Swin-V2G**: Using Transformers, developed by Liu et al. (Microsoft, 2021): `https://arxiv.org/pdf/2111.09883v1.pdf`

6

Understanding Text with Natural Language Processing

Natural Language Processing (**NLP**) is the field of machine learning that deals with the understanding of language, in the form of text data. It is one of the fields that has seen a strong evolution in the last few years, achieving great results in the areas of sentiment analysis, chatbots, text summarization, and machine translation. NLP is at the core of the assistants developed by Amazon (Alexa), Google, and Apple (Siri), as well as modern assistants such as ChatGPT and Llama 2.

In this chapter, we will learn how to use GluonNLP, an MXNet Gluon library specific to NLP, how to build our own networks, and how to use its Model Zoo API for several applications of pre-trained models.

Specifically, we will cover the following topics:

- Introducing NLP networks
- Classifying news highlights with topic modeling
- Analyzing sentiment in movie reviews
- Translating text from Vietnamese to English

Technical requirements

Apart from the technical requirements specified in the *Preface*, the following technical requirements apply to this chapter:

- Ensure that you have completed *Recipe 1, Installing MXNet, Gluon, GluonCV and GluonNLP*, from *Chapter 1, Up and Running with MXNet*
- Ensure that you have completed *Recipe 4, Toy dataset for text classification: Load, manage, and visualize a spam email dataset*, from *Chapter 2, Working with MXNet and Visualizing Datasets: Gluon and DataLoader*

The code for this chapter can be found at the following GitHub URL: `https://github.com/PacktPublishing/Deep-Learning-with-MXNet-Cookbook/tree/main/ch06`.

Furthermore, you can access each recipe directly from Google Colab; for example, use the following link for the first recipe of this chapter: `https://colab.research.google.com/github/PacktPublishing/Deep-Learning-with-MXNet-Cookbook/blob/main/ch06/6_1_Introducing_NLP_Networks.ipynb`.

Introducing NLP networks

In the previous chapters, we saw how different architectures, such as **Multi-Layer Perceptrons** (**MLPs**) and **Convolutional Neural Networks** (**CNNs**), deal with numerical data and images, respectively. In this recipe, we will analyze the most important architectures to process natural language expressed as text data.

The most important characteristic of natural language is that it is a list of words of variable length, and the order of those words matters; it is a sequence. The previous architectures that we have analyzed are not suited for variable-length data inputs and also do not exploit the relationships among words effectively.

In this recipe, we will introduce neural networks that have been developed to process sequences of words:

1. We will start by applying the network introduced in the previous chapter, that is, CNNs for text processing, called **TextCNNs**.
2. Afterward, we will introduce **Recurrent Neural Networks** (**RNNs**) and their vanilla implementation. Then, we will continue with an improved version known as **Long Short-Term Memory** (**LSTM**).
3. Then, as we did with computer vision, we will introduce **GluonNLP Model Zoo**, one of the most value-adding features of MXNet. We will leverage these libraries MXNet and GluonNLP to understand and implement **transformers** and their self-attention mechanisms, and how these networks deal with sequences of variable length.

Getting ready

As in previous chapters, in this recipe, we will be using a little bit of matrix operations and linear algebra, but it will not be hard at all.

How to do it...

In this recipe, we will be doing the following:

1. Applying CNNs for text processing (TextCNNs)
2. Introducing RNNs
3. Improving RNNs with LSTM

4. Introducing GluonNLP Model Zoo
5. Paying attention to Transformers

Let's go through each of these network architectures in detail next.

Applying CNNs for text processing

CNNs were introduced in the previous chapter and are typically used for working with images. However, with some slight changes, CNNs can work very efficiently with text data.

Images are 2D data and, as shown in *Chapter 5, Analyzing Images with Computer Vision*, we worked with two layers on this data:

- 2D convolutions layers
- Max pooling layers

These operations are slightly modified to work with text data, which can be seen as a 1D sequence. Therefore, for 1D convolution layers, we have the following:

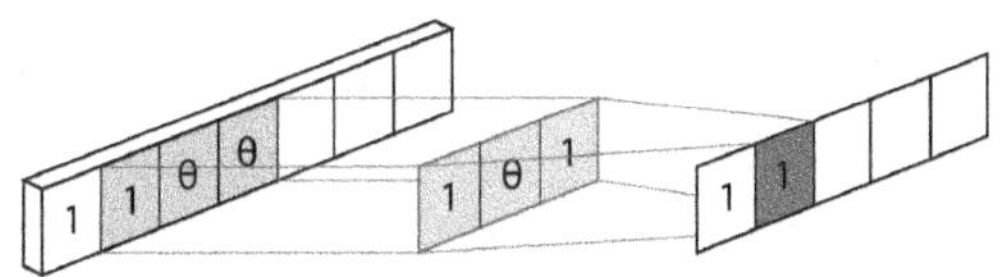

Figure 6.1 – 1D convolution

As can be seen in *Figure 6.1*, the sequence combined with the kernel varies over time, yielding a new sequence as output. Please note how the number of words that are analyzed at the same time is the kernel size (3 in the preceding figure).

For the max pooling layers, as we only have one dimension, which corresponds to time, these layers are known as **max-over-time pooling** layers:

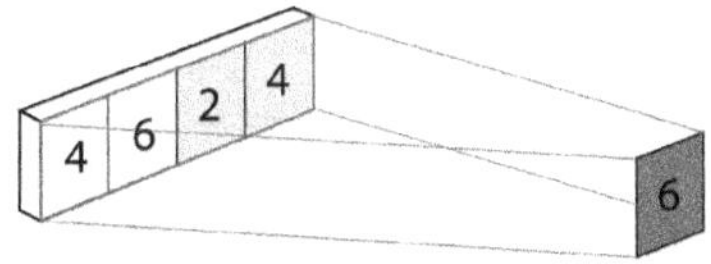

Figure 6.2 – Max-over-time pooling

As can be seen from *Figure 6.2*, the maximum value from the sequence is selected.

Using MXNet Gluon, we can define the 1D convolution and max-over-time layers as follows:

```
# TextCNN Components
# 1D Convolution
```

```
conv1d = mx.gluon.nn.Conv1D(3, 100, activation='relu')
# Max-Over-Time Pooling
max_over_time_pooling = mx.gluon.nn.GlobalMaxPool1D()
```

An example of how to work with these layers can be found in the GitHub code.

As with the CNN architecture shown in the previous chapter (see *Figure 5.6*), typically, after a feature learning phase, we have a classifier. As an example application, this kind of architecture will help us later with sentiment analysis.

Introducing Recurrent Neural Networks (RNNs)

As discussed in the recipe introduction, RNNs are architectures that deal with sequences of data of variable length. For NLP, these data points are sentences, composed of words, but they can also be utilized for sequences of images (video), for example.

RNN's history is a series of step-by-step attempts to improve the recurrent processing of different inputs of a sequence. There have been several important contributions, the most notable being Hopfield (1982), Jordan (1986), and Elman (1990).

The idea behind RNNs is that once the output is processed from the input, that output is fed again to the model (in a recurrent manner), in combination with the new incoming input. This mechanism allows the model to have memory (can access past data) and process the new input, taking into account that past information as well. This basic architecture is typically referred to as a **vanilla RNN**.

Please note that, as introduced in *Recipe 4, Understanding text datasets – loading, managing, and visualizing the Enron Email dataset*, in *Chapter 2, Working with MXNet and Visualizing Datasets: Gluon and DataLoader*, the inputs to NLP networks, including RNNs, are not the words obtained from the dataset, but numerical representations of those words, such as one-hot encoding or word embeddings.

If a sequence of data is modeled as successive inputs, x(1), x(2), ... x(t), the architecture of an RNN can be visualized as follows:

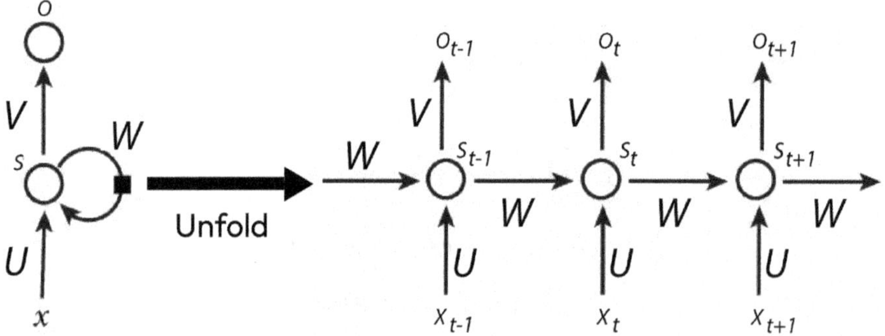

Figure 6.3 – RNN architecture

In *Figure 6.3*, we can see how every input *x(t)* is processed over time, and how the processed input (hidden state) is looped back for the next iterations. Let's take a deeper look at what is happening at each step, as in the preceding figure, activation functions and biases are not shown for simplicity. The actual equations of a vanilla RNN cell are as follows:

$$\begin{aligned}
\boldsymbol{a}^{(t)} &= \boldsymbol{b} + \boldsymbol{W}\boldsymbol{h}^{(t-1)} + \boldsymbol{U}\boldsymbol{x}^{(t)} \\
\boldsymbol{h}^{(t)} &= \tanh(\boldsymbol{a}^{(t)}) \\
\boldsymbol{o}^{(t)} &= \boldsymbol{c} + \boldsymbol{V}\boldsymbol{h}^{(t)} \\
\hat{\boldsymbol{y}}^{(t)} &= \mathrm{softmax}(\boldsymbol{o}^{(t)})
\end{aligned}$$

Figure 6.4 – Equations for a vanilla RNN cell

As can be seen in *Figure 6.4*, for each step, the input, *x(t)*, is multiplied by a weight matrix, *U*, and a bias vector, *b*, is added, which yields the value *a(t)*, assuming there was no previous input, *h(t - 1) = 0*. The state value, *h(t)*, is computed as the output of the activation function, *tanh*, of that value. When there is a previous input, the previous state value, *h(t - 1)*, is multiplied by a weight matrix, *W*, and added to the computations of values *a(t)* and *h(t)*.

The state value, *h(t)*, is then multiplied by a weight matrix, *V*, and a bias vector, *c*, is added which yields the value *o(t)*. The output of the cell is computed as the output of the activation function, *softmax*, of that value, yielding the final output value *y(t)*.

These cells can be stacked together to produce a complete RNN.

Using MXNet and Gluon, we can easily create our own custom RNN networks:

```
# RNNs MXNet Implementation Example
class RNNModel(mx.gluon.Block):
    """
    A basic RNN Model
    """
    def __init__(self, num_hidden, num_layers, embed_size, **kwargs):
        super(RNNModel, self).__init__(**kwargs)
        self.rnn = mx.gluon.rnn.RNN(
            num_hidden,
            num_layers,
            input_size=embed_size)
    def forward(self, inputs, hidden):
        output, hidden = self.rnn(inputs, hidden)
        return output, hidden
```

We can use this code to define a custom RNN:

```
# RNN with 3 hidden cells, 1 layer and expecting inputs with 20
embeddings
rnn = RNNModel(3, 1, 20)
rnn.collect_params().initialize(mx.init.Xavier(), ctx=ctx)
```

Furthermore, we use the following to process a sequential input:

```
rnn_hidden = hidden_initial
outputs = []
for index in range(3):
    rnn_output, rnn_hidden = rnn(inputs[index], rnn_hidden)
    outputs.append(rnn_output)
```

The previous code shown runs a custom RNN. Feel free to play with the notebook available in the GitHub repository accompanying this book.

To conclude, one advantage of RNNs is that the information present in previous inputs is kept stored in the states transferred along time steps. However, this information is constantly multiplied by different weight matrices and passed through non-linear functions (*tanh*), and the outcome is that, after several time steps, the state information is modified and does not work anymore as memory; the information stored has changed too much. Long-term memory storage is a problem for RNNs.

Training RNNs

In the previous chapters, we saw how to train networks using supervised learning, by computing the loss function between the expected outputs, the **ground truth**, and the actual outputs of the network. This error could then be back-propagated from the outer layers of the network to the inner layers of the network and update the weights of these layers.

With RNNs, at each time step, the network is shown a sequence of inputs and the expected sequence in the outputs. Errors are computed for each time step and back-propagated from the outer layers of the network to the inner layers of the network. This variation, suitable for RNNs, is called **Back-Propagation Through Time (BPTT)**.

As with other networks, computing the different gradients involves the iterative multiplication of matrices. This operation is exponential, which means that, after several occurrences, the values will either shrink or blow up. This leads to a problem we have already discussed: **vanishing gradients** and **exploding gradients**. These issues make the training of RNNs very unstable. Another important drawback of BPTT is that as it is a sequential computation, it cannot be parallelized.

Improving RNNs with Long Short-Term Memory (LSTM)

LSTMs were introduced by Hochreiter and Schmidhuber in 1997, as a mechanism to solve the problems described previously (lack of long-term memory and vanishing/exploding gradients). In LSTMs, instead of having the previous state multiplied and passed through the non-linear function, the connection is much more straightforward. To provide this mechanism, each LSTM cell receives two inputs from the previous cell: the **hidden state** (**ht**) and the **cell state** (**ct**):

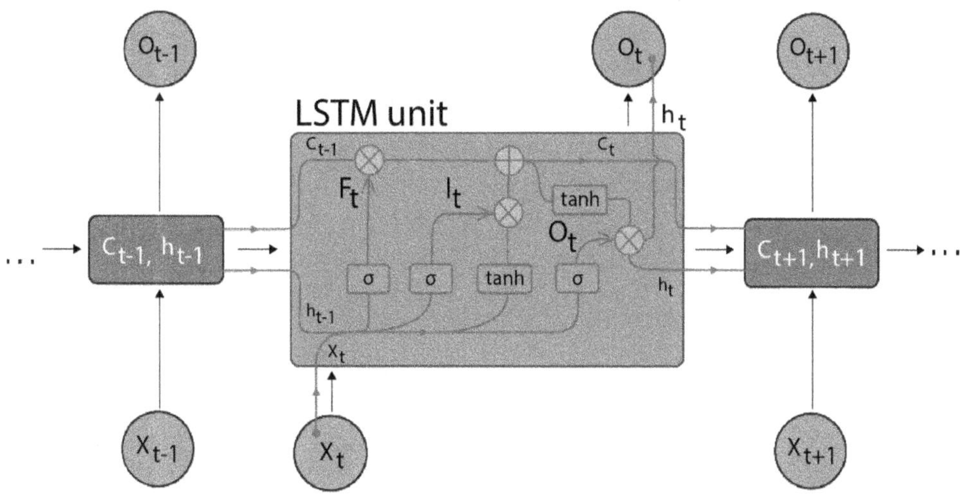

Figure 6.5 – LSTM network

The key components of LSTMs are called gates, which define how a certain input is modified to become a part of the outputs. These vectors have values between 0 and 1 and help activate/deactivate the information from the input. They are, therefore, a sigmoid operation (depicted in *Figure 6.5* as σ) of a weighted sum of the inputs, followed by a multiplication operation. To emphasize, each of the three gates present in an LSTM cell allows how much of the input or the state passes through each of the outputs. Taking this into account, the following equations define the LSTM behavior:

$$
\begin{aligned}
i_t &= sigmoid(W_{ii}x_t + b_{ii} + W_{hi}h_{(t-1)} + b_{hi}) \\
f_t &= sigmoid(W_{if}x_t + b_{if} + W_{hf}h_{(t-1)} + b_{hf}) \\
g_t &= \tanh(W_{ig}x_t + b_{ig} + W_{hc}h_{(t-1)} + b_{hg}) \\
o_t &= sigmoid(W_{io}x_t + b_{io} + W_{ho}h_{(t-1)} + b_{ho}) \\
c_t &= f_t * c_{(t-1)} + i_t * g_t \\
h_t &= o_t * \tanh(c_t)
\end{aligned}
$$

Figure 6.6 – Equations for an LSTM cell

The equations in *Figure 6.6* can be explained as follows:

- **Input gate (it)**: This is the gate that decides how the information from the previous state and the current input will be updated.
- **Forget gate (ft)**: This is the gate that decides how the information from the previous state and the current input will become part of the long-term memory (cell state). This is how much of the current step we want to forget.
- **Memory cell candidate (gt)**: This is the computation that decides how the information from the previous state and the current input will become part of the memory cell. It must allow for positive and negative values; therefore, *tanh* is the activation function selected.
- **Output gate (ot)**: This is the gate that decides how the information from the previous state and the current input will become part of the output.
- **Memory cell (ct)**: This is the computation that combines the previous state (c_{t-1}) and the current memory cell candidate (g_t) into the new cell state
- **Output state (ht)**: This is the computation that combines the memory cell with the output gate value.

Using MXNet and Gluon, we can easily create our own custom LSTM networks:

```
# LSTMs MXNet Implementation Example
class LSTMModel(mx.gluon.Block):
    """
    A basic LSTM Model
    """
    def __init__(self, num_hidden, num_layers, embed_size, **kwargs):
        super(LSTMModel, self).__init__(**kwargs)
        self.lstm = mx.gluon.rnn.LSTM(
            num_hidden,
            num_layers,
            input_size=embed_size)
    def forward(self, inputs, hidden):
        output, hidden = self.lstm(inputs, hidden)
        return output, hidden
```

We can use the following code to define a custom RNN:

```
# LSTM with 3 hidden cells, 1 layer and expecting inputs with 20
embeddings
lstm = LSTMModel(3, 1, 20)
 lstm.collect_params().initialize(mx.init.Xavier(), ctx=ctx)
```

Furthermore, we can use the following to process a sequential input:

```
lstm_hidden = [hidden_initial, state_initial]
 outputs = []
for index in range(3):
    lstm_output, lstm_hidden = lstm(inputs[index], lstm_hidden)
    outputs.append(lstm_output)
```

The code shown here runs a custom LSTM network. Feel free to play with the notebook available in the GitHub repository accompanying this book.

To conclude, LSTMs allow RNNs to be trained more optimally and have been implemented to solve a large number of tasks in NLP, such as sentiment analysis and language modeling.

Introducing GluonNLP Model Zoo

One of the best features that MXNet GluonCV provides is its large pool of pre-trained models, readily available for its users to use and deploy in their own applications. This model library is called **Model Zoo**.

In Model Zoo, models have been pre-trained for the following tasks:

- Language models
- Sentiment analysis
- Machine translation
- Sentence classification
- Question answering
- Named entity recognition
- Joint intent classification and slot labeling

In this chapter, we will examine in detail the pre-trained models included for sentiment analysis and machine translation.

Paying attention with Transformers

Although LSTMs have been proven to work well for a lot of applications, they also have significant drawbacks, including taking longer and requiring more memory to train, as well as being sensitive to random initialization. New architectures have been developed that overcome these limitations. One of the most important examples is **Transformers**.

Transformers were introduced by Google Brain in 2017. It is a novel approach of an encoder-decoder architecture (as seen in *Recipe 4, Segmenting objects in images with PSPNet and DeepLab-v3*, in *Chapter 5, Analyzing Images with Computer Vision*) with a repurposed mechanism to process sequences of data, called **attention**. The largest improvement of this architecture is that it does not depend on processing the data sequentially. All the data can be processed in parallel, allowing for faster training and inference. This improvement allowed a very large amount of text, the **corpus**, to be processed, yielding **Large Language Models** (**LLMs**) such as **Bidirectional Encoder Representations from Transformers** (BERT).

The architecture of Transformers can be represented as follows:

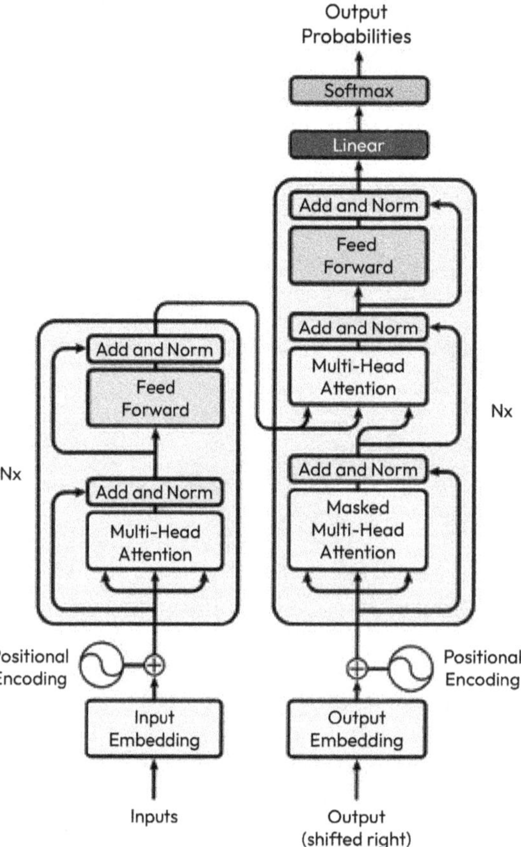

Figure 6.7 – Transformer architecture

In *Figure 6.7*, we can distinguish several components:

- **Input and output preprocessing**: Embeddings and positional encodings are computed before inputting the data into the network.
- **Encoder-decoder architecture**: The left part corresponds to the encoder and the right part corresponds to the decoder. Feed-forward processing, residual connections, and normalization are parts of this component.

- **Attention heads**: The sequence inputs the path from the encoder to the decoder and the sequence outputs are all processed via this mechanism.

Let's see each of these components in more detail.

Input and output preprocessing

In the original paper, the inputs and outputs, before being fed into the network, are processed through **learned embeddings**. These embedding vectors typically have a size of 512 elements.

Furthermore, as we will see afterward, these embeddings are passed through a *softmax* function, combining several pieces of information into one number, but also losing the positional information in the process. To maintain the positional information through the whole encoding and decoding, for both inputs and outputs, the input embeddings are added as a vector with information on position. These are called **positional encodings**.

Encoder-decoder architecture

As noted in the original *Google Brain's original Transformer* paper: `https://arxiv.org/pdf/1706.03762.pdf` the encoder and decoder are composed of six identical layers (in *Figure 6.5*, $N = 6$ in the left diagram and right diagram, respectively). Each encoder layer has two components, a multi-head self-attention mechanism followed by a fully connected feed-forward network. For each decoded layer, another attention component is added, a masked multi-head self-attention mechanism. Attention heads will be explained in the following section.

Residual connections are also added, similar to what we saw for ResNets in *Recipe 2, Classifying images with MXNet - GluonCV Model Zoo, AlexNet, and ResNet*, in *Chapter 5, Analyzing Images with Computer Vision*.

This information together is normalized using layer normalization, which is similar to batch normalization, introduced in *Recipe 3, Training for regression models*, in *Chapter 3, Solving Regression Problems*. The most important difference is that, as described in the paper introducing `layer normalization` (`https://arxiv.org/abs/1607.06450`), with layer normalization, all the hidden units in a layer share the same normalization terms, but different input data can have different normalization terms. Layer normalization has been proven to work better than batch normalization for processing sequences.

By combining all these layers across the different encoder and decoder steps (including the embeddings), the vectors will all have a dimension of 512.

Attention heads

In Transformers, how each word in a sequence is connected to each of the words in another sequence is done via the attention mechanism. For example, if we have an input sequence with N words in English (*I love you very much*) and its translation to French has M words (*Je t'aime beaucoup*), the attention matrix of weights between these two sequences will have *NxM* dimensions. Connecting sequences using this mechanism has a strong advantage over a recurrent mechanism (such as the one used in RNNs), which is parallelization. Attention is a matrix operation and therefore can be optimally parallelized.

	I	love	you	very	much
je		0.02	0.06	0.01	0.01
t'	0.11	0.01	0.80	0.03	0.05
aime	0.03		0.03	0.01	0.01
beaucoup	0.02	0.02	0.02	0.41	0.53

Figure 6.8 – Example of attention matrix

In the Transformer paper, three of the matrices shown in *Figure 6.9* were introduced, *Query (Q), Key (K), and Value (V)*. The following is an explanation of each of these matrices:

- **Query**: This is the input representation of each input word
- **Key and value**: Similar to a **hash map** that maps keys into values, these matrices are used for indexing (key) and providing a representation (value) of the word (different from the query)

The combination of the operations carried out by these three matrices is called the dot-product attention function and can be described as follows:

$$\text{Attention}(Q, K, V) = \text{softmax}(\frac{QK^T}{\sqrt{d_k}})V$$

Figure 6.9 – Attention function

When the output is a representation of the input sequence, this mechanism is called a **self-attention head**. In the architecture diagram shown earlier in *Figure 6.7*, the two attention mechanisms closer to the input sequence and the output sequence (lower part of the diagram) are self-attention mechanisms, as the sequence outputted by the function is the same as the sequence that is inputted. When this is not the case, such as in the attention mechanism that connects the encoder to the decoder, this is known as a **cross-attention head**. The self-attention head for the output vectors is masked, meaning that only past information is available in the training of the network. This allows Transformer models to generate text from a limited input (auto-regressive models such as GPT-3 or BLOOM).

Instead of processing all the input data in parallel, in Google Brain's Attention is All You Need original paper (`https://arxiv.org/abs/1706.03762`), eight attention heads are used in parallel. As the output is expected to have the same dimensionality (512), each head works with a reduced vector (with a dimensionality of 64). This is described in the paper as follows:

"Multi-head attention allows the model to jointly attend to information from different representation subspaces at different positions."

Reducing the dimensionality allows for the total computation cost to be similar to using a complete (full-dimensionality) attention head.

Implementation in GluonNLP

GluonNLP has its own implementation of a Transformer model and therefore, getting our encoder and decoder is straightforward, as follows:

```
# Transformers MXNet Implementation Example
# Transformer with 6 layers (encoder and decoder), 2 parallel heads,
and expecting inputs with 20 embeddings
transformer_encoder, transformer_decoder, _ = nlp.model.transformer.
get_transformer_encoder_decoder(
    num_layers=6,
    num_heads=2,
    units=20)
transformer_encoder.collect_params().initialize(mx.init.Xavier(),
ctx=ctx)
 transformer_decoder.collect_params().initialize(mx.init.Xavier(),
ctx=ctx)
```

Now, we can use the encoder to process the inputs; however, with Transformers, we can process the whole input at the same time:

```
encoded_inputs, _ = transformer_encoder(inputs[0])
```

Large-scale Transformers are the current state of the art for most NLP tasks, such as topic modeling, sentiment analysis, or question answering, and the encoder and decoder architectures are also used separately for different tasks.

How it works...

In this recipe, we have introduced several networks to work with NLP using MXNet, Gluon, and GluonNLP:

- RNNs
- LSTM
- Transformers

We have reviewed how each of these architectures works and analyzed its advantages and disadvantages, as well as how each one has improved on the previous one, exploring concepts such as sequences, BPTT, memory cells, and attention.

In the following recipes, we will explore how to use these architectures to solve practical problems such as topic modeling, sentiment analysis, and machine translation.

There's more...

Some of the concepts explored in this recipe are too advanced to cover in detail in this book. I strongly suggest taking a look at the following references if you would like to get a deeper understanding:

- **TextCNNs (Paper):** `https://aclanthology.org/D14-1181.pdf`
- **RNNs (Intuitive explanation):** `https://towardsdatascience.com/a-battle-against-amnesia-a-brief-history-and-introduction-of-recurrent-neural-networks-50496aae6740`
- **RNNs (Backpropagation Through Time):** `https://d2l.ai/chapter_recurrentneural-networks/bptt.html`
- **Vanishing/exploding gradients research papers (Learning Long-Term Dependencies with Gradient Descent is Difficult):** `http://www.comp.hkbu.edu.hk/~markus/teaching/comp7650/tnn-94-gradient.pdf`
- **Vanishing/exploding gradients research papers: (On the difficulty of training Recurrent Neural Networks):** `https://arxiv.org/pdf/1211.5063.pdf`
- **Vanishing/exploding gradients research papers (The exploding gradient problem demystified - definition, prevalence, impact, origin, tradeoffs,and solutions):** `https://arxiv.org/pdf/1712.05577.pdf`
- **LSTMs (Paper):** `http://www.bioinf.jku.at/publications/older/2604.pdf`
- **LSTMs (Intuitive explanation):** `http://colah.github.io/posts/2015-08-Understanding-LSTMs/`
- **Transformers (Original paper – Attention Is All You Need):** `https://arxiv.org/pdf/1706.03762.pdf`
- **Transformers (Intuitive explanation):** `https://towardsdatascience.com/transformers-89034557de14`
- **Transformers (Layer Normalization):** `https://arxiv.org/pdf/1607.06450.pdf`
- **State-of-the-art models in NLP (GPT-3 paper):** `https://arxiv.org/pdf/2005.14165.pdf`
- **State-of-the-art models in NLP (BLOOM (open source alternative):** `https://huggingface.co/blog/bloom`

In this chapter, we are going to analyze in detail the following tasks: topic modeling, sentiment analysis, and text translation. However, MXNet GluonNLP Model Zoo contains lots of models pre-trained for a large number of tasks. You are encouraged to explore the different examples included at `https://nlp.gluon.ai/model_zoo/index.html`.

Classifying news highlights with topic modeling

In this recipe, we are going to study one of the most interesting tasks in NLP, topic modeling. In this task, the user must find the number of topics given a set of documents. Sometimes, the topics (and the number of topics) are known beforehand and the supervised learning techniques that we have seen in previous chapters can be applied. However, in a typical scenario, topic modeling datasets do not provide ground truth and are therefore unsupervised learning problems.

To achieve this, we will use a pre-trained model from GluonNLP Model Zoo and apply its word embeddings to feed a clustering algorithm, which will yield the clustered topics. We will apply this process to a new dataset: *1 Million News Headlines*.

Getting ready

As in previous chapters, in this recipe, we will be using a little bit of matrix operations and linear algebra, but it will not be hard at all.

Furthermore, we will be working with text datasets. Therefore, we will revisit some concepts already seen in *Recipe 4, Understanding text datasets – loading, managing, and visualizing the Enron Email dataset*, from *Chapter 2, Working with MXNet and Visualizing Datasets: Gluon and DataLoader*.

How to do it...

In this recipe, we will be carrying out the following steps:

1. Exploring the *1 Million News Headlines* dataset
2. Applying word embeddings
3. Clustering the topics using K-means
4. Putting everything together

Let's go through each of these steps in the following subsections.

Exploring the 1 Million News Headlines dataset

This dataset is one of the most well-known datasets used for topic modeling. It contains 19 years of noteworthy news headlines published by the **Australian Broadcasting Corporation** (**ABC**) website, from 2003 to 2021 (inclusive). The topic of each news headline is not included.

As expected from a real-world dataset, the corpus contains a large number of words. Therefore, before going further, we will proceed to clean the data and follow the process described in *Recipe 4, Understanding text datasets – loading, managing, and visualizing the Enron Email dataset*, from *Chapter 2, Working with MXNet and Visualizing Datasets: Gluon and DataLoader*:

- Tokenizing
- Removing stop words
- Stemming
- Lemmatization

Furthermore, this dataset contains more than 1 million headlines (1.2 million, actually). In order to be able to process them in a time-efficient manner, we will work with a reduced subset of 5%:

```
reduced_number_headlines = int(0.05 * number_headlines)
 print(reduced_number_headlines)
62209
```

If we analyze this subset to compute the number of words each headline has, we can plot the following histogram:

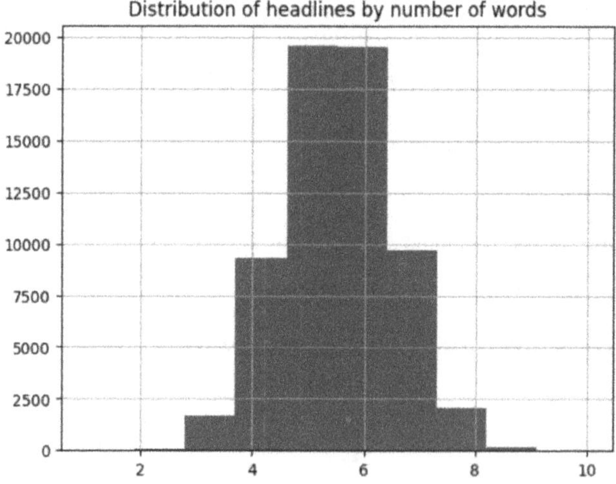

Figure 6.10 – Distribution of headlines by number of words

As can be seen in *Figure 6.10*, most of the headlines have between 4 and 8 words.

Applying word embeddings

In *Recipe 4, Understanding text datasets – loading, managing, and visualizing the Enron Email dataset*, from *Chapter 2, Working with MXNet and Visualizing Datasets: Gluon and DataLoader*, we introduced two embedding models from GluonNLP Model Zoo: Google's **word2vec** and **GloVe** from Stanford University. For this use case, we are going to work with word2vec, because it was trained on a dataset called Google News, composed of 3 million words and phrases from a corpus of 100 billion words. As the corpus is composed of news information, this embedding model is very well suited for our use case.

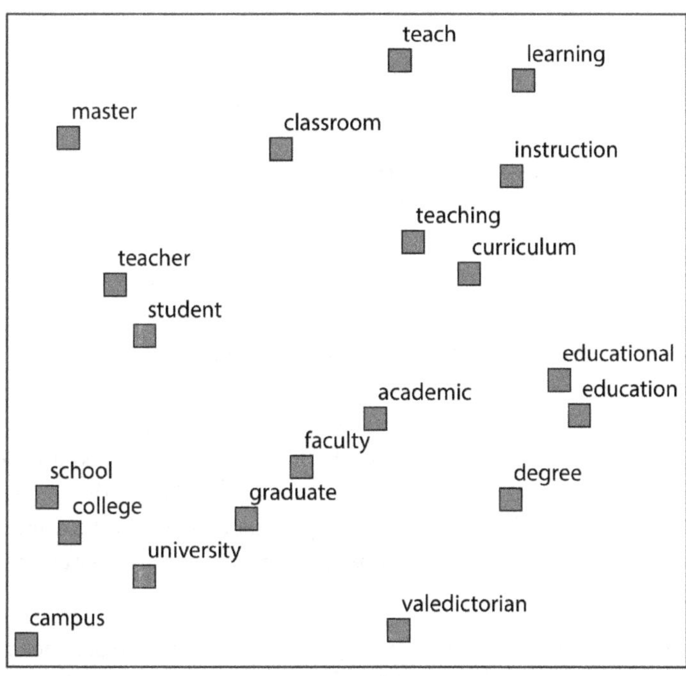

Figure 6.11 – 2D representation of word2vec embeddings

By using this model, each word is transformed into a vector with 300 components. However, for our application with headlines (full sentences), we are more interested in a representation of the complete headline and not just its independent words. A simple but effective method to accomplish this for our application is to compute the average vector of each preprocessed word.

Clustering the topics using K-means

With our headline embeddings ready, the last step to classify our headlines is to cluster them. There are many clustering algorithms, such as **expectation maximization clustering** and **mean shift clustering**. However, for our application, we will use my favorite one, **K-means**, which is implemented in a Python package we have already talked about, **scikit-learn**.

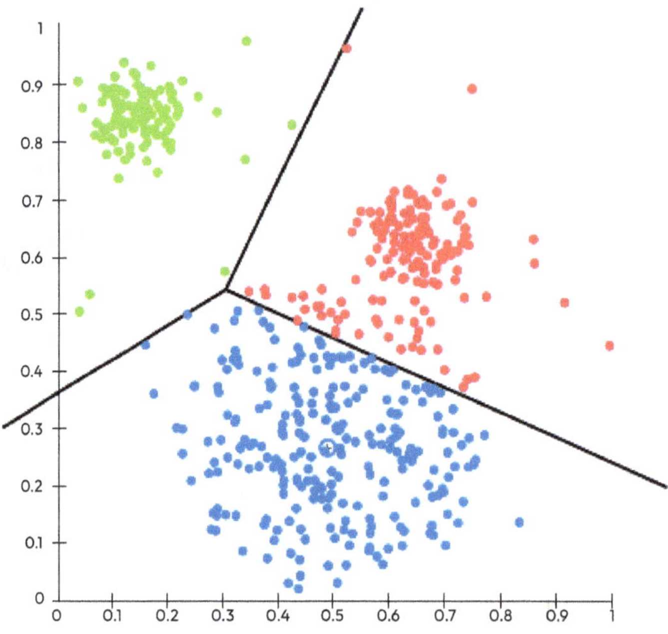

Figure 6.12 – K-means visualization

The intuitive idea behind K-means is that given a number of clusters, *K*, it will assign the clusters' centroids randomly, and add each newly seen vector to the closest centroid (assignment step). Then, it will compute the new centroid as the mean of the vectors that belong to the cluster (update step) and iterate. This process is repeated until there is no significant change in the centroids' positions. Therefore, the full dataset can be iterated several times. For large datasets, other criteria can be added for convergence and stopping criteria.

One important drawback of K-means is that the number of clusters is an input parameter, and therefore, it requires some intuition or knowledge about the dataset. In practice, knowing this information beforehand is difficult. Therefore, I strongly recommend running the algorithm for several different values of *K*. For our use case, the number of clusters chosen is 4.

Putting everything together

After this three-step process (cleaning the data, embedding, and clustering), we are ready to analyze some results.

The most interesting output of topic modeling is identifying which headline topics are associated with each cluster. A helpful approach to doing this is to visualize the most important words for each cluster and come up with the connecting topic. Therefore, we can plot the following for the first identified cluster (cluster 0):

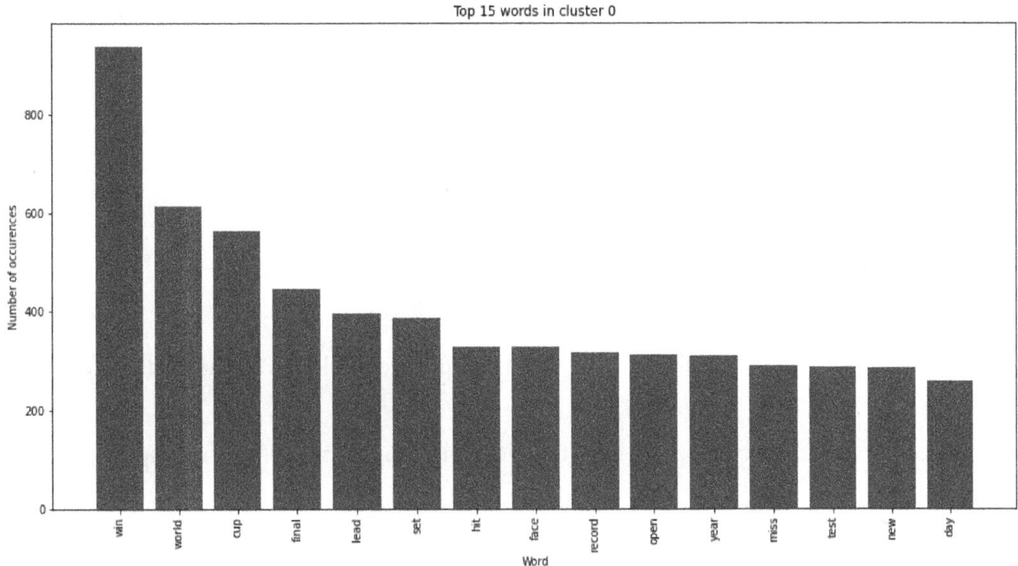

Figure 6.13 – Top 15 words in order of importance for the first cluster

In *Figure 6.13*, we can see that the most important words are `win`, `world`, `cup`, and `final`. All these words are sports-related, and therefore the topic for cluster 0 is *sports*.

We can put together the most important words per cluster:

```
Cluster 0 : ['win', 'world', 'cup', 'final', 'lead', 'set', 'hit',
'face', 'record', 'open', 'year', 'miss', 'test', 'new', 'day']
Cluster 1 : ['govt', 'iraq', 'urg', 'nsw', 'polic', 'continu', 'say',
'australia', 'vic', 'consid', 'qld', 'iraqi', 'forc', 'secur', 'sar']
Cluster 2 : ['plan', 'council', 'new', 'govt', 'fund', 'say', 'boost',
'group', 'water', 'concern', 'health', 'report', 'claim', 'seek',
'warn']
Cluster 3 : ['polic', 'man', 'kill', 'charg', 'court', 'murder',
'crash', 'death', 'attack', 'woman', 'face', 'arrest', 'probe', 'car',
'dead']
```

From the preceding information, we can conclude the topics for each cluster:

- `Cluster 0`: Sports
- `Cluster 1`: Global affairs
- `Cluster 2`: Economy
- `Cluster 3`: Crime/current happenings

With this information, now we can take any of the headlines that were not used during the clustering process and predict their topic:

```
import random
random_index = random.randint(0, len(headlines_full)) random_headline
= headlines_full["headline_text"][random_index] print(random_index,
random_headline)
```

This piece of code will yield a similar statement to the following:

```
771004 waratahs count cost with loss to cheetahs
```

The preceding sentence is a clear reference to sports; therefore, we would expect the predicted cluster to be `0`.

In order to predict the cluster group, we need to follow following steps: Define the function, run it and verify the result.

1. Let's code a prediction function that will perform the necessary cleaning and preprocessing, embedding, and clustering steps:

```
def predict(cluster_km, headline):    """    This function
predicts the cluster  of a headline via K-Means
    """
    # Cleaning
    headline_clean = clean_text(headline)
    headline_pre = process_words(headline_clean)
    # Embeddings
    bag_of_words_list = headline_pre.split()
    number_of_words = len(bag_of_words_list)
    # Process 1st word (to be able to concatenate)
    word_embeddings_array = w2v[bag_of_words_list[0]].reshape(1,
embedding_features)
    # To manage headlines with just 1 meaningful word
    word_index = -1
    for word_index, word in enumerate(bag_of_words_list[1:]):
        word_embeddings = w2v[word].reshape(1, embedding_
features)
        word_embeddings_array = mx.nd.concat(word_embeddings_
array,
 word_embeddings, dim=0)
    assert(number_of_words == word_index + 2)
    average_embedding_headline_pre = mx.nd.mean(word_embeddings_
array, axis=0).reshape(1, embedding_features)
    # Clustering
```

```
    selected_cluster = cluster_km.predict(average_embedding_
headline_pre.asnumpy())
    return selected_cluster
```

2. Let's run this function with the following code:

```
predicted_cluster = predict(cluster_km, random_headline)
print(predicted_cluster)
```

3. The output is the actual predicted cluster:

```
[0]
```

Great work!

How it works...

In this recipe, we have explored the NLP task known as topic modeling. This task tries to come up with the topics associated with a given set of documents. Typically, no answer is given (no ground truth), and so this task is better solved via unsupervised learning. We attempted to solve this task with ABC's *1 Million News Headlines* dataset.

We followed a three-step approach:

1. Data processing and cleaning
2. Word embeddings
3. Clustering

For the first step, we followed a typical pipeline for any NLP problem:

1. Data cleaning
2. Tokenizing
3. Removing stop words
4. Stemming
5. Lemmatization

For the second step, we applied Google's word2vec to compute embeddings for each word, and each headline embedding was computed as the average of the embeddings of each one of its words.

In the third step, we explored the unsupervised learning algorithm K-means, selected four clusters, and computed its centroids. We generated the following topic clusters sports, global affairs, economy and crime, and current happenings.

With this information, we selected a random headline and accurately predicted the topic it was related to.

There's more...

Unsupervised learning is a very wide topic and an active field of research. To learn more, a good starting point is its Wikipedia article: `https://en.wikipedia.org/wiki/Unsupervised_learning`.

Apart from the *1 Million News Headlines* dataset, another well-known reference dataset for topic modeling is the 20 Newsgroups dataset. I recommend working with the larger 6 Newsgroups choice as many Newsgroups had a lot of themes in common. More information can be found at `http://qwone.com/~jason/20Newsgroups/`.

One simplification we followed during the processing of embeddings is that the computation of our headline embedding was done by averaging each of the corresponding word embeddings. However, there are other approaches, known as document embeddings or sentence embeddings, with models such as **Doc2Vec** or **Sentence-BERT**, which can be more useful for other applications. An analysis comparing some of these approaches can be found at `https://www.analyticsvidhya.com/blog/2020/08/top-4-sentence-embedding-techniques-using-python/`.

For a detailed explanation of how the K-means algorithm works, it is suggested you review `https://towardsdatascience.com/k-means-clustering-explained-4528df86a120`.

When predicting the topic of a given headline, K-means is equivalent to another algorithm, called 1-nearest neighbor, which is the specific case of K-nearest neighbor with K = 1. More information regarding this supervised learning algorithm can be found at `https://en.wikipedia.org/wiki/K-nearest_neighbors_algorithm`.

Analyzing sentiment in movie reviews

Sentiment analysis is the use of several different techniques, including NLP, to identify the emotional state associated with human-generated information, text in our case. In this recipe, we are going to perform sentiment analysis on real-world movie reviews. We will classify the reviews into two sentiments: positive or negative.

To achieve this, we will use several pre-trained models from GluonNLP Model Zoo, and apply its word embeddings to feed a classifier, which will output the predicted sentiment. We will apply this process to a new dataset: **IMDb Movie Reviews**.

Getting ready

As in previous chapters, in this recipe, we will be using a little bit of matrix operations and linear algebra, but it will not be hard at all.

Furthermore, we will be classifying text datasets. Therefore, we will revisit some concepts already seen in *Recipe 4, Understanding text datasets – loading, managing, and visualizing the Enron Email dataset*, from *Chapter 2, Working with MXNet and Visualizing Datasets: Gluon and DataLoader*.

How to do it...

In this recipe, we will be carrying out the following steps:

1. Exploring the *IMDb Movie Reviews* dataset
2. Combining TextCNNs with word embeddings
3. Introducing BERT
4. Putting everything together

Exploring the IMDb Movie Reviews dataset

This dataset was collected in 2011 by researchers from Stanford University. It is split into a training set and a test set, with each of the sets having 25,000 reviews. They included at most 30 reviews per movie. The sentiments are quite polar, with negative reviews with values between [1, 4] and positive reviews with values between [7, 10].

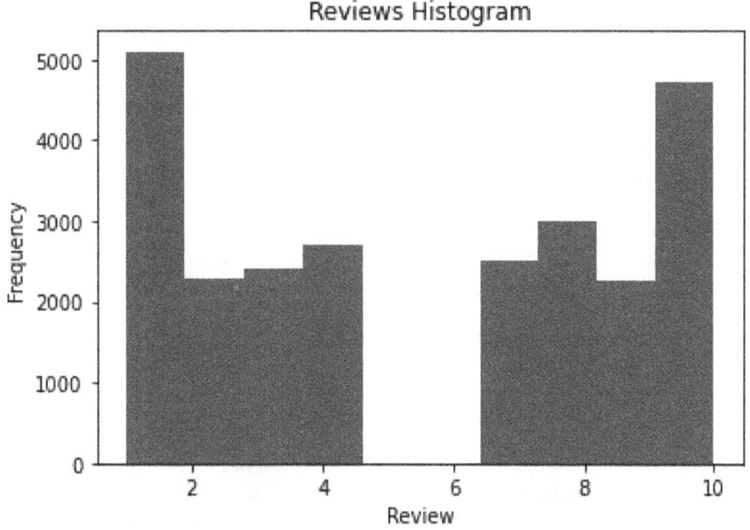

Figure 6.14 – Histogram of movie reviews (imbalanced dataset)

For our analysis, we will simplify the sentiment values to a binary sentiment classification task. Therefore, negative reviews are assigned a 0 value and positive reviews a 1 value. As a by-product of this simplification, the dataset becomes balanced.

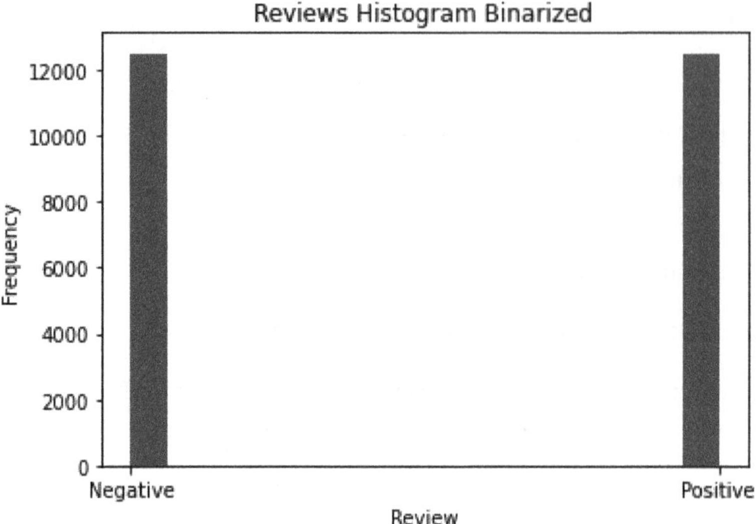

Figure 6.15 – Histogram of binarized movie reviews (balanced dataset)

Another point to take into account, noted by the paper's authors, is that as nuances in the language can contain information regarding the sentiment, the preprocessing of the reviews must not include usual stop words and stemming. We took this remark into account in our preprocessing:

```
def process_words_basic(
    text,
    lemmatizer = lemmatizer):
    words = nltk.tokenize.word_tokenize(text)
    filtered_words_post = []
    for word in words:
        if word.isalpha():
            filtered_words_post.append(lemmatizer.lemmatize(word))
    return filtered_words_post
```

The file can be accessed from `https://github.com/PacktPublishing/Deep-Learning-with-MXNet-Cookbook/blob/main/ch06/utils.py`.

This function is applied to all samples in the dataset.

Combining TextCNNs with word embeddings

After processing the dataset, we are now ready to use it with any architecture of our choice. In the first recipe of this chapter, we showed how we can use CNNs with sequences. In order to provide language information, TextCNNs can use pre-trained token representations as input. For this recipe, we will use two word embeddings that will generate inputs for our model:

- **word2vec**: These embeddings were introduced in *Recipe 4, Understanding text datasets - loading, managing, and visualizing the Enron Email dataset*, from *Chapter 2, Working with MXNet and Visualizing Datasets: Gluon and DataLoader*
- **BERT**: A language model introduced by Google in 2018

Introducing BERT

RNNs and Transformers can work with large sequences of text. However, one of the largest disadvantages is that the data is processed in a single direction, from left to right. BERT provides a mechanism so that every word representation (token) can jointly use information from both directions, to the left and to the right of that specific word.

Another distinctive characteristic of BERT is that its attention mechanisms are solely based on self-attention layers; no cross-attention layers are used.

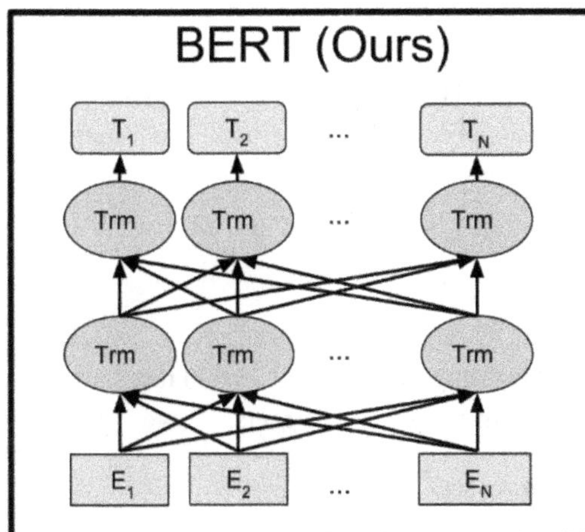

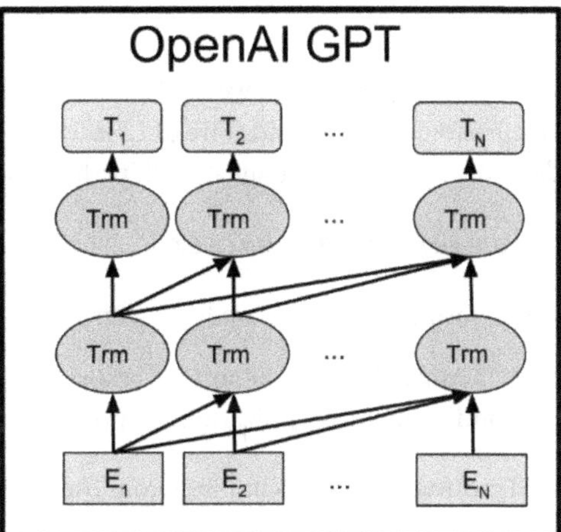

Figure 6.16 – BERT architecture: Comparison of BERT bidirectional approach with Transformers, such as GPT-1; left-to-right-only approach

BERT was trained in an unsupervised manner, using two task objectives:

- **Masked language model**: The word under analysis is not shown, and therefore the model needs to understand its meaning from context alone
- **Next-sentence prediction**: Given two sentences, this task predicts whether they have a connection (one could happen after the other in a longer text) or they are not related

This training methodology, combined with the BERT architecture, proved to be very successful, beating state-of-the-art on 11 NLP tasks at the time the paper was published.

GluonNLP provides two pre-trained BERT models with the following features:

- `BERT_12_768_12`: 12 layers, 768-dimensional embedding vectors, 12 self-attention heads. This model is known as **BERT base**.
- `BERT_24_1024_16`: 24 layers, 1,024-dimensional embedding vectors, 16 self-attention heads. This model is known as **BERT large**.

For our experiments, we will use the BERT base model, which can be easily loaded with the following code statement:

```
bert_model, vocab = nlp.model.get_model(
    'bert_12_768_12',
    dataset_name='book_corpus_wiki_en_uncased',
    use_classifier=False,
    use_decoder=False,
    ctx=ctx)
```

With the preceding function, we can easily obtain a BERT model (`bert_model`) and its vocabulary (`vocab`), based on the architecture of 12 layers, 768-dimensional embedding vectors, 12 self-attention heads, and a dataset from English Wikipedia (`book_corpus_wiki_en_uncased`).

Putting everything together

Let's summarize all the steps we have seen so far.

Our *IMDb Movie Reviews* dataset is composed of 25,000 training samples and 25,000 test samples. For cost and compute optimization purposes, we work with the following datasets:

- **Training set**: 5,000 samples (from the original training set)
- **Validation set**: 1,250 samples (from the original training set; no overlap with our 5,000-sample training set)
- **Test set**: 25,000 samples

We use two embedding models as inputs to TextCNN:

- **word2vec**: Vectors with 300 components
- **BERT base**: Vectors with 768 components

The TextCNN architecture is very similar for both approaches. The kernel sizes for TextCNN are 3, 4, and 5, that is, analyzing 3, 4, and 5 words at the same time, and the number of channels is equivalent to the embedding components. Furthermore, as we have a binary output (*negative* or *positive*), the classifier is a fully connected layer with one sigmoid output (the sigmoid activation function is included in the loss function due to computational optimizations).

For the training, equivalent parameters are used for both embedding models:

- **Optimizer**: Adam
- **Learning rate**: 10^{-3}
- **Loss function**: Sigmoid cross-entropy
- **Epochs**: 5
- **Batch size**: 4

With these parameters, we have the following results using a word2vec embedding model:

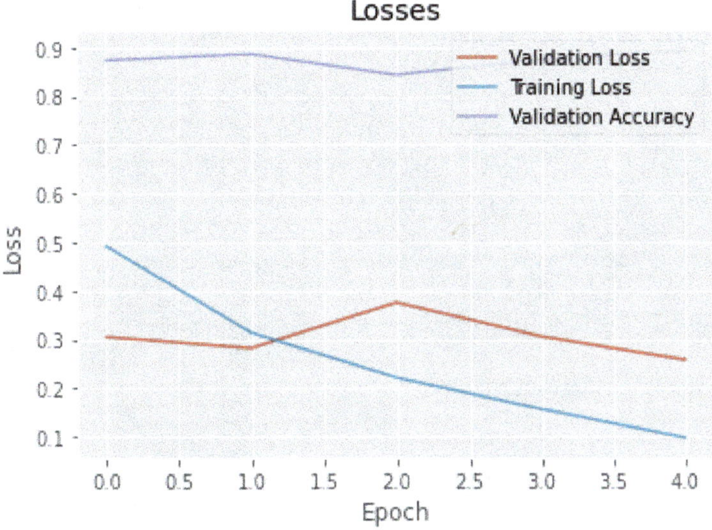

Figure 6.17 – Training loss/validation loss and accuracy using word2vec

In *Figure 6.17*, we can see how the training improved with the epochs, yielding the best validation accuracy of 0.89 at the end of the training process.

We can check the results qualitatively, selecting a movie review from our test set (unseen samples) and seeing the output of our sentiment analysis algorithm. The example movie review is from `https://ieee-dataport.org/open-access/imdb-movie-reviews-dataset` with license CC BY 4.0):

```
I went and saw this movie last night after being coaxed to by a few
friends of mine. I'll admit that I was reluctant to see it because
from what I knew of Ashton Kutcher he was only able to do comedy. I
was wrong. Kutcher played the character of Jake Fischer very well, and
Kevin Costner played Ben Randall with such professionalism. The sign
of a good movie is that it can toy with our emotions. This one did
exactly that. The entire theater (which was sold out) was overcome by
```

```
laughter during the first half of the movie, and were moved to tears
during the second half. While exiting the theater I not only saw many
women in tears, but many full grown men as well, trying desperately
not to let anyone see them crying. This movie was great, and I suggest
that you go see it before you judge.
```

We can format this input as embeddings expected by our TextCNN network:

```
# Formatting single input as expected for the network
seq_output, _ = process_dataset_sample(test_dataset[0][0])
 seq_output_reshaped = mx.nd.array(seq_output, ctx=ctx).expand_
dims(axis=0)
```

We can pass it through our best model from training:

```
# Retrieve best model from training
text_cnn.load_parameters(model_file_name)
review_sentiment = text_cnn(seq_output_rehaped)
# We can omit sigmoid processing, outputs of the network
# with positive values are positive reviews
if review_sentiment >= 0:
    print(review_sentiment, "The review is positive")
else:
    print(review_sentiment, "The review is negative")
```

These commands yield the following output:

```
[[2.5862172]]
 <NDArray 1x1 @gpu(0)> The review is positive
```

As can be seen in the preceding output, our algorithm has classified the review correctly as *positive*.

However, for a more thorough and formal analysis, we can quantitatively process the full test set and compute the final accuracy:

```
Final Test Accuracy: 0.87724
```

However, that number alone does not provide the full information about *Type I* and *Type II* errors. Therefore, we can also display the results as a confusion matrix (introduced in *Recipe 4, Evaluating classification models*, in *Chapter 4, Solving Classification Problems*):

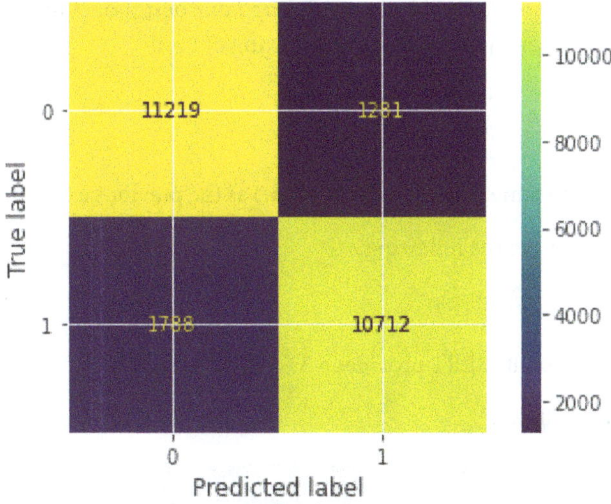

Figure 6.18 – Confusion matrix using word2vec

When following the same approach, but this time using our BERT model for embeddings, we have the following results:

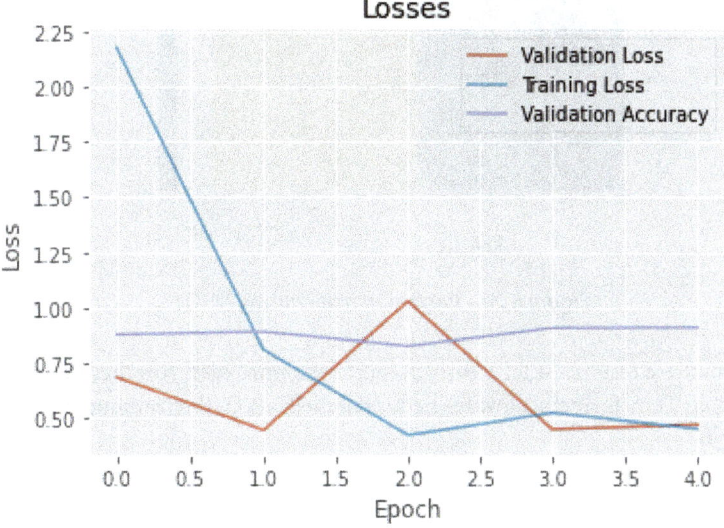

Figure 6.19 – Training loss/validation loss and accuracy using BERT

In *Figure 6.19*, we can see how the training improved with the epochs, yielding the best validation accuracy of 0.91 at the end of the training process. This figure is higher than with word2vec, as expected, as BERT is able to build more contextual relationships between the words in the reviews.

We can also pass the same review from the test set through our best model from training, using the same code statements as previously, obtaining the following output:

```
[[15.462966]]
 <NDArray 1x1 @gpu(0)> The review is positive
```

This experiment produces the same result (positive review) as the previous experiment with *word2vec*.

For the test set accuracy, we have the following:

```
Final Test Accuracy: 0.90848
```

Compared to the word2vec result, BERT provides a 3% higher accuracy.

The confusion matrix is as follows:

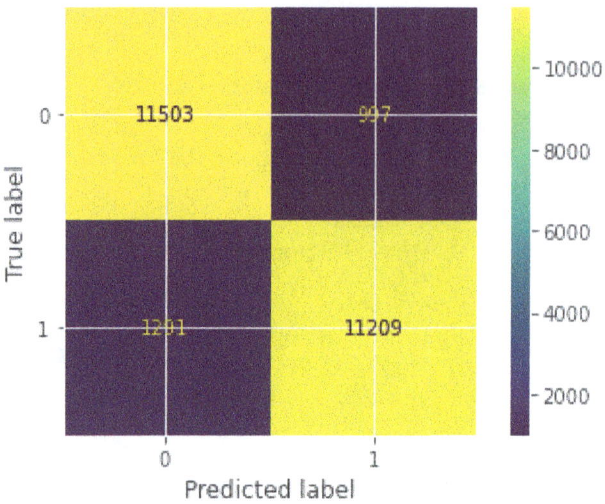

Figure 6.20 – Confusion matrix using BERT

As we can see from these results, BERT clearly outperforms word2vec. Another important advantage to mention is that, as Transformers allow for better parallelization, the training process is also faster.

How it works...

In this recipe, we tackled the sentiment analysis problem. We analyzed an architecture, TextCNN, to solve this task and explored how it can be applied to different word embedding models.

We explored a new dataset, *IMDb Movie Reviews*, and made adequate transformations so that we could work with the dataset in a constrained computation environment and simplify it to a binary classification task.

We introduced BERT, a new word embedding model introduced by Google in 2018, and compared it to a previously explored model, word2vec, understanding its differences, advantages, and constraints. We understood the two most important advantages of BERT: using bidirectional information for each word and masking each word in training so that the information about each word is purely based on its context.

We ran experiments to compare these two word embedding approaches and saw that despite both approaches solving the problem rather well (the test accuracy for word2vec and BERT being 88% and 91%, respectively), BERT performed better.

There's more...

Sentiment analysis is a well-researched task in the literature. To learn more, it is recommended to read this: `https://www.datarobot.com/blog/introduction-to-sentiment-analysis-what-is-sentiment-analysis/`.

The paper introducing the *IMDb Movie Reviews* dataset, which also proposed a model for sentiment analysis, can be found here: *Learning Word Vectors for Sentiment Analysis*, `https://ai.stanford.edu/~ang/papers/acl11-WordVectorsSentimentAnalysis.pdf`.

BERT was introduced in the paper `https://arxiv.org/pdf/1810.04805.pdf`. However, a more intuitive explanation can be found here: `https://huggingface.co/blog/bert-101`. Reading the preceding article is strongly encouraged due to its analysis of how data can embed biases in our models.

BERT is very powerful and can be complemented with even better language models, such as RoBERTa (improved version) or DistilBERT (smaller model with similar performance), and lots of models fine-tuned for specific tasks. A list of the pre-trained models available in MXNet GluonNLP can be found at `https://nlp.gluon.ai/model_zoo/bert/index.html`.

Translating text from Vietnamese to English

Translating text automatically (machine translation) has been a very interesting and useful use case for NLP since its inception, as breaking language barriers has lots of applications, including chatbots and automated subtitles in multiple languages.

Before deep learning, machine translation was typically approached as a statistical problem. Even after deep learning, it was not until Google, in 2016, applied deep learning to machine translation that the area of **Neural Machine Translation** (**NMT**) was born. This model set the foundation for translating tasks now available in LLMs, such as **OpenAI GPT** and **Google Bard**.

In this recipe, we will apply these techniques to translate sentences from Vietnamese to English, using pre-trained models from GluonNLP Model Zoo.

Getting ready

As in previous chapters, in this recipe, we will be using a little bit of matrix operations and linear algebra, but it will not be hard at all.

Furthermore, we will be classifying text datasets. Therefore, we will revisit some concepts already seen in *Recipe 4, Understanding text datasets – loading, managing, and visualizing the Enron Email dataset*, from *Chapter 2, Working with MXNet and Visualizing Datasets: Gluon and DataLoader*.

How to do it...

In this recipe, we will be carrying out the following steps:

1. Exploring the *IWSLT2015* dataset
2. Evaluating machine translators (BLEU)
3. Introducing the GNMT model and exploring Transformers for this task
4. Putting everything together

Let's look at these steps in detail next.

Exploring the IWSLT2015 dataset

The **International Workshop on Spoken Language Translation** (**IWSLT**) is a yearly scientific workshop focused on all forms of translation (not necessarily machine translation). They have generated several very important datasets and benchmarks that have helped the field of machine translation evolve. In 2015, an English-Vietnamese dataset was published, composed of a training set of 130,000+ sentence pairs and validation/test sets with 1,000+ sentence pairs. This dataset is publicly available with MXNet GluonNLP and can be easily retrieved, as follows:

```
# IWSLT2015 Dataset (Train, Validation and Test)
# Dataset Parameters
src_lang, tgt_lang = "vi", "en"
src_max_len, tgt_max_len = 50, 50
iwslt_train_text = nlp.data.IWSLT2015("train",
                                      src_lang=src_lang,
                                      tgt_lang=tgt_lang)
iwslt_val_text   = nlp.data.IWSLT2015("val",
                                      src_lang=src_lang,
                                      tgt_lang=tgt_lang)
iwslt_test_text  = nlp.data.IWSLT2015("test",
```

```
                                src_lang=src_lang,
                                tgt_lang=tgt_lang)
iwslt_src_vocab = iwslt_train_text.src_vocab
iwslt_tgt_vocab = iwslt_train_text.tgt_vocab
```

This version of the dataset provides the following data:

```
Length of train set: 133166
Length of val set   : 1553
Length of test set  : 1268
```

The preprocessing is similar to previous pipelines we have already seen, and includes the following steps:

1. Sentence clipping (to define maximum values)
2. Tokenizing
3. Adding **End-of-Sentence** (**EOS**) tokens to the source sentence (Vietnamese) and **Beginning-of-Sentence** (**BOS**) and EOS tokens to the target sentence (English)

Furthermore, to optimize training, a bucketing process is applied, where sentences are grouped by similar length:

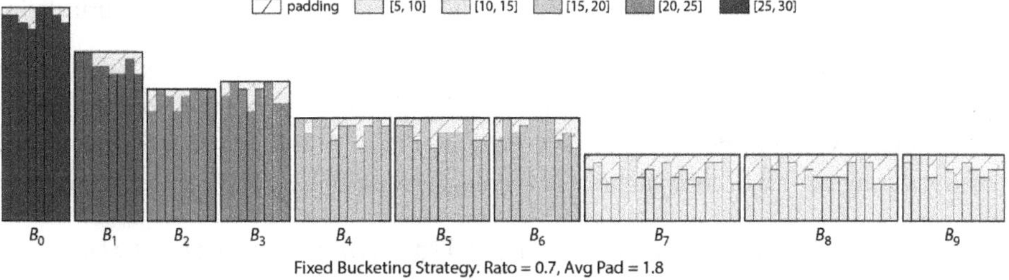

Figure 6.21 – Fixed bucket sampler

The example in *Figure 6.21* shows this strategy with 10 buckets, yielding a minimal amount of padding (required so that all sentences in 1 batch can be processed in parallel). The size of the buckets is also exponentially increased. Use MXNet GluonNLP, as follows:

```
# Bucket scheme
bucket_scheme = nlp.data.ExpWidthBucket(bucket_len_step=1.2)
```

In the preceding example, the size (width) of each bucket is augmented by 20% (1.2 increments).

Evaluating machine translators (BLEU)

Evaluating how successful machine translation systems are is very difficult. For example, using a single number to measure the quality of a translation is inherently subjective. For our use case, we will work with a widely used metric called **BiLingual Evaluation Understudy** (**BLEU**).

With BLEU, several reference translations are provided, and it tries to measure how close the automated translation is to its reference translations. To do so, it compares the different N-grams (size 1 to 4) of the automated translation to the N-grams of the reference translations.

Input: Bud Powell etait un pianiste de legende. **Reference**: Bud Powell was a legendary pianist.	**sentence BLEU** (0-100)
Candidate 1: Bud Powell was a legendary pianist.	100
Candidate 2: Bud Powell was a historic piano player.	46.7
Candidate 3: Bud Powell was a New Yorker.	54.1

Figure 6.22 – BLEU metric

As can be seen in *Figure 6.22*, BLEU tries to minimize the subjectivity associated with translations.

Another metric is **Perplexity**, which defines approximately how "surprised" the model is to see a translated word. When the model is not surprised, it means it is performing well; therefore, for Perplexity, a lower value is better. Computing Perplexity is much faster than BLEU, and so it is used more as a checking metric during per-batch computations in training, leaving BLEU for per-epoch computations.

Introducing the GNMT model and exploring Transformers for this task

As mentioned, the largest improvement in the field of machine translation was introduced by Google in 2016 with their **Google Neural Machine Translator** (**GNMT**) model (`https://ai.googleblog.com/2016/09/a-neural-network-for-machine.html`).

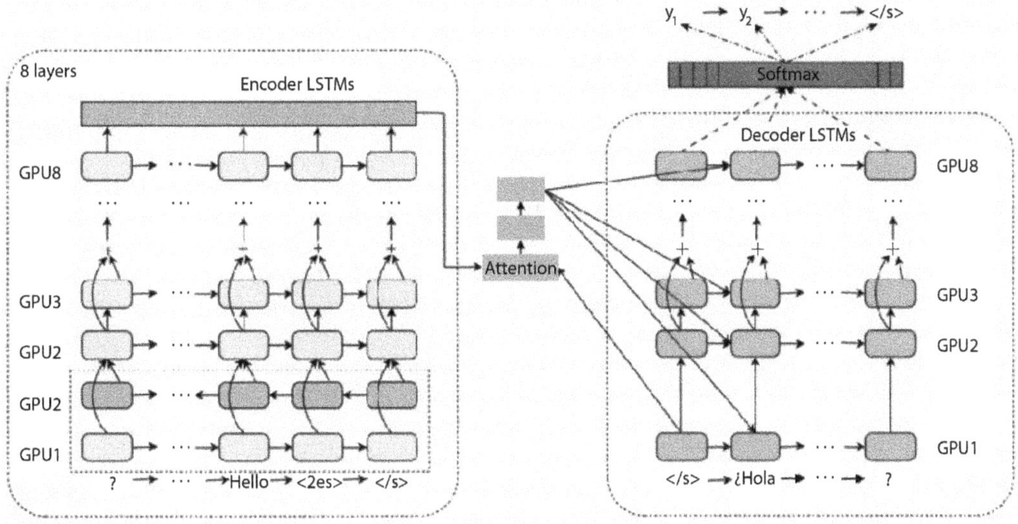

Figure 6.23 – GNMT architecture

GNMT is a pioneer of transformers and makes use of attention with an encoder-decoder architecture as well. The encoder and the decoder are LSTM RNNs, with eight layers in the encoder and another eight layers in the decoder. The attention mechanism implemented in the model is a cross-attention layer.

At the end of the model, a beam-search sampler is chained to generate new translations to maximize the trained conditional probability of the translations. As in the paper, in our implementation, the scoring function includes a length penalty so that all words in the translation are covered.

We'll compare GNMT with Transformers for our use case of Vietnamese-to-English machine translation. For our application, these are the most important parameters for each model:

- **GNMT**:
    - Number of layers for the encoder: 2
    - Number of layers for the decoder: 2
    - Number of units: 512
- **Transformer**:
    - Number of layers for the encoder: 4
    - Number of layers for the decoder: 4
    - Number of units: 512

In the next sections, we will compare both architectures for the same translation task.

Putting everything together

Let's summarize all the steps we have seen so far.

Our *IWSLT2015* Vietnamese-to-English dataset is composed of 133,000+ training samples and 1,000+ validation/test samples. We'll work with the complete datasets.

We'll use two models for our machine translation use case:

- GNMT
- Transformer

For the training, equivalent parameters are used for both architectures:

- **Optimizer**: Adam
- **Learning rate**: 10^{-3}, with a learning rate schedule of a step decay, halving the learning rate every epoch after half training
- **Loss function**: Masked softmax cross-entropy, similar to the cross-entropy loss functions we have already explored with the added feature that when predictions are longer than their valid length, the redundant words are masked out
- **Epochs**: 12
- **Batch size**: 128

With these parameters, we have the following evolution in the training using the GNMT model:

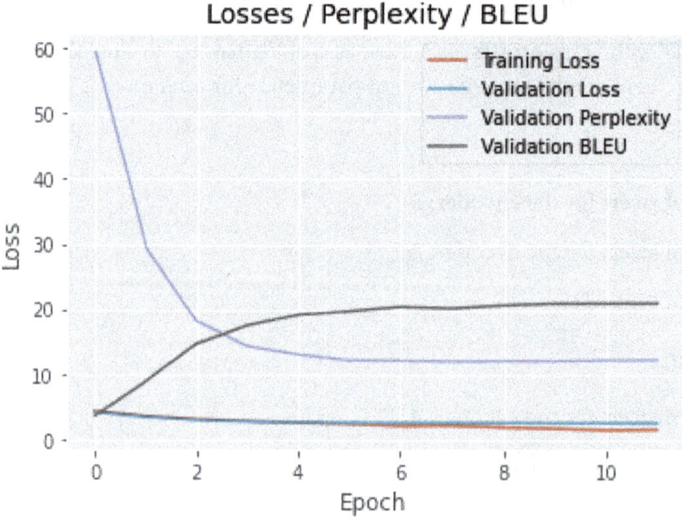

Figure 6.24 – GNMT training evolution (training loss and validation loss, Perplexity, and BLEU)

Furthermore, for the best iteration, the loss, perplexity, and BLEU score (multiplied by 100) obtained in the test set are as follows:

```
Best model test Loss=2.3807, test ppl=10.8130, test bleu=23.15
```

Current state-of-the-art models can yield above 30 points in their BLEU score, but this score is certainly very high.

Qualitatively, we can also check how well our model is performing with a sentence example. In our case, we chose `I like to read books`, and this can be verified with the following code:

```
print("Qualitative Evaluation: Translating from Vietnamese to
English")
expected_tgt_seq = "I like to read books."
 print("Expected translation:")
 print(expected_tgt_seq)
# From Google Translate
src_seq = "Tôi thích đọc sách kỹ thuật."
 print("In Vietnamese (from Google Translate):")
 print(src_seq)
translation_out = nmt.utils.translate(
    gnmt_translator,
    src_seq,
    iwslt_src_vocab,
    iwslt_tgt_vocab,
    ctx)
print("The English translation is:")
 print(" ".join(translation_out[0]))
```

These code statements will give us the following output:

```
Qualitative Evaluation: Translating from Vietnamese to English
Expected translation:
 I like to read books.
 In Vietnamese (from Google Translate):
 Tôi thích đọc sách kỹ thuật.
 The English translation is:
 I like to read books .
```

As can be seen from the results, the text has been correctly translated from Vietnamese to English.

Now, we are going to repeat the same exercises with our Transformer model. With the parameters defined previously for its training, we have the following evolution in the training:

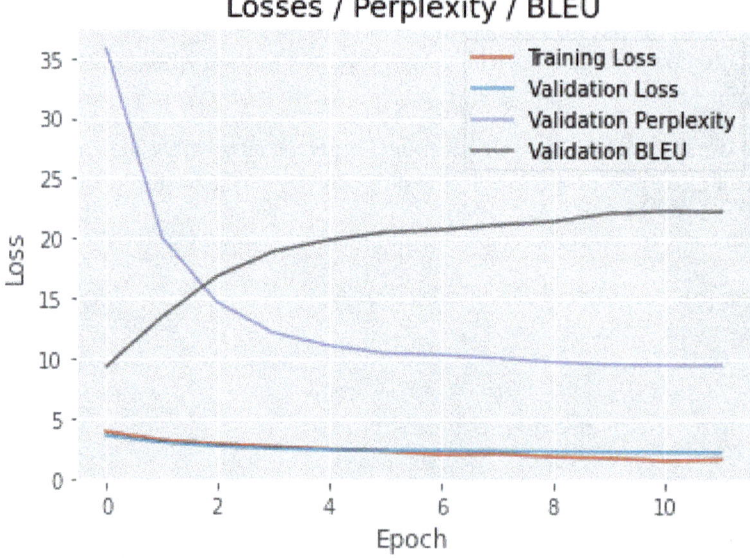

Figure 6.25 – Transformer training evolution (training loss and validation loss, Perplexity, and BLEU)

Furthermore, for the best iteration, the loss, perplexity, and BLEU score (multiplied by 100) obtained in the test set are as follows:

```
Best model test Loss=2.1171, test ppl=8.3067, test bleu=24.16
```

As we can see, the Transformer architecture yields around ~0.015 higher BLEU score points.

As done for GNMT, we can also check how well our model is performing qualitatively with the same sentence example and code. The output is as follows:

```
Qualitative Evaluation: Translating from Vietnamese to English
Expected translation:
 I like to read books.
 In Vietnamese (from Google Translate):
 Tôi thích đọc sách kỹ thuật.
 The English translation is:
 I like to read books .
```

As can be seen from the results, the text has been correctly translated from Vietnamese to English.

How it works...

In this recipe, we solved one of the most useful tasks in NLP, machine translation. We introduced a new architecture, GNMT, the precursor of the Transformer, and compared both models.

We explored a new dataset, *IWSLT2015*, which, among other language pairs, supports translations between Vietnamese and English. We introduced the Perplexity and BLEU metrics, which are widely used to evaluate translation models.

We ran experiments to compare these two models and saw that, despite both approaches solving the problem rather well (the BLEU scores in the test set for GNMT and the Transformer were 23.15 and 24.34, respectively), the Transformer performed better.

There's more...

Machine translation is a difficult problem to tackle. Two very good official guides from MXNet GluonNLP where this problem is solved are the following:

- **Official machine translation tutorials of MXNet GluonNLP:** `https://nlp.gluon.ai/examples/machine_translation/index.html`
- **AMLC19-GluonNLP:** `https://github.com/eric-haibin-lin/AMLC19-GluonNLP`

This recipe used code from the previous references. I would like to kindly thank the contributors.

The IWSLT conference takes place every year. For more info, please visit their official site: `https://iwslt.org/`.

We introduced two new metrics for translation problems, Perplexity and BLEU. Work is actively being carried out to improve these metrics, with new metrics being developed recently, such as **SacreBLEU**. Some references that tackle this very important topic are as follows:

- **Perplexity**: `http://blog.echen.me/2021/12/23/a-laymans-introduction-to-perplexity-in-nlp/`
- **BLEU**: `https://towardsdatascience.com/bleu-bilingual-evaluation-understudy-2b4eab9bcfd1`
- **Improving BLEU with SacreBLEU**: `https://aclanthology.org/W18-6319.pdf`

We also discussed GNMT for the first time, which was one of the first real-world systems that used deep learning for translation (NMT), developed by Google in 2016. The blog post where this was announced is worth reading: `https://ai.googleblog.com/2016/09/a-neural-network-for-machine.html`.

There are many translation models that have used the *IWSLT2015* dataset. The results can be found at `https://paperswithcode.com/sota/machine-translation-on-iwslt2015-english-1`.

Furthermore, in this recipe, we analyzed language-to-language translation, which has been the de facto approach of the industry for a long time, using English as a bridge language for multilingual translation. This is an active area of research, and recently, Meta, formerly known as Facebook, has developed the **No Language Left Behind** (**NLLB-200**) model. More information about this breakthrough can be found at `https://ai.facebook.com/blog/nllb-200-high-quality-machine-translation/`.

7
Optimizing Models with Transfer Learning and Fine-Tuning

As models grow in size (the depth and number of processing modules per layer), training them grows exponentially as more time is spent per epoch, and typically, more epochs are required to reach optimum performance.

For this reason, **MXNet** provides state-of-the-art pre-trained models via **GluonCV** and **GluonNLP** libraries. As we have seen in previous chapters, these models can help us solve a variety of problems when our final dataset is similar to the one the selected model has been pre-trained on.

However, sometimes this is not good enough, and our final dataset might have some nuances that the pre-trained model is not picking up. In these cases, it would be ideal to combine the stored knowledge of the pre-trained model with our final dataset. This is called transfer learning, where the knowledge of our pre-trained model is transferred to a new task (final dataset).

In this chapter, we will learn how to use GluonCV and GluonNLP, which are MXNet Gluon libraries that are specific to **Computer Vision** (**CV**) and **Natural Language Processing** (**NLP**), respectively. We will also learn how to retrieve pre-trained models from their model zoos, and how to optimize our own networks by transferring the learnings from these pre-trained models.

Specifically, we will cover the following topics in our recipes:

- Understanding transfer learning and fine-tuning
- Improving performance for classifying images
- Improving performance for segmenting images
- Improving performance for translating English to German

Technical requirements

Apart from the technical requirements specified in the *Preface*, the following technical requirements apply:

- Ensure that you have completed the first recipe, *Installing MXNet, Gluon, GluonCV and GluonNLP*, from *Chapter 1, Up and Running with MXNet.*
- Ensure that you have completed *Chapter 5, Analyzing Images with Computer Vision*, and *Chapter 6, Understanding Text with Natural Language Processing.*

The code for this chapter can be found at the following GitHub URL: `https://github.com/PacktPublishing/Deep-Learning-with-MXNet-Cookbook/tree/main/ch07`.

Furthermore, you can access each recipe directly from Google Colab; for example, the first recipe of this chapter can be found here: `https://colab.research.google.com/github/PacktPublishing/Deep-Learning-with-MXNet-Cookbook/blob/main/ch07/7_1_Understanding_Transfer_Learning_and_Fine_Tuning.ipynb`.

Understanding transfer learning and fine-tuning

In the previous chapters, we saw how we could leverage MXNet, GluonCV, and GluonNLP to retrieve pre-trained models in certain datasets (such as ImageNet, MS COCO, and IWSLT2015) and use them for our specific tasks and datasets.

In this recipe, we will introduce a methodology called **transfer learning**, which will allow us to combine the information from pre-trained models (on general knowledge datasets) and the information from the new domain (the dataset from the task we want to solve). There are two main significant advantages to this approach. On the one hand, pre-training datasets are typically large-scale (ImageNet-22k has 14 million images), and using a pre-trained model saves us that training time. On the other hand, we use our specific dataset not only for evaluation but also for training the model, improving its performance in the desired scenario. As we will discover, there is not always an easy way to achieve this, as it requires the capability to obtain a sizable dataset, or even one right way, as it might not yield the expected results. We will also explore the optional next step after transfer learning, called fine-tuning, where we will try to use our specific dataset to modify the model parameters even further. We will put both techniques to the test.

Getting ready

As for previous chapters, in this recipe, we will be using some matrix operations and linear algebra, but it will not be hard at all.

How to do it...

In this recipe, we will be looking at the following steps:

1. Introducing transfer learning
2. Describing the advantages of transfer learning and when to use it
3. Understanding the fundamentals of representation learning
4. Focusing on practical applications

Let's dive into each of these steps.

Introducing transfer learning

In the previous chapters, we learned how to train deep learning neural networks from scratch, exploring problems in CV and NLP. As introduced in *Chapter 3*, *Solving Regression Problems*, deep learning neural networks try to imitate the biological networks in our brains. One interesting point of view is that when we (and our brains) learn new tasks, we leverage previous knowledge we have acquired in a very strong way. For example, a very good tennis player will become a relatively good player at squash with a few hours of play. Transfer learning is a field of study that contains different techniques to achieve similar results as in this example.

Traditional ML
vs
Transfer Learning
• Isolated, single task learning:
• Knowledge is not retained or accumulated. Learning is performed w.o. considering past learned knowledge in other tasks
• Learning of a new tasks relies on the previous learned tasks:
• Learning process can be faster, more accurate and/or need less training data
Dataset 1
Learning System Task 1
Dataset 2
Learning System Task 2
Dataset 1
Learning System Task 1
Knowledge
Dataset 2
Learning System Task 2

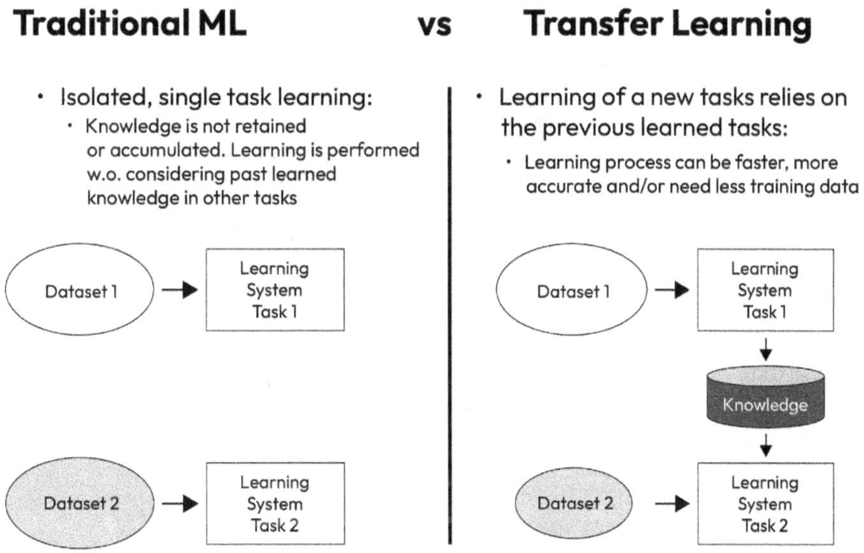

Figure 7.1 – Comparison between traditional machine learning (ML) and transfer learning

In *Figure 7.1*, we can see a comparison between both paradigms where, in transfer learning, the approach to solving Task 2 leverages the knowledge acquired while solving Task 1. This implies, however, that to solve a single desired task (Task 2), we are training the model twice (for Task 1 and for Task 2 later). In practice, as we will see in the next steps, we will work with pre-trained models from MXNet's GluonCV and GluonNLP model zoos, and therefore, we will only have to train the model once, for Task 2.

Describing the advantages of transfer learning and when to use it

There are several reasons why using transfer learning offers advantages:

- **Faster**: As we leverage pre-trained models from model zoos, the training will converge much faster than training from scratch, requiring much fewer epochs and less time.
- **More general**: Typically, pre-trained models have been trained in large-scale datasets (such as ImageNet); therefore, the parameters (weights) learned are generalistic and can then be reused for a large number of tasks. It is an objective that outputs from the feature extraction part of the pre-trained model (also known as **representations**), learned by training using large-scale datasets that are general and domain-invariant (can be reused).
- **Requires less data**: To adapt a pre-trained model for a given new task, the amount of data required is much less than for training that specific model architecture from scratch. This is because representations can be reused (as mentioned in the previous point).
- **More environmentally friendly**: As the training time, datasets, and compute requirements for transfer learning are much lower than training from scratch, less pollution is required to train a model.
- **Performance improvements**: It has been proven (for example, in `https://www.cv-foundation.org/openaccess/content_cvpr_2014/papers/Oquab_Learning_and_Transferring_2014_CVPR_paper.pdf`) that using transfer learning with small-scale datasets yields strong performance improvements, and on large-scale datasets, the same performance point is achieved much faster than training from scratch.

In *Figure 7.2*, different methods to compute representations are analyzed, and although specialized networks can reach better performance, this is only possible if large-scale datasets, high-end compute resources, and longer training times are given.

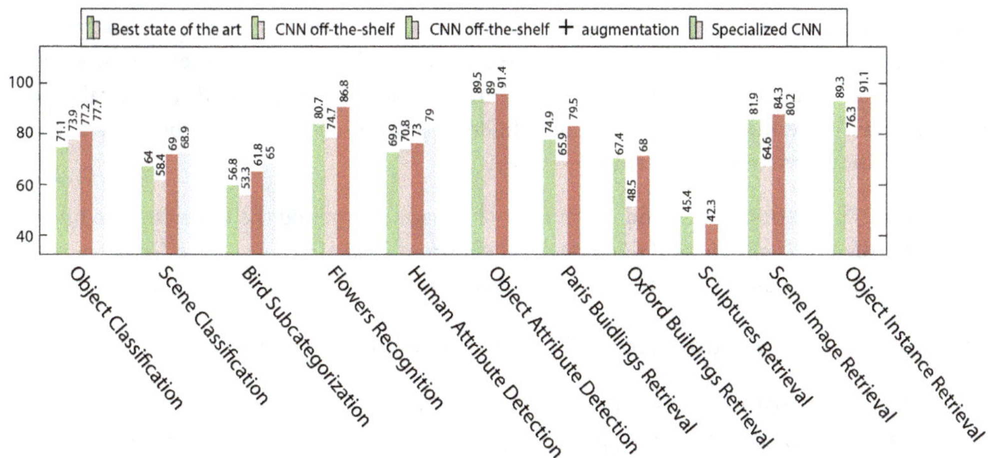

Figure 7.2 – Comparing different approaches for representations

In a more general setting, there are different ways to achieve transfer learning, as shown in the following figure:

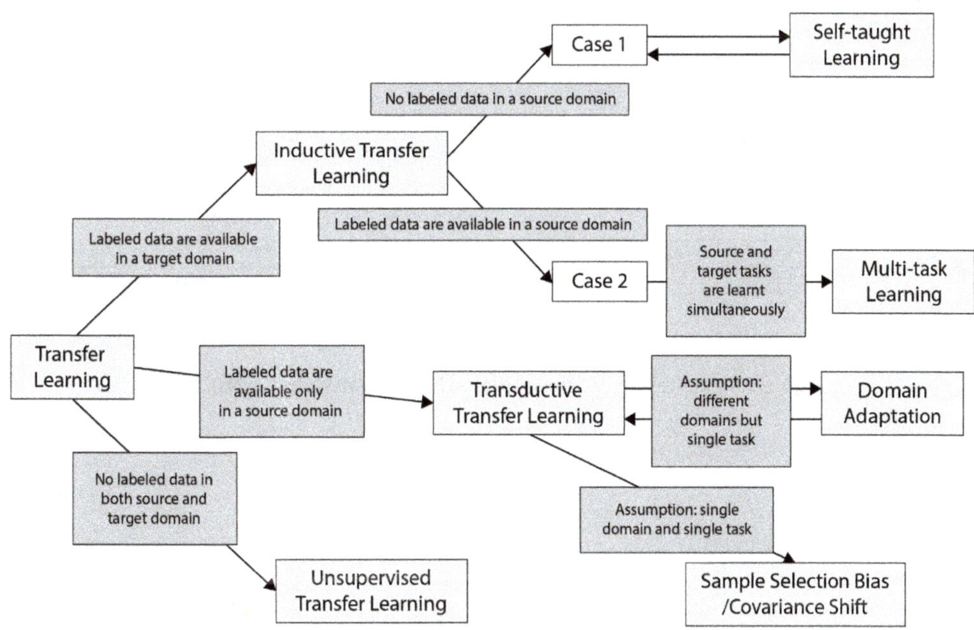

Figure 7.3 – Different types of transfer learning

In *Figure 7.3*, we can see the different types of transfer learning, depending on the similarity of the source and target domain and the availability of source and target data. In this chapter, we will explore the usual setting of having a pre-trained model in a similar domain to our intended task (equal source and target domain), and the tasks will be slightly different, with some amount of labeled data in the target domain (**inductive transfer learning**).

Andrew Ng, chief scientist of Baidu and co-founder of Google Brain, said the following in a tutorial in NIPS 2016 called *Nuts and Bolts of Building AI Applications Using Deep Learning*: "*In the next few years, we'll see a lot of concrete value driven through transfer learning*," and he was right.

Understanding the fundamentals of representation learning

In this section, we will answer the question, from a more theoretical point of view, about how to use transfer learning and why it works. In *Chapter 5, Analyzing Images with Computer Vision*, and *Chapter 6, Understanding Text with Natural Language Processing*, we introduced the concept of **representations** for features in images using GluonCV and for words/sentences in text using GluonNLP.

We can revisit, in *Figure 7.4*, the usual architecture of a CNN architecture:

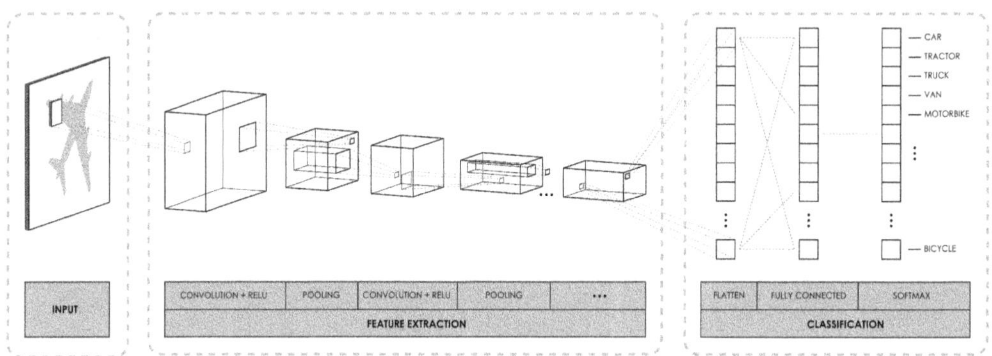

Figure 7.4 – Refresher of Convolutional Neural Networks (CNNs)

In *Figure 7.5*, we can revisit the usual Transformer architecture:

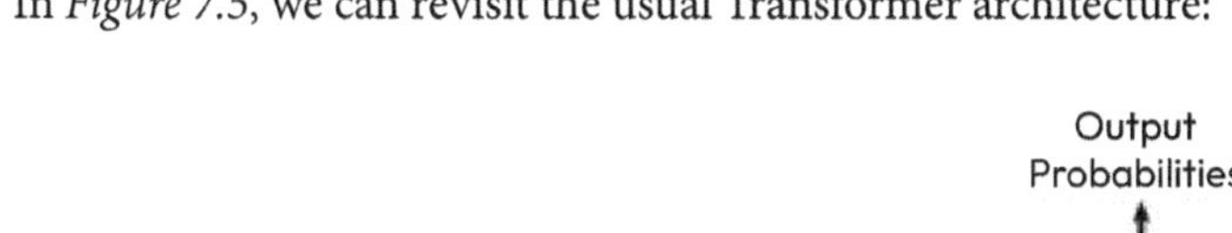

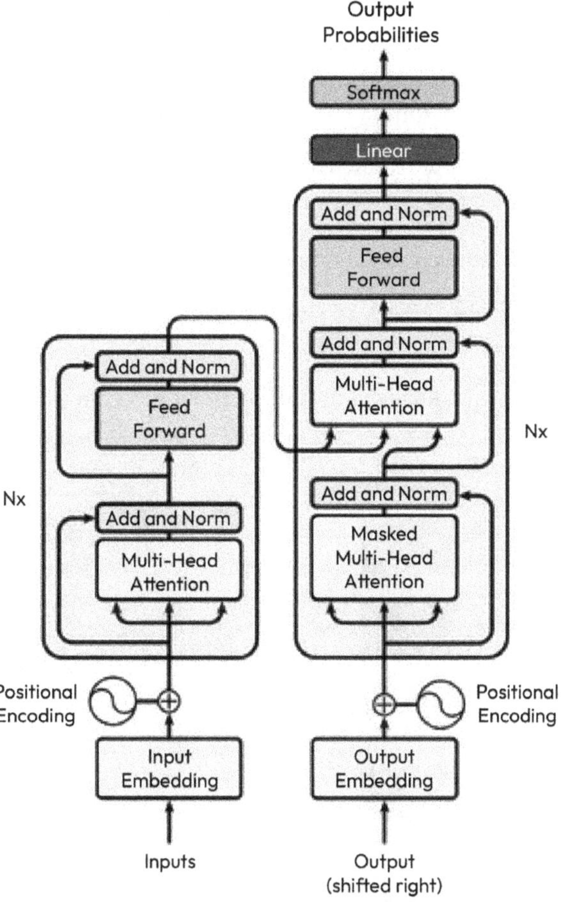

Figure 7.5 – Refresh of the Transformer architecture (encoder on the left, and decoder on the right)

The underlying idea is common in both fields; for example, the feature extractor part of CNNs and the encoder in Transformers are representations, and the training of these network sections is called **representation learning**, an active field of study due to the capability of being able to train these networks in both supervised and unsupervised settings.

The main idea behind transfer learning is to transfer the representations learned in a task to a different task; therefore, we will typically follow the next steps:

1. Retrieve a pre-trained model from MXNet's Model Zoo (GluonCV or GluonNLP).
2. Remove the last layers (typically, a classifier). Keep the parameters in the rest of the layers frozen (not updatable during training).

3. Add new layers (a new classifier) corresponding to the new task
4. Train the updated model (only the new layers, not frozen, will be updated during training) with the target data.

If we have enough labeled data for the task that we want to solve (target task), another step (that can be done after the previous step or substituting it) is called **fine-tuning**.

Fine-tuning takes into account that the representations originally learned might not fit perfectly with the target task and, therefore, could also improve with updating. In this scenario, the steps are as follows:

1. Unfreeze the weights of the representation network.
2. Retrain the network with target data, typically with a smaller learning rate as the representations should be close (same domain).

Both processes (transfer learning and fine-tuning) are summarized visually in *Figure 7.6*.

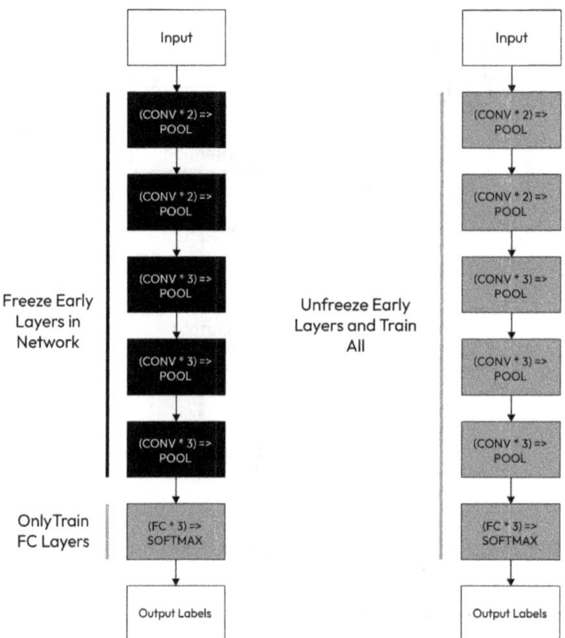

Figure 7.6 – Transfer learning and fine-tuning

Both processes can be applied sequentially, with adequate **hyperparameters** for each one.

Focusing on practical applications

In this section, we will use what we have learned so far about representation learning and we will apply it to a practical example: detecting cats and dogs.

To do this, we will retrieve a model from the **GluonCV Model Zoo**; we will remove the classifier (last layers) and keep the feature extraction stage. We will then analyze how the representations of the cats and dogs have been learned. To load the model, we can use this code snippet:

```
alexnet = gcv.model_zoo.get_model("resnet152_v2", pretrained=True,
ctx=ctx)
```

In the previous code snippet, for the `pretrained` parameter, we have assigned the value of `True`, indicating that we want the pretrained weights to be retrieved (and not only the architecture of the model).

When trained correctly, CNNs learn hierarchical representations of the features of the images in the training dataset, with each progressive layer learning more and more complex patterns. Therefore, when an image is processed (when processing on successive layers), the network can compute more complex patterns associated with the network.

Now, we can use a new MXNet library, MXBoard (see the recipe for installation instructions), with this model to evaluate the different steps that a dog image goes through and see some examples of how a pre-trained model computes its representations:

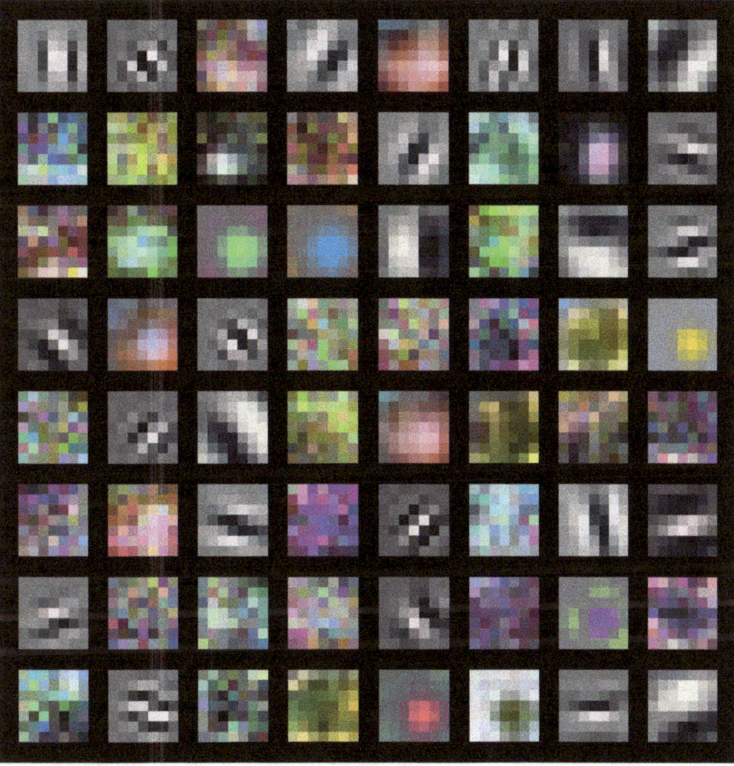

Figure 7.7 – Cat and dog representations – convolutional filters

In *Figure 7.7*, we can see the convolutional filters corresponding to the first convolutional layer of a ResNet152 pre-trained network (on ImageNet). Please note how these filters focus on simple patterns such as specific shapes (vertical and horizontal lines, circles, and so on) and specific colors (red blobs).

Let's analyze the results with a specific image:

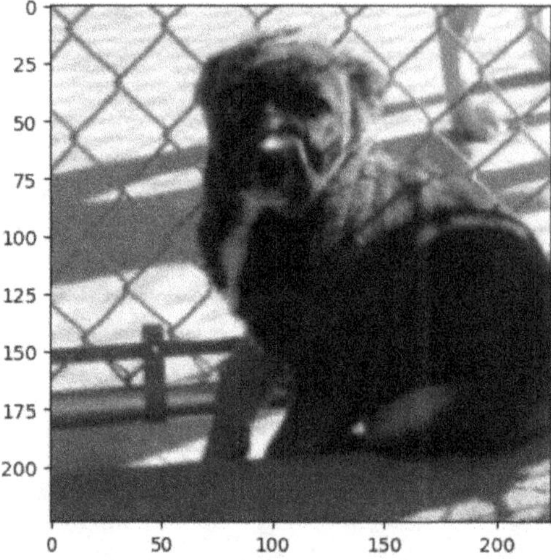

Figure 7.8 – Example image of a dog

We select an image from our *Dogs vs. Cats* dataset, such as the dog depicted in *Figure 7.8*. When passing this image through our network, we will find results similar to the following:

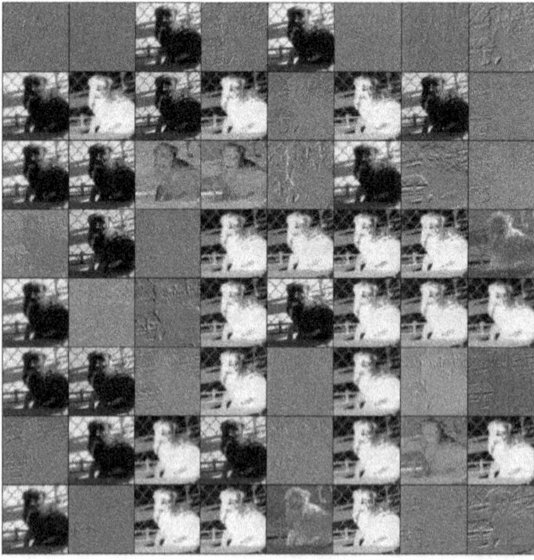

Figure 7.9 – Output from convolutional filters

In *Figure 7.9*, we can see the output of the filters in *Figure 7.7* for our dog example. Note how different outputs highlight simple shapes such as the eyes or the legs (larger values, closer to white).

Finally, as the image traverses the network, its features are more and more compressed, yielding (for ResNet152) a final vector of 2,048 elements. This vector can be computed easily with networks retrieved using MXNet's Model Zoo:

```
resnet152.features(summary_image.as_in_context(ctx))
```

This code excerpt provides the following output:

```
[[2.5350871e-04 2.8519407e-01 1.6196619e-03 ... 7.2884483e-05
  2.9618644e-07 7.8995163e-03]]
 <NDArray 1x2048 @gpu(0)>
```

As we can see, we have a `2048` element.

How it works...

In this recipe, we introduced the concepts of transfer learning and fine-tuning. We explained when it made sense to use these two different techniques and their advantages.

We also explored when these techniques can be useful and their connections to representation learning, explaining how representations play a significant role in the knowledge being transferred when using these techniques. We used a new library, **MXBoard**, to produce visualizations for the representations.

Moreover, we intuitively and practically showed how to apply these techniques to CV and NLP tasks and computed a representation for a specific example.

There's more...

Transfer learning, including fine-tuning, is an active field of study. In this recipe, we have only covered the most useful scenario for deep learning, inductive transfer learning. For a more comprehensive but still easy-to-read introduction, I recommend reading *Transfer learning: a friendly introduction*, which can be found at: `https://journalofbigdata.springeropen.com/articles/10.1186/s40537-022-00652-w`.

Moreover, the concept of transferring knowledge from one system to another is not new, and there are references to concepts such as **learning to learn** and **knowledge transfer** as early as 1995, when a **Neural Information Processing Systems** (**NeurIPS**) workshop on the topic was presented. A summary of the workshop can be found here: `http://socrates.acadiau.ca/courses/comp/dsilver/nips95_ltl/nips95.workshop.pdf`.

Furthermore, as introduced 21 years later in the same venue, Andrew Ng was able to correctly foresee the importance of transfer learning. His 2016 NeurIPS tutorial can be found here (jump to 1h 37m for the transfer learning quote): `https://www.youtube.com/watch?v=F1ka6a13S9I`.

Improving performance for classifying images

After introducing transfer learning and fine-tuning in the previous recipe, in this one, we will apply it to **image classification**, a CV task.

In the second recipe, *Classifying images with MXNet – GluonCV Model Zoo, AlexNet, and ResNet*, in *Chapter 5, Analyzing Images with Computer Vision*, we saw how we could use GluonCV to retrieve pre-trained models and use them directly for an image classification task. In the first instance, we looked at training them from scratch, effectively only leveraging past knowledge by using the architecture of the pre-trained model, without leveraging any past knowledge contained in the pre-trained weights, which were re-initialized, deleting any past information. Afterward, the pre-trained models were used directly for the task, effectively also leveraging the weights/parameters of the model.

In this recipe, we will combine the weights/parameters of the model with the target dataset, applying the techniques introduced in this chapter, transfer learning and fine-tuning. The dataset used for the pre-training was **ImageNet-1k** (source task) and we will run several experiments to train and evaluate our models in a new (target) task, using the `Dogs vs Cats` dataset.

Getting ready

As for previous chapters, in this recipe, we will be using some matrix operations and linear algebra, but it will not be hard at all.

Furthermore, we will be working with text datasets; therefore, we will revisit some concepts already seen in the second recipe, *Classifying images with MXNet: GluonCV Model Zoo, AlexNet, and ResNet*, in *Chapter 5, Analyzing Images with Computer Vision*.

How to do it...

In this recipe, we will be looking at the following steps:

1. Revisiting the *ImageNet-1k* and *Dogs vs. Cats* datasets
2. Training a **ResNet** model from scratch with *Dogs vs Cats*
3. Using a pre-trained ResNet model to optimize performance via transfer learning from *ImageNet-1k* to *Dogs vs Cats*
4. Fine-tuning our pre-trained ResNet model on *Dogs vs Cats*

Let's look at these steps in detail next.

Revisiting the ImageNet-1k and Dogs vs Cats datasets

ImageNet-1k and *Dogs vs Cats* are both image classification datasets; however, they are quite different. *ImageNet-1k* is a large-scale dataset containing ~1.2 million images labeled into 1,000 classes and has been used extensively in research and academia for benchmarking. *Dogs vs Cats* is a small-scale dataset containing 1,400 images depicting either a dog or a cat, and its fame is mostly due to a Kaggle competition launched in 2013.

MXNet GluonCV does not provide methods to directly download any of the datasets. However, we do not need the *ImageNet-1k* dataset (its size is ~133 GB), only the pre-trained parameters for our chosen model. The pre-trained models can be downloaded directly from the MXNet GluonCV Model Zoo, we have seen examples in previous chapters and we will use them again in this one.

Here are some examples from *ImageNet-1k*:

Figure 7.10 – ImageNet-1k examples

The source of the preceding figure is `https://cs.stanford.edu/people/karpathy/cnnembed/`.

For *Dogs vs Cats*, all the information on how to retrieve the dataset can be found in the second recipe, *Classifying images with MXNet: GluonCV Model Zoo, AlexNet, and ResNet*, in *Chapter 5*, *Analyzing Images with Computer Vision*. Taking that recipe's code as a reference, we can display some examples:

Figure 7.11 – Dogs vs Cats dataset

In *Figure 7.10* and *Figure 7.11*, we can see how some images from *ImageNet-1k* resemble some of the images from *Dogs vs Cats*.

Training a ResNet model from scratch with Dogs vs Cats

As described in the second recipe, *Classifying images with MXNet: GluonCV Model Zoo, AlexNet, and ResNet*, in *Chapter 5*, *Analyzing Images with Computer Vision*, we will be using **softmax cross-entropy** as the loss function and **accuracy** and the **confusion matrix** as evaluation metrics.

We have the following evolution in the training using the ResNet model:

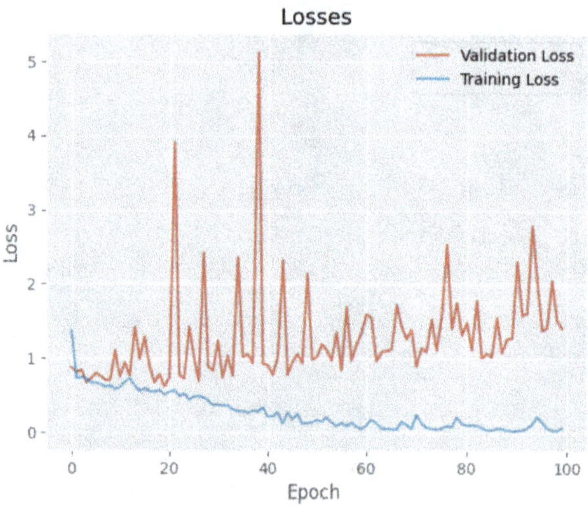

Figure 7.12 – ResNet training evolution (training loss and validation loss, and validation accuracy) – training from scratch

Furthermore, for the best iteration, the `accuracy` value obtained in the test set is as follows:

```
('accuracy', 0.75)
```

The confusion matrix is as follows:

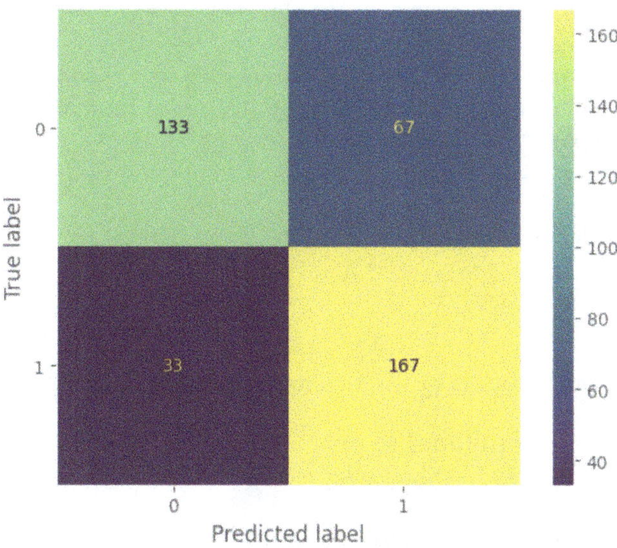

Figure 7.13 – Confusion matrix in Dogs vs Cats for a ResNet model trained from scratch

Both the accuracy value obtained (75%) and *Figure 7.13* show quite average performance after training for several epochs (100, in this example). You are encouraged to run your own experiments trying out different hyperparameters.

Qualitatively, we can also check how well our model is performing with an example image. In our case, we chose the following:

Figure 7.14 – Qualitative example of Cats vs Dogs, specifically a cat

We can check the output of our model by running this image through it with the following code snippet:

```
# Qualitative Evaluation
# Qualitative Evaluation
# Expected Output
print("Expected Output:", example_label)
# Model Output
example_output = resnet50_ft(example_image_preprocessed)
 class_output = np.argmax(example_output, axis=1).asnumpy()[0]
 print("Class Output:", class_output)
assert class_output == 0 # Cat 0
```

These code statements will give us the following output:

```
Expected Output: 0
Class Output: 0
```

As can be seen from the results, the image has been correctly classified as a cat.

Using a pre-trained ResNet model to optimize performance via transfer learning from ImageNet-1k to Dogs vs Cats

In the previous recipe, we trained a new model from scratch using our dataset. However, this has two important drawbacks:

- A large amount of data is required for training from scratch.
- The training process can take a long time due to the large size of the dataset and the number of epochs needed for the model to learn the task.

Therefore, in this recipe, we will follow a different approach: we will use pre-trained models from MXNet GluonCV to solve the task. These models have been trained in *ImageNet-1k*, a dataset that contains the classes we are interested in (cats and dogs); therefore, we can use those learned representations and easily transfer them to *Dogs vs Cats* (same domain).

For a ResNet model, use the following:

```
# ResNet50 from Model Zoo (This downloads v1d)
 resnet50 = gcv.model_zoo.get_model("resnet50_v1d", pretrained=True,
ctx=ctx)
```

As we can see in the previous code snippet, following the discussion in this chapter's first recipe, *Understanding transfer learning and fine-tuning*, for the `pretrained` parameter, we have assigned the value of `True`, indicating that we want the pre-trained weights to be retrieved (and not only the architecture of the model).

In order to adequately evaluate the improvements that transfer learning brings, we are going to evaluate our pre-trained model directly (the source task is *ImageNet-1k*) before applying transfer learning to *Dogs vs Cats* and after applying it. Therefore, using our pre-trained model as is, we obtain the following:

```
('accuracy', 0.925)
```

The confusion matrix is as follows:

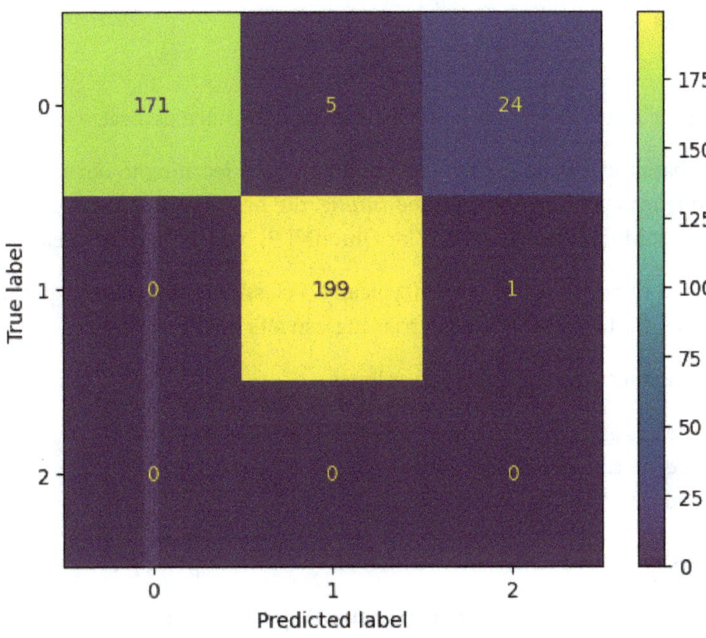

Figure 7.15 – Confusion matrix in Dogs vs Cats for a pre-trained ResNet model

As we can see, our pre-trained Transformer model is already showing good performance values as it is the same domain; however, simply using a pre-trained model does not yield better performance than training from scratch. The great advantage of using pre-trained models is the time savings, as loading one just takes a few lines of code.

We can also check how well our model is performing qualitatively with the same image example. Note how the code slightly differs from the previous qualitative image excerpt, as now we need to convert ImageNet classes (the output of our ResNet50 pre-trained model) to our classes (`0` for cats and `1` for dogs). The new code is given as follows:

```
# Qualitative Evaluation
# Expected Output
print("Expected Output:", example_label)
# Model Output
```

```
example_output = resnet50(example_image_preprocessed)
class_output = model.CLASSES_DICT[np.argmax(example_output, axis=1).
asnumpy()[0]]
print("Class Output:", class_output)
assert class_output == 0 # Cat
```

These code statements will give us the following output:

```
Expected Output: 0
Class Output: 0
```

As can be seen from the result, the image has been correctly classified as a cat.

Now that we have a baseline for comparison, let's apply transfer learning to our task. From the first recipe, *Understanding transfer learning and fine-tuning*, the first step was to retrieve a pre-trained model from the MXNet Model Zoo (GluonCV or GluonNLP), which we have already done.

The second step was to remove the last layers (typically, a classifier), keeping the parameters in the rest of the layers frozen (not updatable during training), so let's do it!

We can replace the classifier with the following snippet:

```
# Replace the classifier (with gradients activated)
 resnet50_tl.fc = mx.gluon.nn.Dense(2)
 resnet50_tl.fc.initialize(ctx=ctx)
```

We can freeze the ResNet feature extraction layers with the following snippet:

```
for param in resnet50_tl.collect_params().values():
param.grad_req = 'null'
```

We can replace the classifier with the following snippet:

```
# Replace the classifier (with gradients activated)
 resnet50_tl.fc = mx.gluon.nn.Dense(2)
 resnet50_tl.fc.initialize(ctx=ctx)
```

Now, we can apply the usual training process with *Dogs vs Cats*, and we have the following evolution in the training using the ResNet model:

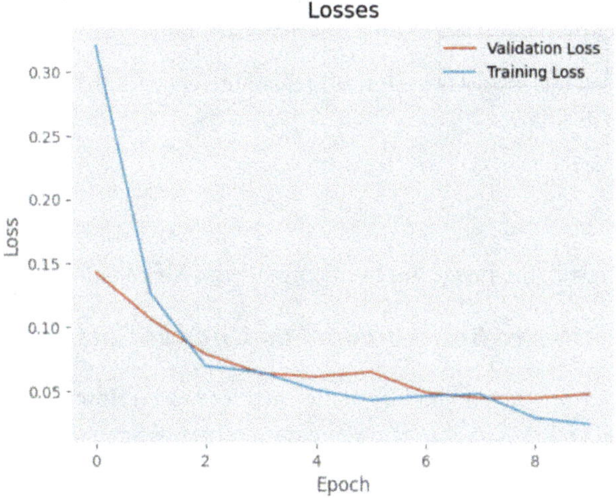

Figure 7.16 – ResNet training evolution (training loss and validation loss) – transfer learning

Furthermore, for the best iteration, the accuracy obtained in the test set is as follows:

```
('accuracy', 0.985)
```

The confusion matrix is as follows:

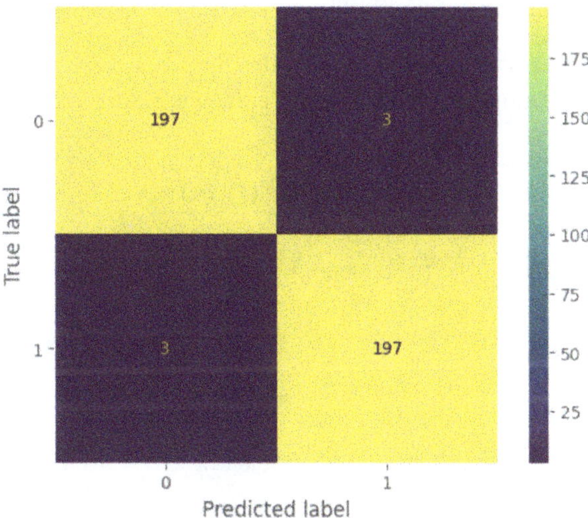

Figure 7.17 – Confusion matrix in Dogs vs Cats for a ResNet model with transfer learning

Compared with our previous experiment of training from scratch, this experiment yields much higher performance, and it took us literally minutes to get this model to start working well for us in our intended task, whereas the training required for the previous experiment took hours and required several tries to tune the hyperparameters, which can then turn into several days of effort in total.

We can also check how well our model is performing qualitatively with the same image example and code. The output is given as follows:

```
Expected Output: 0
Class Output: 0
```

As can be seen from the result, the image has been correctly classified as a cat.

Fine-tuning our pre-trained ResNet model on Dogs vs Cats

In the previous recipe, we *froze* the parameters in the encoder layers. However, as the dataset we are currently working with (*Dogs vs Cats*) has enough data samples, we can *unfreeze* those parameters and train the model, effectively allowing the new training process to update the representations (with transfer learning, we were working directly with the representations learned for *ImageNet-1k*). This process is called fine-tuning.

There are two variants of fine-tuning:

- Applying transfer learning by freezing the layers and unfreezing them afterward (fine-tuning after transfer learning)
- Directly applying fine-tuning without the preliminary step of freezing the layers (fine-tuning directly)

Let's compute both experiments and draw conclusions by comparing the results.

For the first experiment, we can take the network obtained in the previous recipe, unfreeze the layers, and restart the training. In MXNet, to unfreeze the encoder parameters, we can run the following snippet:

```
# Un-freeze weights
for param in resnet50_ft.collect_params().values():
    if param.name in updated_params:
        param.grad_req = 'write'
```

Now, we can apply the usual training process with *Dogs vs Cats*, and we have the following evolution in the training using the ResNet model:

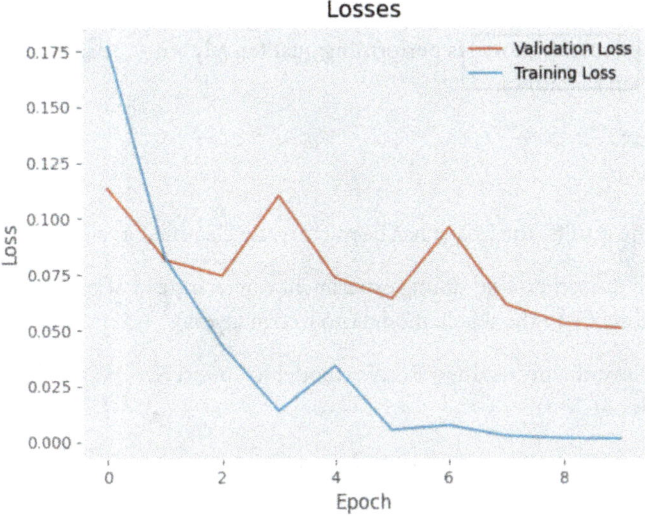

Figure 7.18 – ResNet training evolution (training loss and validation loss) – fine-tuning after transfer learning

Furthermore, for the best iteration, the accuracy obtained in the test set is as follows:

```
('accuracy', 0.90255)
```

The confusion matrix is as follows:

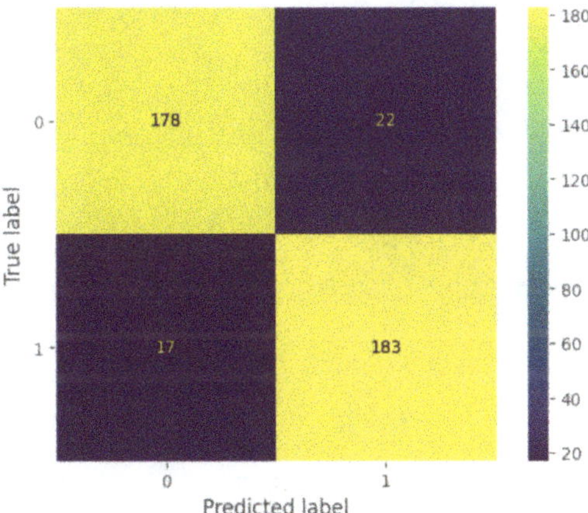

Figure 7.19 – Confusion matrix in Dogs vs Cats for a ResNet model with fine-tuning after transfer learning

Compared with our previous experiment of transfer learning, this experiment yields worse performance. This is due to a combination of the size of the dataset and the hyperparameters chosen. You are encouraged to try your own experiments.

We can also check how well our model is performing qualitatively with the same image example and code. The output is given as follows:

```
Expected Output: 0
Class Output: 0
```

As can be seen from the results, the image has been correctly classified as a cat.

Let's continue now with the second fine-tuning experiment where, instead of applying transfer learning, we apply fine-tuning directly to the whole model (no frozen layers).

We need to again retrieve the pre-trained ResNet model for *ImageNet-1k*, with the following code snippet for MXNet GluonCV:

```
# ResNet50 from Model Zoo (This downloads v1d)
resnet50 = gcv.model_zoo.get_model("resnet50_v1d", pretrained=True,
ctx=ctx)
```

And now, without freezing, we can apply the training process, which will update all layers of our ResNet model, giving the following loss curves:

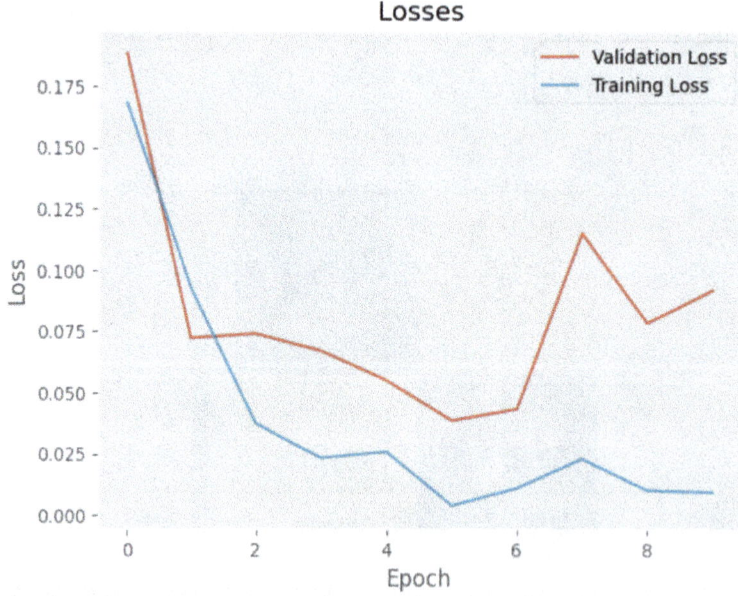

Figure 7.20 – ResNet training evolution (training loss and validation loss) – fine-tuning without freezing

Furthermore, for the best iteration, the accuracy obtained in the test set is as follows:

```
('accuracy', 0.98)
```

The value is similar to the previous experiment with transfer learning. For the confusion matrix, we have the following:

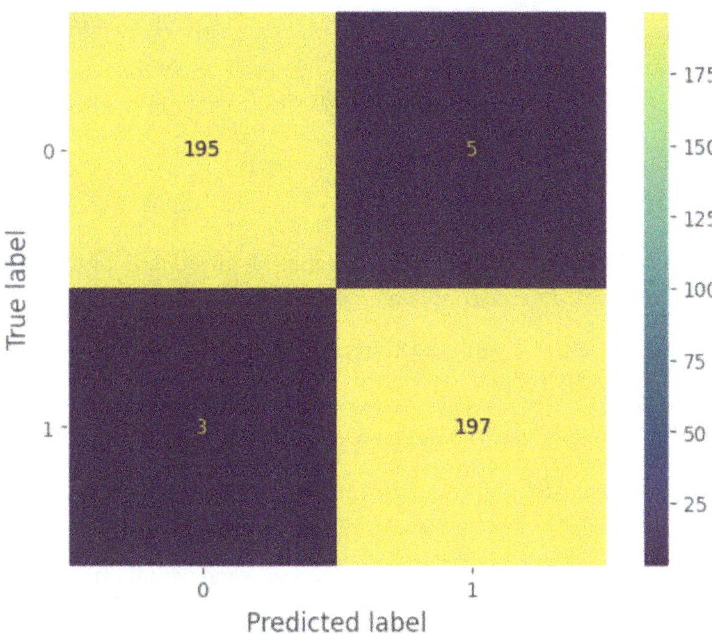

Figure 7.21 – Confusion matrix in Dogs vs Cats for a ResNet model with fine-tuning without freezing

As mentioned, compared with our previous fine-tuning experiment, we can see how this experiment yields higher performance. Empirically, this has been proven to be a repeatable result and has been indicated to be because initially freezing the encoder allows for the decoder to learn (using the encoder representations) the new task at hand. From an information flow point of view, in this step, there is a knowledge transfer from the feature extraction stage to the classifier. In the secondary step when the feature extraction stage is unfrozen, the learned parameters from the classifier perform auxiliary transfer learning – this time, from the classifier to the feature extraction stage.

We can also check how well our model is performing qualitatively with the same image example and code. The output is given as follows:

```
Expected Output: 0
Class Output: 0
```

As can be seen from the results, the image has been correctly classified as a cat.

How it works...

In this recipe, we applied the techniques of transfer learning and fine-tuning, introduced at the beginning of the chapter, to the task of image classification, which was also presented previously, in the second recipe, *Classifying images with MXNet: GluonCV Model Zoo, AlexNet, and ResNet*, in *Chapter 5, Analyzing Images with Computer Vision*.

We revisited two known datasets, *ImageNet-1k* and *Dogs vs Cats*, which we intended to combine using knowledge transfer based on the former dataset and refining that knowledge with the latter. Moreover, this was achieved by leveraging the tools that MXNet GluonCV provided:

- A pre-trained ResNet model for *ImageNet-1k*
- Tools for easy-to-use access to *Dogs vs Cats*

Furthermore, we continued using the loss functions and metrics introduced for image classification, softmax cross-entropy, accuracy, and the confusion matrix.

Having all these tools readily available within MXNet and GluonCV allowed us to run the following experiments with just a few lines of code:

- Training a model from scratch in *Dogs vs Cats*
- Using a pre-trained model to optimize performance via transfer learning from *ImageNet-1k* to *Dogs vs Cats*
- Fine-tuning our pre-trained model on *Dogs vs Cats* (with and without freezing layers)

After running the different experiments, we obtained an effective tie between transfer learning and fine-tuning directly (accuracies of 0.985 and 0.98, respectively). The actual results obtained when running these experiments might differ based on model architecture, datasets, and hyperparameters chosen, so you are encouraged to try out different techniques and variations.

There's more...

Transfer learning, including fine-tuning, is an active field of research. A recent paper published in 2022 explores the latest advances in image classification. The paper is titled *Deep Transfer Learning for Image Classification: A survey*, and can be found here: `https://www.researchgate.net/publication/360782436_Deep_transfer_learning_for_image_classification_a_survey`.

For a more general approach to CV use cases, a recent paper was published, *Transfer Learning Methods as a New Approach in Computer Vision Tasks with Small Datasets*, where the problem of small datasets is evaluated, and these techniques are applied to solve medical imaging tasks. It can be found here: `https://www.researchgate.net/publication/344943295_Transfer_Learning_Methods_as_a_New_Approach_in_Computer_Vision_Tasks_with_Small_Datasets`.

Improving performance for segmenting images

In this recipe, we will apply transfer learning and fine-tuning to **semantic segmentation**, a CV task.

In the fourth recipe, *Segmenting objects in images with MXNet: PSPNet and DeepLab-v3*, in *Chapter 5, Analyzing Images with Computer Vision*, we saw how we could use GluonCV to retrieve pre-trained models and use them directly for a semantic segmentation task, effectively leveraging past knowledge by using the architecture and the weights/parameters of the pre-trained model.

In this recipe, we will continue leveraging the weights/parameters of the model, obtained for a task consisting of classifying images among a set of 21 classes using semantic segmentation models. The dataset used for the pre-training was *MS COCO* (source task) and we will run several experiments to evaluate our models in a new (target) task, using the *Penn-Fudan Pedestrian* dataset. In these experiments, we will also include knowledge from the target dataset to improve our semantic classification performance.

Getting ready

As for previous chapters, in this recipe, we will be using some matrix operations and linear algebra, but it will not be hard at all.

Furthermore, we will be working with text datasets; therefore, we will revisit some concepts already seen in the fourth recipe, *Segmenting objects in images with MXNet: PSPNet and DeepLab-v3*, in *Chapter 5, Analyzing Images with Computer Vision.*

How to do it...

In this recipe, we will be looking at the following steps:

1. Revisiting the *MS COCO* and *Penn-Fudan Pedestrian* datasets
2. Training a **DeepLab-v3** model from scratch with *Penn-Fudan Pedestrian*
3. Using a pre-trained DeepLab-v3 model to optimize performance via transfer learning from *MS COCO* to *Penn-Fudan Pedestrian*
4. Fine-tuning our pre-trained DeepLab-v3 model on *Penn-Fudan Pedestrian*

Let's look at these steps in detail next.

Revisiting the MS COCO and Penn-Fudan Pedestrian datasets

MS COCO and *Penn-Fudan Pedestrian* are both object detection and semantic segmentation datasets; however, they are quite different. *MS COCO* is a large-scale dataset containing ~150k images labeled into 80 classes (21 main ones) and has been used extensively in research and academia for benchmarking. *Penn-Fudan Pedestrian* is a small-scale dataset containing 170 images of 423 pedestrians. For this recipe, we will focus on the semantic segmentation task.

MXNet GluonCV does not provide methods to directly download any of the datasets. However, we do not need the *MS COCO* dataset (its size is ~19 GB), only the pre-trained parameters for our chosen model.

Here are some examples from *MS COCO*:

Figure 7.22 – MS COCO example

For *Penn-Fudan Pedestrian*, all the information on how to retrieve the dataset can be found in the fourth recipe, *Segmenting objects in images with MXNet: PSPNet and DeepLab-v3*, in *Chapter 5*, *Analyzing Images with Computer Vision*. Taking that recipe's code as a reference, we can display some examples:

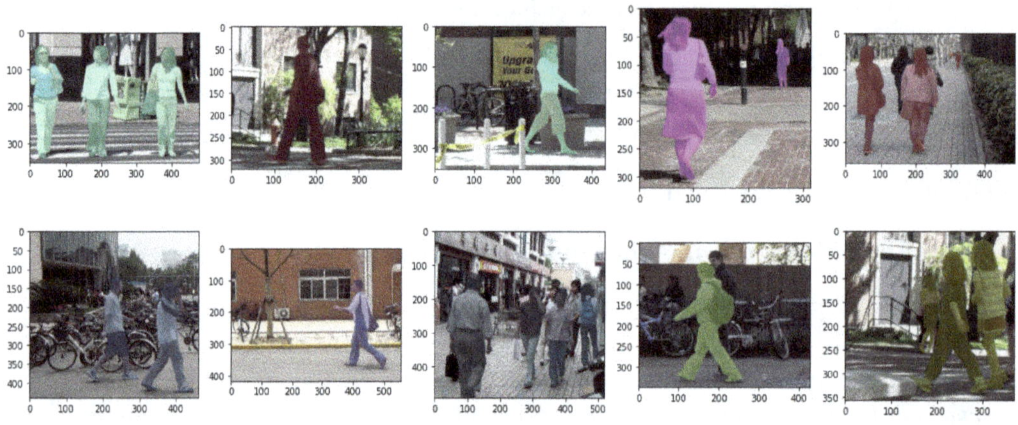

Figure 7.23 – Penn-Fudan Pedestrian dataset examples

From *Figures 7.22* and *7.23*, we can see how some images from *MS COCO* resemble some of the images from *Penn-Fudan Pedestrian*.

Training a DeepLab-v3 model from scratch with Penn-Fudan Pedestrian

As described in the fourth recipe, *Segmenting objects in images with MXNet: PSPNet and DeepLab-v3*, in *Chapter 5*, *Analyzing Images with Computer Vision*, we will be using softmax cross-entropy as the loss function and pixel accuracy and **Mean Intersection over Union** (**mIoU**) as evaluation metrics.

By following the code in our recipe, we have the following evolution while training from scratch our *DeepLab-v3* model:

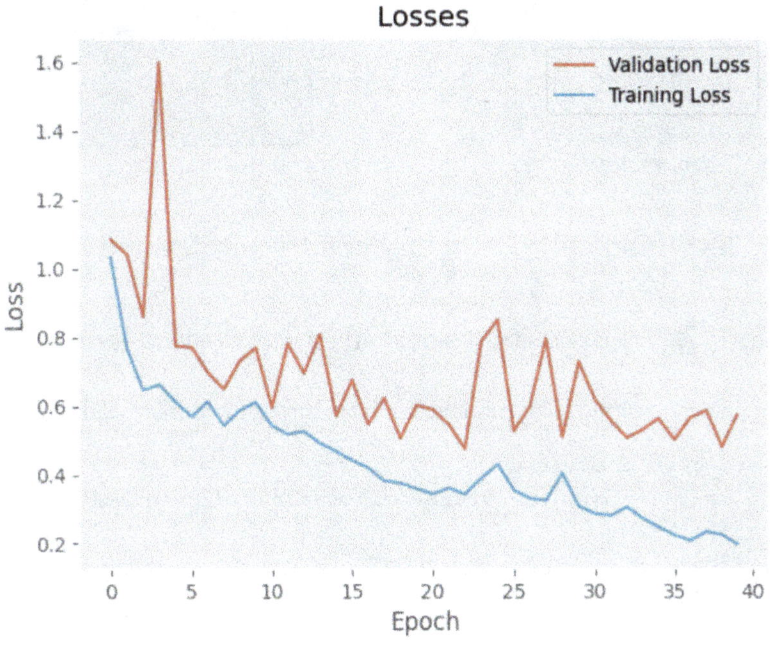

Figure 7.24 – DeepLab-v3 training evolution (training loss and validation loss) – training from scratch

Furthermore, for the best iteration, the pixel accuracy and mIoU values obtained in the test set are as follows:

```
PixAcc:  0.8454046875
mIoU  :  0.6548404063890942
```

Even after training for 40 epochs, the evaluation values obtained do not show strong performance (an mIoU value of only 0.65).

Qualitatively, we can also check how well our model is performing with an example image. In our case, we chose the following:

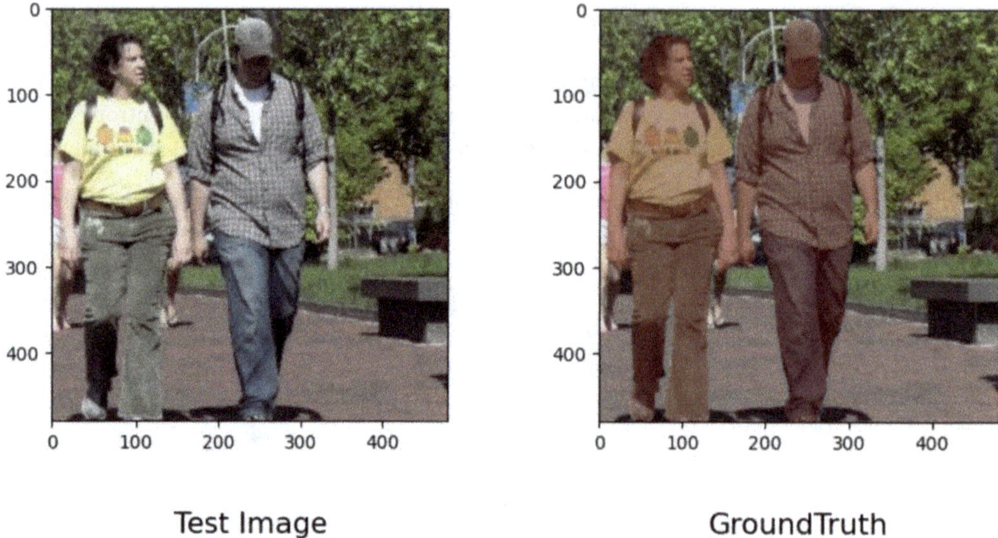

Figure 7.25 – Image example of Penn-Fudan Pedestrian for qualitative results

We can check the output of our model by running this image through it with the following code snippet:

```
# Compute and plot prediction
transformed_image = gcv.data.transforms.presets.segmentation.test_
transform(test_image, ctx)
 output = deeplab_ts(transformed_image)
 filtered_output = mx.nd.argmax(output[0], 1)
 masked_output = gcv.utils.viz.plot_mask(test_image, filtered_output)
 axes = fig.add_subplot(1, 2, 2)
 axes.set_title("Prediction", fontsize=16, y=-0.3)
 axes.axis('off')
 axes.imshow(masked_output);
```

The preceding code snippet shows the ground truth segmentations and the prediction from our model:

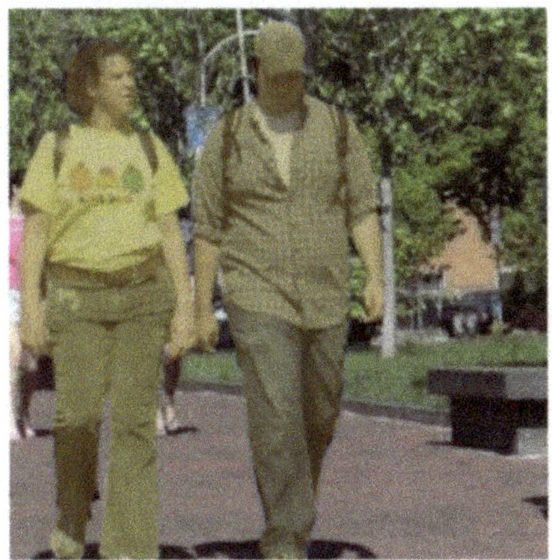

GroundTruth   Prediction

Figure 7.26 – Ground truth and prediction from DeepLab-v3 trained from scratch

As can be seen from the results, the pedestrians have only started to be correctly segmented. To improve the results, we will need to train for more epochs and/or adjust the hyperparameters. However, a better, faster, and simpler approach would be to use transfer learning and fine-tuning.

Using a pre-trained DeepLab-v3 model to optimize performance via transfer learning from MS COCO to Penn-Fudan Pedestrian

In the previous recipe, we trained a new model from scratch using our dataset. However, this has three important drawbacks:

- A large amount of data is required for training from scratch.
- The training process can take a very long time due to the large size of the dataset and the number of epochs needed for the model to learn the task.
- The compute resources required might be expensive or difficult to procure.

Therefore, in this recipe, we will follow a different approach. We will use pre-trained models from the MXNet GluonCV Model Zoo to solve the task. These models have been trained in *MS COCO*, a dataset that contains the classes we are interested in (`person` in this case); therefore, we can use those learned representations and easily transfer them to *Penn-Fudan Pedestrian* (same domain).

For a DeepLab-v3 model, we have the following:

```
# DeepLab-v3 from Model Zoo
deeplab_pt
gcv.model_zoo.get_model('deeplab_resnet101_coco'
pretrained=True, ctx=ctx)
```

As we can see in the preceding code snippet, following the discussion in this chapter's first recipe, *Understanding Transfer-Learning and Fine-Tuning*, for the `pretrained` parameter, we have assigned the value of `True`, indicating that we want the pretrained weights to be retrieved (and not only the architecture of the model).

In order to evaluate adequately the improvements that transfer learning brings, we are going to directly evaluate our pre-trained model in our target task (the task source is *MS COCO*) before applying transfer learning to *Penn-Fudan Pedestrian* and after applying it. Therefore, using our pre-trained model as is, we obtain the following:

```
PixAcc:  0.9640322916666667
mIoU  :  0.476540873665686
```

As we can see, our pre-trained Transformer model is already showing good performance values as it is in the same domain. Moreover, the great advantage of using pre-trained models is the time savings, as loading a pre-trained model just takes a few lines of code.

We can also check how well our model is performing qualitatively with the same image example and code. The output is given as follows:

GroundTruth

Prediction

Figure 7.27 – Ground truth and prediction from a DeepLab-v3 pre-trained model

As can be seen from the results, the pedestrians have been correctly segmented. Please note a side advantage of using pre-trained models in *Figure 7.27*: in the ground truth image, the people in the background were not segmented, but the pre-trained model correctly picked them up (which might explain the low mIoU values).

Now that we have a baseline for comparison, let's apply transfer learning to our task. In the first recipe, *Understanding transfer learning and fine-tuning*, the first step was to retrieve a pre-trained model from the MXNet Model Zoo (GluonCV or GluonNLP), which we have already done.

The second step is to remove the last layers (typically, a classifier), keeping the parameters in the rest of the layers frozen (not updatable during training), so let's do it!

We can freeze the *DeepLab-v3* feature extraction layers with the following snippet:

```
for param in deeplab_tl.collect_params().values():
param.grad_req = 'null'
```

Furthermore, we will also need to replace the segmentation task head. Previously, it supported 21 classes from *MS COCO*. For our experiments, two classes are enough, `background` and `person`. This is done with the following snippet:

```
# Replace the last layers
deeplab_tl.head = gcv.model_zoo.deeplabv3._DeepLabHead(2)
deeplab_tl.head.initialize(ctx=ctx)
deeplab_tl.head.collect_params().setattr('lr_mult', 10)
```

Now, we can apply the usual training process with *Penn-Fudan Pedestrian*, and we have the following evolution in the training using the *DeepLab-v3* model:

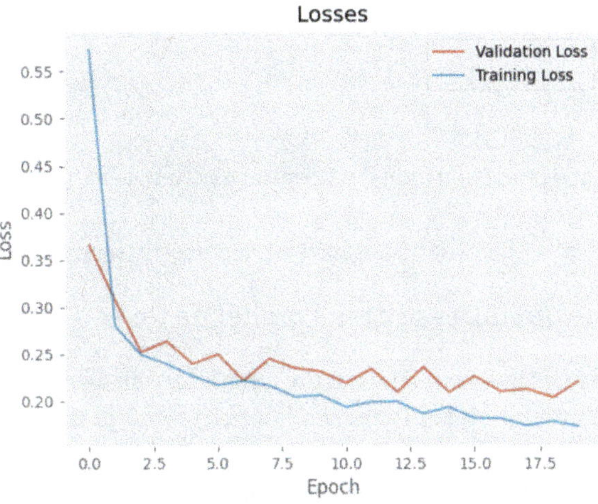

Figure 7.28 – DeepLab-v3 training evolution (training loss and validation loss) – transfer learning

Furthermore, for the best iteration, the evaluation metrics obtained in the test set are as follows:

```
PixAcc:  0.9503427083333333
mIoU  :  0.8799470898171042
```

Compared with our previous experiments of training from scratch and pre-training, this experiment yields slightly better performance, and it took us literally minutes to get this model to start working for us in our intended task, whereas the training required for the training from scratch experiment took hours and required several tries to tune the hyperparameters, which turned into several days of effort in total.

We can also check how well our model is performing qualitatively with the same image example and code. The output is given as follows:

GroundTruth Prediction

Figure 7.29 – Ground truth and prediction from the DeepLab-v3 pre-trained model with transfer learning

As can be seen from the results, the pedestrians have been correctly segmented.

Fine-tuning our pre-trained DeepLab-v3 model on Penn-Fudan Pedestrian

In the previous recipe, we *froze* the parameters in the encoder layers. However, with the dataset we are currently working with (*Penn-Fudan Pedestrian*), we can *unfreeze* those parameters and train the model, effectively allowing the new training process to update the representations (with transfer learning, we were working directly with the representations learned for *MS COCO*). As introduced in this chapter, this process is called fine-tuning.

There are two variants of fine-tuning:

- Apply transfer learning by freezing the layers and unfreezing them afterward.
- Directly apply fine-tuning without the preliminary step of freezing the layers.

Let's compute both experiments and draw conclusions by comparing the results.

For the first experiment, we can take the network obtained in the previous recipe, unfreeze the layers, and restart the training. In MXNet, to unfreeze the encoder parameters, we can run the following snippet:

```
for param in deeplab_ft.collect_params().values():
param.grad_req = 'write'
```

Now, we can apply the usual training process with *Penn-Fudan Pedestrian*, and we have the following evolution in the training using the *DeepLab-v3* model:

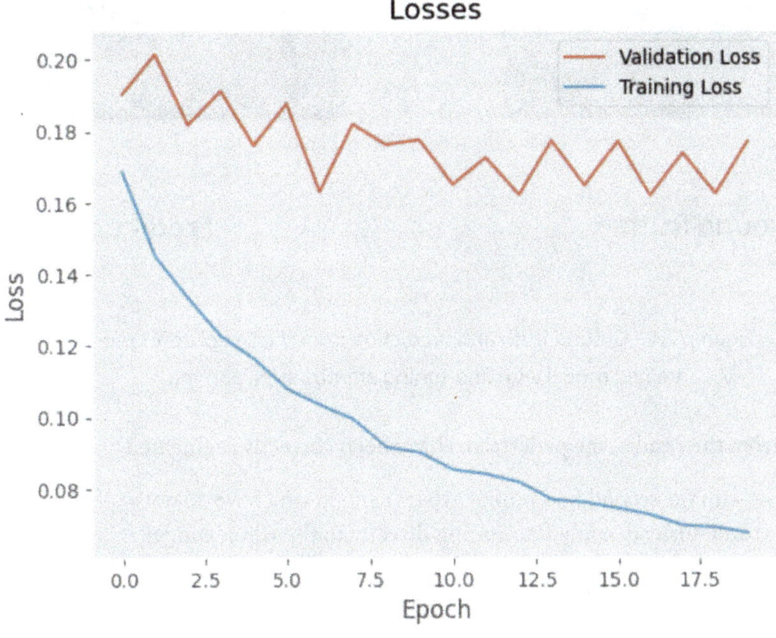

Figure 7.30 – DeepLab-v3 training evolution (training loss and validation loss) – fine-tuning after transfer learning

Furthermore, for the best iteration, the evaluation metrics obtained in the test set are as follows:

```
PixAcc:  0.9637550347222222
mIoU  :  0.9091450223893902
```

Compared with our previous experiment in transfer learning, this experiment yields ~3% better performance in mIoU, a very good increase taking into account the low training time invested.

We can also check how well our model is performing qualitatively with the same image example and code. The output is given as follows:

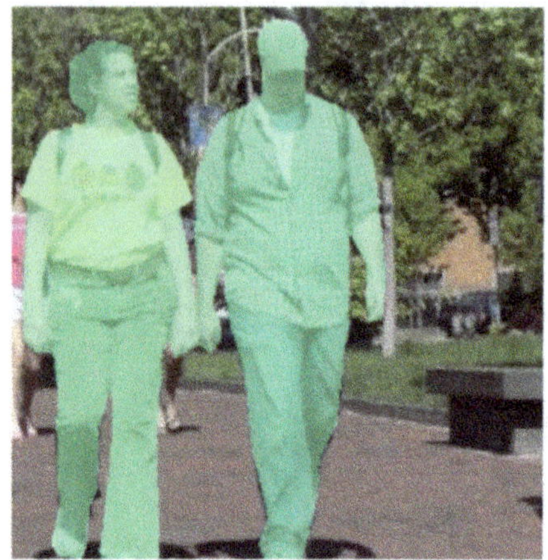

GroundTruth Prediction

Figure 7.31 – Ground truth and prediction from the DeepLab-v3 pre-trained model with fine-tuning after transfer learning

As can be seen from the results, the pedestrians have been correctly segmented.

Let's continue now with the second fine-tuning experiment, in which we do not apply transfer learning (no frozen layers) and, instead, apply fine-tuning directly to the whole model.

We need to retrieve the pre-trained *DeepLab-v3* model for *MS COCO*, with the following code snippet for MXNet GluonCV:

```
# DeepLab-v3 from Model Zoo
 deeplab_ft_direct = gcv.model_zoo.get_model("deeplab_resnet101_coco",
pretrained=True, ctx=ctx)
```

And now, without freezing, we can apply the training process, which will update all layers of our *DeepLab-v3* model:

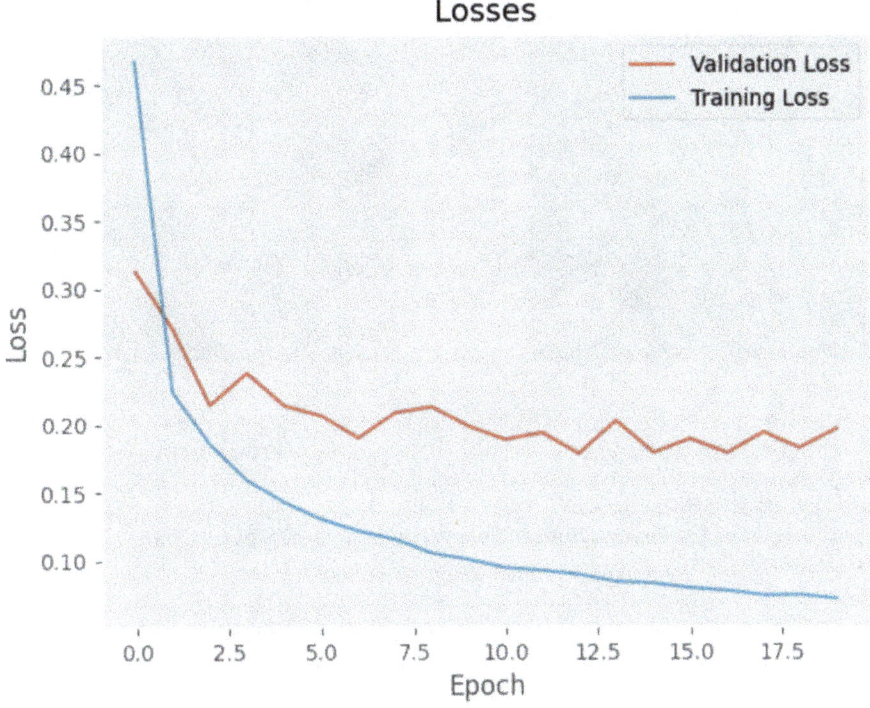

Figure 7.32 – DeepLab-v3 training evolution (training loss and validation loss) – fine-tuning without freezing

Furthermore, for the best iteration, the evaluation metrics obtained in the test set are as follows:

```
PixAcc:  0.9639182291666667
mIoU  :  0.9095065032946663
```

Compared with our previous fine-tuning experiment, we can see how these experiments yield very similar performance. Empirically, it has been proven that this fine-tuning experiment can also yield slightly lower results because initially freezing the encoder allows for the decoder to learn (using the encoder representations) the new task at hand. From a point of view, in this step, there is a knowledge transfer from the encoder to the decoder. In a secondary step, when the encoder is unfrozen, the learned parameters from the decoder perform auxiliary transfer learning, this time from the decoder to the encoder.

We can also check how well our model is performing qualitatively with the same image example and code. The output is given as follows:

GroundTruth Prediction

Figure 7.33 – Ground truth and prediction from the DeepLab-v3 pre-trained model with fine-tuning without freezing

As can be seen from the results, the pedestrians have been correctly segmented, although, as mentioned, if we look at the person on the right, the arm closer to the person on the left could be segmented better. As discussed, sometimes this version of fine-tuning yields slightly lower results than other approaches.

How it works...

In this recipe, we applied the techniques of transfer learning and fine-tuning, introduced at the beginning of the chapter, to the task of image classification, which was also presented previously, in the fourth recipe, *Segmenting objects in images with MXNet: PSPNet and DeepLab-v3*, in *Chapter 5, Analyzing Images with Computer Vision.*

We revisited two known datasets, *MS COCO* and *Penn-Fudan Pedestrian*, which we intended to combine using knowledge transfer based on the former dataset and refining that knowledge with the latter. Moreover, MXNet GluonCV provided the following:

- A pre-trained *DeepLab-v3* model for *MS COCO*
- Tools for easy-to-use access to *Penn-Fudan Pedestrian*

Furthermore, we continued using the loss functions and metrics introduced for semantic segmentation, softmax cross-entropy, pixel accuracy, and mIoU.

Having all these tools readily available within MXNet and GluonCV allowed us to run the following experiments with just a few lines of code:

- Training a model from scratch with *Penn-Fudan Pedestrian*
- Using a pre-trained model to optimize performance via transfer learning from *MS COCO* to *Penn-Fudan Pedestrian*
- Fine-tuning our pre-trained model on *Penn-Fudan Pedestrian* (with and without freezing layers)

After running the different experiments and taking into account the qualitative results and the quantitative results, transfer learning (with a pixel accuracy of 0.95 and mIoU of 0.88) has been the best experiment for our task. The actual results obtained when running these experiments might differ based on model architecture, datasets, and hyperparameters chosen, so you are encouraged to try out different techniques and variations.

There's more...

Transfer learning, including fine-tuning, is an active field of research. A recent paper published in 2022 explores the latest advances in image classification. The paper is titled *Deep Transfer Learning for Image Classification: A survey*, and can be found here: `https://www.researchgate.net/publication/360782436_Deep_transfer_learning_for_image_classification_a_survey`.

An interesting paper that combines transfer learning and semantic segmentation is *Semantic Segmentation with Transfer Learning for Off-Road Autonomous Driving*, in which a change of domain is also studied by the usage of synthetic data. It can be found here: `https://www.researchgate.net/publication/333647772_Semantic_Segmentation_with_Transfer_Learning_for_Off-Road_Autonomous_Driving`.

A more general overview is given in this paper: *Learning Transferable Knowledge for Semantic Segmentation with Deep Convolutional Neural Network*, accepted for **Computer Vision and Pattern Recognition** (**CVPR**) symposium in 2016. It can be found here: `https://openaccess.thecvf.com/content_cvpr_2016/papers/Hong_Learning_Transferrable_Knowledge_CVPR_2016_paper.pdf`.

Improving performance for translating English to German

In the previous recipes, we have seen how we can leverage pre-trained models and new datasets for transfer learning and fine-tuning applied to CV tasks. In this recipe, we will follow a similar approach, but with an NLP task, translating from English to German.

In the fourth recipe, *Translating text from Vietnamese to English*, in *Chapter 6*, *Understanding Text with Natural Language Processing*, we saw how we could use GluonNLP to retrieve pre-trained models and use them directly for a translation task, training them from scratch, effectively only leveraging past knowledge by using the architecture of the pre-trained model.

In this recipe, we will also leverage the weights/parameters of the model, obtained for a task consisting of translating text from English to German using **machine translation** models. The dataset that we will use for pre-training will be *WMT2014* (task source), and we will run several experiments to evaluate our models in a new (target) task, using the dataset *WMT2016* dataset (with a ~20% increased vocabulary of words and sentences for German-English pairs).

Getting ready

As for previous chapters, in this recipe, we will be using some matrix operations and linear algebra, but it will not be hard at all.

Furthermore, we will be working with text datasets; therefore, we will revisit some concepts already seen in the fourth recipe, *Understanding text datasets – load, manage, and visualize Enron Emails dataset* from *Chapter 2, Working with MXNet and Visualizing Datasets: Gluon and DataLoader*.

How to do it...

In this recipe, we will be looking at the following steps:

1. Introducing the *WMT2014* and *WMT2016* datasets
2. Training a Transformer model from scratch with *WMT2016*
3. Using a pre-trained Transformer model to optimize performance via transfer learning from *WMT2014* to *WMT2016*
4. Fine-tuning our pre-trained Transformer model on *WMT2016*

Let's look at these steps in detail next.

Introducing the WMT2014 and WMT2016 datasets

WMT2014 and *WMT2016* are multi-modal (multi-language) translation datasets, including Chinese, English, and German corpus. *WMT2014* was first introduced in 2014 in *Proceedings of the Ninth Workshop on Statistical Machine Translation*, as part of the evaluation campaign of the translation models. This workshop was upgraded to its own conference in 2016, and *WMT2016* was introduced as part of the evaluation campaign of translation models in *Proceedings of the First Conference on Machine Translation*. Both datasets are very similar, retrieving information from news sources, and the largest difference is the corpus (the size of the vocabulary used for both). WMT2014 is about ~140k distinct words, whereas WMT2016 is slightly larger with ~150k words, and specifically for German-English pairs, an increase of ~20% of words and sentences.

MXNet GluonNLP provides ready-to-use versions of these datasets. For our case, we will work with *WMT2016*, which only contains *train* and *test* splits. We will further split the test set to obtain *validation* and *test* splits. Here is the code to load the dataset:

```
# WMT2016 Dataset (Train, Validation and Test)
# Dataset Parameters
src_lang, tgt_lang = "en", "de"
src_max_len, tgt_max_len = 50, 50
 wmt2016_train_data = nlp.data.WMT2016BPE(
     'train',
 src_lang=src_lang,
     tgt_lang=tgt_lang)
wmt2016_val_data = nlp.data.WMT2016BPE(
     'newstest2016',
     src_lang=src_lang,
     tgt_lang=tgt_lang)
wmt2016_test_data = nlp.data.WMT2016BPE(
     'newstest2016',
     src_lang=src_lang,
     tgt_lang=tgt_lang)
```

Here is the code to generate *validation* and *test* splits:

```
 # Split Val / Test sets
val_length = 1500
test_length = len(wmt2016_test_text) - val_length
wmt2016_val_data._data[0] = wmt2016_val_data._data[0][:val_length]
 wmt2016_val_data._data[1] = wmt2016_val_data._data[1][:val_length]
 wmt2016_val_data._length = val_length
wmt2016_val_text._data[0] = wmt2016_val_text._data[0][:val_length]
 wmt2016_val_text._data[1] = wmt2016_val_text._data[1][:val_length]
 wmt2016_val_text._length = val_length
wmt2016_test_data._data[0] = wmt2016_test_data._data[0][-test_length:]
 wmt2016_test_data._data[1] = wmt2016_test_data._data[1][-test_
length:]
 wmt2016_test_data._length = test_length
```

After splitting, our *WMT2016* datasets provide the following data:

```
 Length of train set: 4500966
Length of val set   : 1500
Length of test set : 1499
```

From the large number of instances on each of the datasets, we can confirm that these are suitable for our experiments.

Training a Transformer model from scratch in WMT2016

As described in the fourth recipe, *Translating text from Vietnamese to English*, in *Chapter 6, Understanding Text with Natural Language Processing*, we will be using **Perplexity** for our *per-batch computations* in training, and **BLEU** for *per-epoch computations*, which will show us the evolution of our training process, as part of the typically used training and validation losses. We will also use them for quantitative evaluation, and for qualitative evaluation, we will choose a sentence (feel free to use any other sentence you can come up with).

We have the following evolution in the training using the Transformer model:

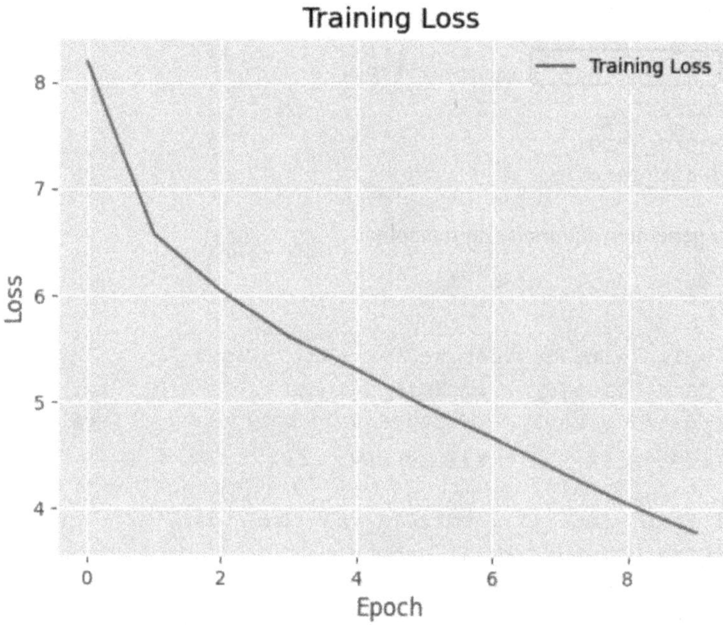

Figure 7.34 – Transformer training evolution (training loss) – training from scratch

Furthermore, for the best iteration, the loss, perplexity, and BLEU score (multiplied by 100) obtained in the test set are as follows:

```
WMT16 test loss: 3.01; test bleu score: 14.50
```

Current **State-of-the-Art** (**SOTA**) models can yield above 30 points in the BLEU score; we reach about halfway with 10 epochs, reaching SOTA performance in ~30 epochs.

Qualitatively, we can also check how well our model is performing with a sentence example. In our case, we chose: `"I learn new things every day"`, and this can be verified with the following code:

```
print("Qualitative Evaluation:  Translating from English to German")
# From Google Translate
expected_tgt_seq = " Ich lerne jeden Tag neue Dinge."
 print("Expected translation:")
 print(expected_tgt_seq)
src_seq = "I learn new things every day."
 print("In English:")
 print(src_seq)
translation_out = nmt.utils.translate(
     transformer_ts_translator,
    src_seq,
     wmt_src_vocab,
    wmt_tgt_vocab,
    ctx)
print("The German translation is:")
 print(" ".join(translation_out[0]))
```

These code statements will give us the following output:

```
Qualitative Evaluation: Translating from English to German
Expected translation:
 Ich lerne jeden Tag neue Dinge.
 In English:
 I learn new things every day.
 The German translation is:
 Ich halte es für so , dass es hier so ist.
```

The German sentence means *I think that's the case here*; therefore, as can be seen from this result, the text has not been correctly translated from English to German, and we would need to invest more time in training to achieve the right results.

Using a pre-trained Transformer model to optimize performance via transfer learning from WMT2014 to WMT2016

In the previous recipe, we trained a new model from scratch using our dataset. However, this has two important drawbacks:

- A large amount of data is required for training from scratch.
- The training process can take a very long time due to the large size of the dataset and the number of epochs needed for the model to learn the task.

Therefore, in this recipe, we will follow a different approach. We will use pre-trained models from MXNet GluonNLP to solve the task. These models have been trained on *WMT2014* a very similar dataset, so the representations learned for this task can be easily transferred to *WMT2016* (same domain).

For a Transformer model, we have the following:

```
wmt_model_name = 'transformer_en_de_512'
wmt_transformer_model_pt, wmt_src_vocab, wmt_tgt_vocab = nlp.model.get_model(
    wmt_model_name,
    dataset_name='WMT2014',
    pretrained=True,
    ctx=ctx)
print('Source Vocab:', len(wmt_src_vocab), ', Target Vocab:', len(wmt_tgt_vocab))
```

The output shows us the size of the vocabulary of the *WMT2014* dataset (the pre-trained English to German translation task):

```
Source Vocab: 36794 , Target Vocab: 36794
```

This is a subset of the whole corpus available for *WMT2014.* As we can also see in the preceding code snippet, following the discussion in this chapter's first recipe, *Understanding transfer learning and fine-tuning*, for the `pretrained` parameter, we have assigned the value of `True`, indicating that we want the pretrained weights to be retrieved (and not only the architecture of the model).

In order to evaluate adequately the improvements that transfer learning brings, we are going to directly evaluate our pre-trained model (task source is *WMT2014*) before applying transfer learning to *WMT2016* and after applying it. Therefore, using our pre-trained model as is, we obtain the following:

```
WMT16 test loss: 1.59; test bleu score: 29.76
```

As we can see, our pre-trained Transformer model is already showing very good performance values as it is the same domain; however, simply using a pre-trained model does not yield SOTA performance, which can be achieved if training from scratch. The great advantage of using pre-trained models is the time and compute savings as loading a pre-trained model just takes a few lines of code.

We can also check how well our model is performing qualitatively with the same sentence example and code. The output is given as follows:

```
Qualitative Evaluation: Translating from English to German
Expected translation:
Ich lerne jeden Tag neue Dinge.
In English:
I learn new things every day.
The German translation is:
Ich lerne neue Dinge, die in jedem Fall auftreten.
```

The German sentence means *I learn new things that arise in every case*; therefore, as can be seen from the results, the text has not yet been correctly translated from English to German, but this time, was much closer than our previous experiment.

Now that we have a baseline for comparison, let's apply transfer learning to our task. In the first recipe, *Understanding transfer learning and fine-tuning*, the first step was to retrieve a pre-trained model from the MXNet Model Zoo (GluonCV or GluonNLP), which we have already done.

The second step is to remove the last layers (typically, a classifier), keeping the parameters in the rest of the layers frozen (not updatable during training), so let's do it!

We can freeze all parameters except the classifier with the following snippet, keeping the parameters frozen (we will unfreeze them in a later experiment):

```
 updated_params = []
for param
wmt_transformer_model_tl.collect_params().values():
    if param.grad_req == "write":
        param.grad_req = "null"
        updated_params += [param.name]
```

Now, we can apply the usual training process with *WMT2016*, and we have the following evolution in the training using the Transformer model:

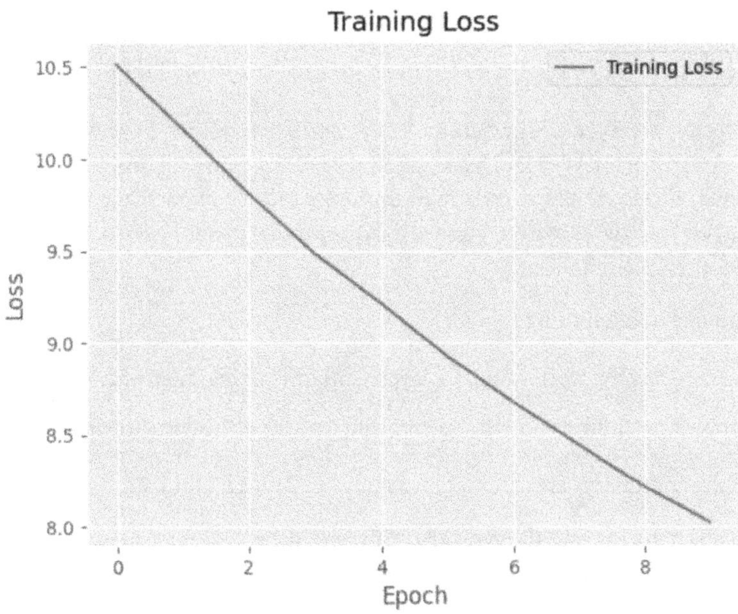

Figure 7.35 – Transformer training evolution (training loss) – transfer learning

Furthermore, for the best iteration, the loss, perplexity, and BLEU score (multiplied by 100) obtained in the test set are as follows:

```
WMT16 test loss: 1.20; test bleu score: 27.78
```

Compared with our previous experiments, this experiment yields slightly lower numerical performance; however, it took us literally minutes to get this model to start working for us in our intended task, whereas training from scratch in our previous experiment took hours and required several tries to tune the hyperparameters, becoming several days of effort in total.

We can also check how well our model is performing qualitatively with the same sentence example and code. The output is given as follows:

```
Qualitative Evaluation: Translating from English to German
Expected translation:
 Ich lerne jeden Tag neue Dinge.
 In English:
 I learn new things every day.
 The German translation is:
 Ich erlerne jedes Mal neue Dinge
```

The German sentence means *I learn new things every time*; therefore, as can be seen from the results, the text has been almost correctly translated from English to German, improving from our previous experiment (pre-trained model), although the (better) quantitative results were suggesting otherwise.

Fine-tuning our pre-trained Transformer model on WMT2016

In the previous recipe, we froze all the parameters except the classifier. However, as the dataset we are currently working with (*WMT2016*) has enough data samples, we can unfreeze those parameters and train the model, effectively allowing the new training process to update the representations (with transfer learning, we were working directly with the representations learned for *WMT2014*). This process, as we know, is called fine-tuning.

There are two variants of fine-tuning:

- Apply transfer learning by freezing the layers and unfreezing them afterward.
- Directly apply fine-tuning without the preliminary step of freezing the layers.

Let's compute both experiments and draw conclusions by comparing the results.

For the first experiment, we can take the network obtained in the previous recipe, unfreeze the layers, and restart the training. In MXNet, to unfreeze the encoder parameters, we can run the following snippet:

```
for param in wmt_transformer_model_ft.collect_params().values():
    if param.name in updated_params:
        param.grad_req = 'write'
```

Now, we can apply the usual training process with *WMT2016*, and we have the following evolution in the training using the Transformer model:

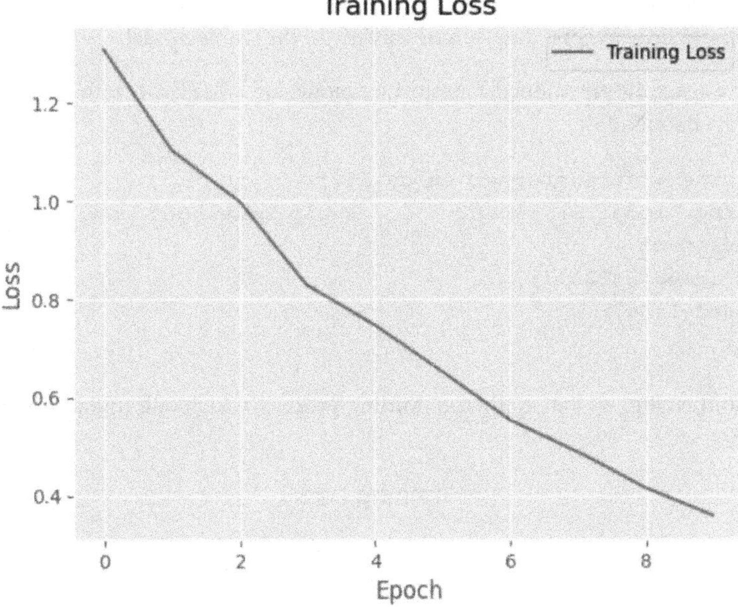

Figure 7.36 – Transformer training evolution (training loss) – fine-tuning after transfer learning

Furthermore, for the best iteration, the loss, perplexity, and BLEU score (multiplied by 100) obtained in the test set are as follows:

```
WMT16 test loss: 1.23; test bleu score: 26.05
```

Compared with our previous experiment in transfer learning, this experiment yields a slightly worse quantitative performance.

Qualitatively, we can also check how well our model is performing with a sentence example. In our case, we chose `"I learn new things every day"`, and the output obtained is as follows:

```
Qualitative Evaluation: Translating from English to German
Expected translation:
 Ich lerne jeden Tag neue Dinge.
 In English:
 I learn new things every day.
 The German translation is:
 Ich lerne jedes Mal Neues.
```

The German sentence means *I learn something new every time*; therefore, as can be seen from the results, the text has been almost correctly translated from English to German.

Let's continue now with the second fine-tuning experiment, where we do not apply transfer learning (no frozen layers), and instead apply fine-tuning directly to the whole model.

We need to retrieve again the pre-trained Transformer model for *WMT2014*, with the following code snippet for MXNet GluonNLP:

```
wmt_model_name = 'transformer_en_de_512'
wmt_transformer_model_ft_direct, _, _ = nlp.model.get_model(
    wmt_model_name,
    dataset_name='WMT2014',
    pretrained=True,
    ctx=ctx)
```

And now, without freezing, we can apply the training process, which will update all layers of our Transformer model:

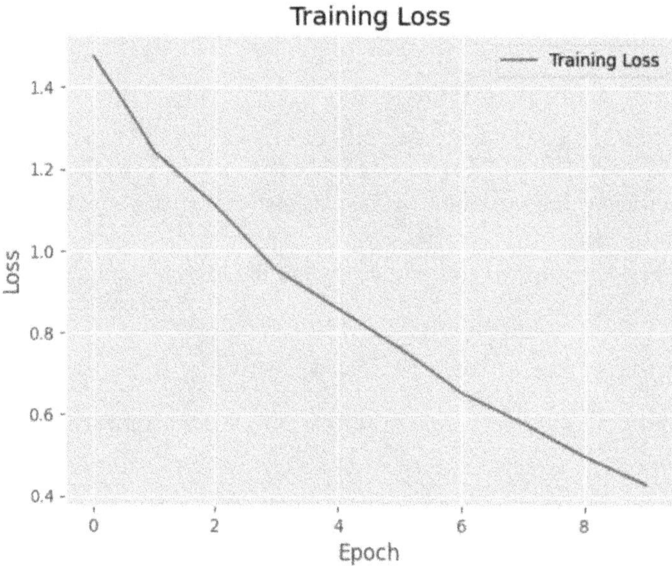

Figure 7.37 – Transformer training evolution (training loss) – fine-tuning without freezing

Furthermore, for the best iteration, the loss, perplexity, and BLEU score (multiplied by 100) obtained in the test set are as follows:

```
WMT16 test loss: 1.22; test bleu score: 26.75
```

Compared with our previous fine-tuning experiment, we can see how this experiment yields a slightly higher performance. Empirically, however, the opposite results were expected (for this experiment to yield a slightly lower performance). This has been proven to be a repeatable result because initially freezing the encoder allows for the decoder to learn (using the encoder representations) the new task at hand. From a point of view, in this step, there is a knowledge transfer from the encoder to the decoder. In a secondary step, when the encoder is unfrozen, the learned parameters from the decoder perform auxiliary transfer learning – this time, from the decoder to the encoder.

Qualitatively, we can also check how well our model is performing with a sentence example. In our case, we chose `"I learn new things every day"`, and the output obtained is as follows:

```
 Qualitative Evaluation: Translating from English to German
Expected translation:
 Ich lerne jeden Tag neue Dinge.
 In English:
 I learn new things every day.
 The German translation is:
 Ich lerne jedes Mal neue Dinge
```

The German sentence means *I learn new things every time*; therefore, as can be seen from the results, the text has been almost correctly translated from English to German.

How it works...

In this recipe, we applied the techniques of transfer learning and fine-tuning, introduced at the beginning of the chapter, to the task of machine translation, which was also presented previously, in the fourth recipe, *Translating text from Vietnamese to English*, in *Chapter 6*, *Understanding Text with Natural Language Processing*.

We explored two new datasets, *WMT2014* and *WMT2016*, which, among other language pairs, support translations between German and English. Moreover, MXNet GluonNLP provided the following:

- A pre-trained Transformer model for *WMT2014*
- A data loader ready to be used with *WMT2016*

Furthermore, we continued using the metrics introduced for machine translation, perplexity, and BLEU.

Having all these tools readily available within MXNet and GluonNLP allowed us to run the following experiments with just a few lines of code:

- Training a model from scratch with *WMT2016*
- Using a pre-trained model to optimize performance via transfer learning from *WMT2014* to *WMT2016*
- Fine-tuning our pre-trained model on *WMT2016* (with and without freezing layers)

We compared the results and derived the best approach for this particular task, which was applying transfer learning and fine-tuning afterward.

There's more...

In this recipe, we introduced two new datasets, *WMT2014* and *WMT2016*. These datasets were introduced as challenges in the **Workshop on Statistical Machine Translation** (**WMT**) conference. The results for 2014 and 2016 are the following:

- **Findings of the 2014 Workshop on Statistical Machine Translation:** `https://aclanthology.org/W14-3302.pdf`
- **Findings of the 2016 Conference on Machine Translation (WMT16):** `https://aclanthology.org/W16-2301.pdf`

Transfer learning, including fine-tuning, for machine translation is an active area of research. A paper published in 2020 explores its applications, titled *In Neural Machine Translation, What Does Transfer Learning Transfer?* and can be found here: `https://aclanthology.org/2020.acl-main.688.pdf`.

For a more general approach to NLP use cases, a recent paper was published, *A Survey on Transfer Learning in Natural Language Processing*, and can be found here: `https://www.researchgate.net/publication/342801560_A_Survey_on_Transfer_Learning_in_Natural_Language_Processing`.

8

Improving Training Performance with MXNet

In previous chapters, we have leveraged MXNet capabilities to solve computer vision and **Natural Language Processing** (**NLP**) tasks. In those chapters, the focus was on obtaining the maximum performance out of pre-trained models, leveraging the *Model Zoos* from `GluonCV` and `GluonNLP`. We trained these models using different approaches: *from scratch*, *transfer learning*, and *fine-tuning*. In this chapter, we will focus on improving the performance of the training process itself and accelerating how we can obtain those results.

To achieve the objective of optimizing the performance of our training loops, MXNet contains different features. We have already briefly used some of those features such as the concept of **lazy evaluation**, which was introduced in *Chapter 1*. We will revisit it in this chapter, in combination with automatic parallelization. Moreover, we will optimize how to access data efficiently, leveraging Gluon DataLoaders in different contexts (CPU, GPU) to perform data transforms.

Moreover, we will explore how to combine multiple GPUs to accelerate training, making use of techniques such as data parallelization for optimal performance. We will also explore how we can use different data types with **Automatic Mixed Precision** (**AMP**), allowing `MXNet` to dynamically optimize the different data formats.

Finally, using problems already explored in the book, we will apply all these techniques with examples. For our computer vision task, we will choose image segmentation, and for our NLP task, we will choose translating text from English to German.

Specifically, this chapter is structured with the following recipes:

- Introducing training optimization features
- Optimizing training for image segmentation
- Optimizing training for translating text from English to German

Technical requirements

Apart from the technical requirements specified in the *Preface*, the following technical requirements apply:

- Ensure that you have completed the *Installing MXNet, Gluon, GluonCV and GluonNLP* recipe from *Chapter 1*.
- Ensure that you have completed *Chapter 5* and *Chapter 6*.
- Ensure that you have completed *Chapter 7*.

The code for this chapter can be found at the following GitHub URL: `https://github.com/PacktPublishing/Deep-Learning-with-MXNet-Cookbook/tree/main/ch08`.

Furthermore, you can access each recipe directly from Google Colab. For example, the code for the first recipe of this chapter can be found here: `https://colab.research.google.com/github/PacktPublishing/Deep-Learning-with-MXNet-Cookbook/blob/main/ch08/8_1_Introducing_training_optimization_features.ipynb`.

Introducing training optimization features

In the previous chapters, we saw how we could leverage *MXNet*, *GluonCV*, and *GluonNLP* to retrieve pre-trained models in certain datasets (such as **ImageNet**, **MS COCO**, or **IWSLT2015**) and use them for our specific tasks and datasets. Furthermore, we used transfer learning and fine-tuning techniques to improve the performance on those tasks/datasets.

In this recipe, we will introduce (and revisit) several concepts and features that will optimize our training loops, after which we will analyze the trade-offs involved.

Getting ready

Similar to the previous chapters, in this recipe, we will be using some matrix operations and linear algebra, but it will not be hard at all, as you will find lots of examples and code snippets to facilitate your learning.

How to do it...

In this recipe, we will work through the following steps:

1. Working with lazy evaluation and automatic parallelization
2. Optimizing DataLoaders: GPU preprocessing and CPU threads
3. Training with `Float32`, `Float16`, and Automatic Mixed Precision
4. Training with multiple GPUs and data parallelization

Let's dive into each of these steps.

Working with lazy evaluation and automatic parallelization

In the *NumPy and MXNet NDArrays* recipe of *Chapter 1*, we introduced lazy evaluation, the strategy that MXNet follows when computing operations. This strategy is optimal for large compute loads, where the actual calculation is deferred until the values are actually needed.

Furthermore, by not executing the computation of the operations until they are needed, MXNet can also parallelize some of those computations, meaning the data involved is not sequentially processed. This process is done automatically and is very useful when sharing data across multiple hardware resources such as CPUs and GPUs.

As a toy example, we can run some experiments with matrix multiplication. Our first experiment will run the generation of four matrices and then a combination of multiplications among them. After each computation, we will force the computation to be finalized (by adding calls to the `wait_to_read()` function). We will compute the results for two configurations. The initial configuration will be to force MXNet to work with one thread (`NaiveEngine`). With this configuration, the computation took this long:

```
Time (s): 134.3672107196594
```

The second configuration tested will be MXNet in its usual, default configuration (`ThreadedEnginePerDevice`, with four CPU threads). With this configuration, the computation took this long:

```
Time (s): 135.26983547210693
```

As we can see, forcing each computation to be finalized before moving to the next one (`wait_to_read()` calls) is counter-productive when working in a multi-threading configuration.

Our second experiment will be very similar; however, this time, we will remove all calls to the `wait_to_read()` function. We will only ensure that all calculations for the multiplication of the matrices are finalized before computing the time taken. For the initial configuration (*NaiveEngine*), the computation takes the following amount of time:

```
Time (s): 134.47382940625321
```

As expected, this is a very similar duration to only working with one thread, as all computations are sequential.

With our second configuration (*ThreadedEnginePerDevice*, with four CPU threads), the computation for this second experiment took the following amount of time:

```
Time (s): 111.36750531196594
```

The results show that when using multiple threads (the default automatic configuration for MXNet), we achieved a ~20% improvement (improvements can be even higher with different workloads more suited for multi-threading).

> **Important Note**
>
> Please note how in the code, we used the `mx.nd.waitall()` function to verify that all computations had been strictly completed before computing the time these operations took.

Optimizing DataLoaders – GPU preprocessing and CPU threads

In the *Understanding image datasets – loading, managing, and visualizing the fashion MNIST dataset* recipe of *Chapter 2*, we introduced **Gluon DataLoader**, an efficient mechanism to generate batch sizes to feed into our models for training and evaluation.

DataLoader has two important roles to play in our data preprocessing. On the one hand, as we have explored in previous chapters, our models are optimized for parallel data processing, meaning that we can ingest several samples (for example, images for an image segmentation task) at the same time in the same *batch* and it will be processed in parallel by the *GPU*. This parameter is called *batch size*. On the other hand, samples typically need to be preprocessed in order to maximize the performance of the model (for example, images are resized and its values allocated to `[0, 1]` from `[0, 255]`). These operations are time-consuming and optimizing them can save large amounts of time and compute.

Let's analyze the effect of the preprocessing of the data in the GPU, compared to the general, default behavior of using the CPU. As a baseline, let's compute how long it takes to just load the dataset using only the CPU. We select the validation split of a segmentation dataset, and the result is as follows:

```
Time (s): 24.319150686264038
```

However, when loading the dataset, we typically apply certain `transform` operations that maximize our network performance. The usual transform operations including image resizing, cropping, transforming to tensors, and normalizing can be defined in MXNet with the following code:

```
input_transform_fn = mx.gluon.data.vision.transforms.Compose([
mx.gluon.data.vision.transforms.Resize(image_size, keep_ratio=True),
mx.gluon.data.vision.transforms.CenterCrop(image_size), mx.gluon.data.
vision.transforms.ToTensor(),
mx.gluon.data.vision.transforms.Normalize([.485, .456, .406], [.229,
.224, .225])
])
```

With these transform operations applied when processing the validation split of a segmentation dataset using only the CPU, the processing time is the following:

```
Time (s): 38.973774433135986
```

As we can see, the processing time has increased by more than 50%, from ~24s to ~39s. However, when we leverage the GPU for the data preprocessing, the processing time is as follows:

```
Time (s): 25.39602303504944
```

As we can see, the GPU-based preprocessing operations have an overhead that is almost negligible (<5%).

Furthermore, performing the preprocessing in the GPU has another advantage: the data can be kept stored in the GPU for our models to process, whereas when preprocessing with the CPU, we need to send a copy of the data to the GPU memory, which can take a significant amount of time. If we actually measure our end-to-end preprocessing pipeline, combining the data preprocessing with the copy operation to the GPU memory, these are the results. With the CPU only, the end-to-end processing time is the following:

```
Time (s): 67.73443150520325
```

As we can see, the copy time is significant, taking the whole pipeline more than 1 minute. However, the result when using the GPU is as follows:

```
Time (s): 23.22727918624878
```

This shows a significant improvement (<40%) in the time it took for the full preprocessing. In summary, this was due to two factors: the fact that the preprocessing operations are faster in the GPU, and secondly, that the data needs to be copied to the GPU at the end of the process so that our models (which are also stored in the GPU) process the data efficiently.

The most important drawback of this approach is the need to keep the full dataset in the GPU. Typically, GPU memory space is optimized for each batch that you use for training or inference, not the whole dataset. This is the reason why this approach typically finishes with the processed data being copied back out of the GPU memory space and into the CPU memory space.

However, there are situations where keeping the data in the GPU memory space might be the right approach – for example, when you are experimenting with datasets, and maybe loading several datasets and testing different preprocessing pipelines. In these situations you want fast turn-around times for your experiments and, therefore, speed is the right variable to optimize for. Moreover, sometimes you are not working with the full training/validation/test splits of a dataset, but just a part of it (again, for example, for experiments). In these cases, optimizing for speed makes sense as well.

For other, more production-oriented environments, the right approach is to preprocess in GPU memory space but keep the data (copying back) in CPU memory space. In this scenario, the results vary slightly:

```
Time (s): 34.58254957199097
```

As we can see, the preprocessing step being done in the GPU is still a significant increase (~50%) in performance, even taking into account the necessary data movements from the CPU to the GPU and back.

Now, we will take a deeper look at how we can leverage two important parameters that Gluon DataLoader takes as input: the number of workers and the batch size. The number of workers is the number of threads that DataLoader will launch in parallel (multi-threading) for data preprocessing. Batch size, as mentioned, is the number of samples that will be processed in parallel.

These parameters are directly related to the number of cores the CPU has, and can be optimized to use the available HW for maximum performance. To find out how many cores our CPU has, Python provides a very simple API:

```
import multiprocessing
multiprocessing.cpu_count()
```

In the environment selected, the number of cores available is shown as follows:

```
4
```

Combining the usage of the CPU and the GPU, we can compute the best performance taking into account different values for the number of workers and the batch size. The results for the selected environment are as follows:

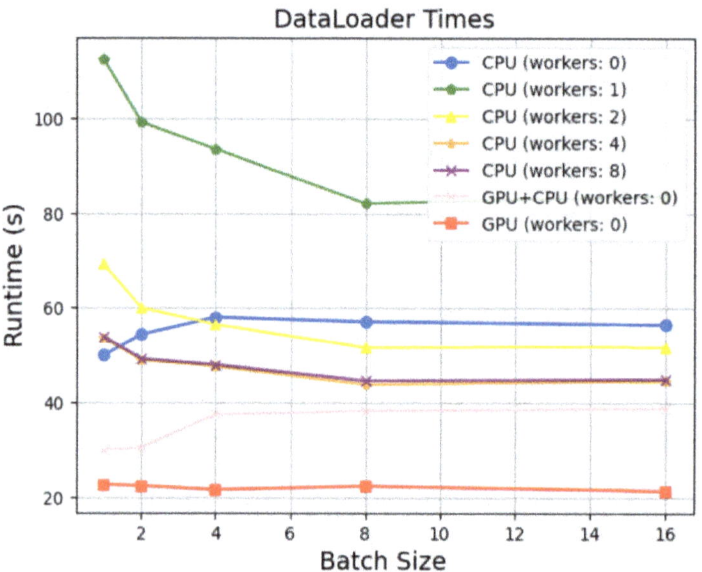

Figure 8.1 – Runtime(s) versus Batch Size for different computation regimes (CPU/GPU and number of workers)

From *Figure 8.1*, we can conclude three important aspects:

- A GPU preprocessing pipeline (data processing plus memory storage) is much faster (+50% runtime improvement), even when copying back the data.
- When combining both the GPU and CPU, as we are only working with one GPU in this environment, it bottlenecks when we copy the data back to the CPU, as it's done per sample (not per batch).

- If working only with a CPU, adding workers improves the processing time. However, the limit is the number of threads. Adding more workers than threads (four in our case) will give no improvement in performance. An increase in the batch size yields better performance until a given number (8 in our case) and won't improve performance further than that.

> **Important Note**
> When working with the GPU, the MXNet Gluon DataLoader only supports the value `0` (zero) for the number of workers.

Training with Float32, Float16, and automatic mixed precision

In the previous recipes, we have seen how to optimize our training loops by using different approaches to optimize the CPU and the GPU for maximum performance for a given model. In this recipe, we will explore how our data inputs, our model parameters, and the different arithmetic calculations around them are computed, and how we can optimize them.

First of all, let's understand how computations work. The default data type for the data inputs and the model parameters is `Float32`, as can be verified (see the recipe code), which yields the following output:

```
Input data type: <class 'numpy.float32'> Model Parameters data type:
<class 'numpy.float32'>
```

This output indicates, as expected, that the data type of our data inputs and our model parameters is `Float32` (single-precision). But what does this mean?

`Float32` indicates two things: on the one hand, that it is a data type that supports decimal numbers using a floating radix point, and on the other hand, that 32 bits are used to store a single number in this format. The most important features of this format are the following:

- The ability to represent large numbers, from 10^{-45} to 10^{+38}
- Variable precision

Using `Float32` as the data type has numerous advantages, mostly connected to its variable precision. However, the training process is an iterative optimization process, for which many of the calculations involved do not require the precision of the `Float32` data type. We could afford, in a controlled way, to trade off some precision if it allowed us to speed up the training process. One of the ways we can achieve that balanced trade-off is with the `Float16` data type (half-precision). Similarly to `Float32`, the most important features of `Float16` are the following:

- The ability to represent large numbers, from 2^{-24} to 2^{+16}
- Variable precision

As an example of the loss of precision, we can display the approximated value of 1/3 in both formats with this code excerpt:

```
a = mx.nd.array([1/3], dtype=np.float32)
 b = a.astype(np.float16)
print("1/3 as Float32: {0:.30f}".format(a.asscalar()))
print("1/3 as Float16: {0:.30f}".format(b.asscalar()))
```

This yields the following:

```
1/3 as Float32: 0.333333343267440795898437500000
1/3 as Float16: 0.333251953125000000000000000000
```

As we can see, none of the representations are exact, with `Float32` yielding higher precision as expected, and `Float16` having more limited accuracy, but potentially enough for some use cases (such as model training, as we will prove shortly).

As mentioned, this loss of accuracy is a trade-off, where we obtain large speed gains in our training loops. To enable `Float16` (half-precision) for our training loops, we need to apply certain changes to our code. First of all, we need to update our model parameters to `Float16`, which we can do with one simple line of code:

```
deeplab_ft_direct_f16.cast('float16')
```

After this, when our model is going to process the data and ground truth, these need to be updated to `Float16` too, so in our training loop, we add the following lines:

```
data   = data.astype('float16', copy=False)
 label = label.astype('float16', copy=False)
```

With these changes, we can now run an experiment to compare the performance of both training loops. For example, we are going to fine-tune a DeepLabv3 pre-trained model in an image segmentation task (see the *Improving performance for segmenting images* recipe of *Chapter 7*). For `Float32`, we obtain the following results:

```
Training time for 10 epochs: 594.4833037853241 / Best validation loss:
0.6800425
```

For `Float16`, these are the results we obtained:

```
Training time for 10 epochs: 199.80901980400085 / Best validation
loss: nan
```

Unfortunately, for `Float16`, although our training time took ~1/3rd than the `Float32` training loop, it did not converge. This is due to several reasons:

- Limited support for large numbers, as any integer larger than `65519` is represented as infinity

- Limited support for small numbers, as any positive decimal number smaller than `1e-7` is represented as `0` (zero)

Thankfully, MXNet offers a solution that automatically combines the best of both worlds:

- Applying `Float32` (single-precision) where it is necessary
- Applying `Float16` (half-precision) where it is not, for runtime optimization

This approach is called **Automatic Mixed Precision** (**AMP**), and in order to enable it, we just need to make a few changes in our code. First of all, before creating our model, we need to initialize the library:

```
amp.init()
```

Then, after initializing our trainer/optimizer, we need to link it with AMP:

```
amp.init_trainer(trainer)
```

And finally, in order to prevent underflow or overflow, we need to enable **loss scaling**, an MXNet feature that allows for the loss computed during the training loop to be adjusted (multiplied or divided) to always be in the range supported by the `Float16` data type. This is done quite conveniently in the training loop:

```
with amp.scale_loss(loss, trainer) as scaled_loss:
mx.autograd.backward(scaled_loss)
```

When we apply these changes and repeat the previous experiment for `Float16` (now with AMP enabled), we obtain the following results:

```
Training time for 10 epochs: 217.64903020858765 / Best validation
loss: 0.7082735
```

As we can see, we obtained very similar results for the validation loss in a much shorter amount of time (~33%).

As the memory footprint of our training loop is now approximately half of what it was before, we can typically either double the size of our model (more layers and larger resolutions), or double our batch size, as the GPU memory consumed will be the same in this case compared to a full `Float32` training loop. Running the same experiment with a double batch size yields the following results:

```
Training time for 10 epochs: 218.82141995429993 / Best validation
loss: 0.18198483
```

As we can see, increasing the batch size has an excellent effect on the performance of our training loop, with a much lower validation loss, and still benefiting from a significantly smaller amount of training time (~33%).

However, typically, as a **Machine Learning Engineer** (**MLE**) or **Data Scientist** (**DS**), we will work with large amounts of data and large models, running training loops expected to last for hours or days. Therefore, it is very common for MLEs/DSs at work to start training loops just before the end of the working day, leaving the training running in the background, and coming back the next working day to analyze and evaluate the results. In such an environment, it is actually a better strategy to optimize performance given an expected training time. With MXNet, we can optimize our training parameters for this as well. For example, we could adjust the training time by doubling the number of epochs. In this case, the experiment yields the following results:

```
Training time for 10 epochs: 645.7392318248749 / Best validation loss:
0.16439788
```

Compared to a vanilla `Float32` training loop, these results are excellent. However, let's not forget that the actual results depend on the specific task, datasets, model, hyperparameters, and so on. You are encouraged to try different options and hyperparameters with toy training loops to find the solution that works best for each case.

Training with multiple GPUs and data parallelization

In this recipe, we will leverage having multiple GPUs in our environment to optimize our training further. MXNet and Gluon allow us to update our training loops to include multiple GPUs very easily.

From a high-level perspective, there are two paradigms to leverage multiple GPUs:

- **Model parallelization**: The model is split into parts and each part is deployed to a specific GPU. This paradigm is very useful when the model does not fit in a single GPU.
- **Data parallelization**: The data batches are split into parts and each part is deployed to a specific GPU that can perform a forward and a backward pass using that data fully.

We will work exclusively with data parallelization as it is the most common use case, yielding high speed-ups, and is also the most convenient given the simplicity of its approach.

In order to apply data parallelization, we will need to make modifications to our training loop, as follows:

1. **Setting the context**: The context is now a list, where each element is a specific GPU context.
2. **Initializing our model in those contexts**: In data parallelization, each GPU will store a copy of all the model parameters.
3. **Adapting hyperparameters**: Batch size is typically set to the largest possible without filling up the GPU memory. When working with several GPUs in parallel, this number can typically be multiplied by the number of GPUs in the context. However, this also has a side effect on the learning rate, which must be multiplied by the same number to keep gradient updates in the same range.
4. **Distributing the data**: Each GPU must have a slice of each batch and run the forward and backward passes with it.

5. **Computing the losses and updating the gradients**: Each GPU will compute the losses associated with their slice of each batch. MXNet automatically combines the losses and computes the gradients that are distributed to each GPU to update their model copy.
6. **Displaying results**: Statistics such as the training loss and the validation loss are typically computed and accumulated during each batch and visualized at the end of each epoch.

Let's see some examples of how to apply each of these steps.

For example, to set the context in an environment with four GPUs is very easy with MXNet and just requires one line of code:

```
ctx_list = [mx.gpu(0), mx.gpu(1), mx.gpu(2), mx.gpu(3)]
```

Initializing the model and custom layers is as easy as that. For our environment, this is how we can initialize a Deeplabv3 network with a `ResNet-101` backbone:

```
deeplab_ft_direct_f32 = gcv.model_zoo.get_model('deeplab_resnet101_
coco', pretrained=True, ctx=ctx_list)
 [...]
deeplab_ft_direct_f32.head.initialize(ctx=ctx_list)
```

To update the hyperparameters, we just need to compute the number of GPUs in the context and update the previously computed batch size and learning rates. For our example, this simply means adding/modifying some lines of code:

```
num_gpus = len(ctx_list)
 [...]
batch_size_per_gpu = 4
batch_size = len(ctx_list) * batch_size_per_gpu
 [...]
trainer = mx.gluon.Trainer(deeplab_ft_direct_f32.collect_params(),
"sgd", {"learning_rate": 0.5})
```

In order to distribute the data evenly across every GPU, MXNet and Gluon have a very convenient function, `split_and_load()`, which automatically allocates the data according to the number of GPUs in the context. For our environment, this is done as follows:

```
data_list   = mx.gluon.utils.split_and_load(data, ctx_list=ctx_list)
 label_list  = mx.gluon.utils.split_and_load(label, ctx_list=ctx_list)
```

To compute the losses and update the gradients, the data distributed in each GPU is processed in parallel using a loop. As MXNet provides automatic parallelization, the calls are not blocking and each GPU computes its outputs and losses independently. Furthermore, MXNet combines those losses to generate the full gradient updates, and redistributes this to each GPU, and all of this is done automatically. We can achieve all this with just a few lines of code:

```
with mx.autograd.record():
outputs = [model(data_slice) for data_slice in data_list]
losses = [loss_fn(output[0], label_slice) for output, label_slice in
zip(outputs, label_list)]
for loss in losses:
loss.backward()
trainer.step(batch_size)
```

Lastly, in order to display the loss computations, each GPU loss needs to be processed and combined. Using automatic parallelization, this can be achieved with just one line of code:

```
current_loss = sum([l.sum().asscalar() for l in losses])
```

With these simple steps, we have been able to modify our training loop to support multiple GPUs, and we are now ready to measure the performance increase of these changes.

As a reminder, using one GPU, we reached the following performance (with a batch size of four):

```
Training time for 10 epochs: 647.753002166748 / Best validation loss:
0.0937674343585968
```

In our environment, with 4 GPUs, we could increase the batch size to 16, the results of which would be as follows:

```
Training time for 10 epochs: 177.23532104492188 / Best validation
loss: 0.082047363743186
```

As expected, we have been able to reduce the time spent in training to ~25% (the expected reduction when going from 1 GPU to 4 GPUs, with some expected loss due to the data distribution) while maintaining our validation scores (even slightly improving them).

How it works...

In this recipe, we take a deeper look at how MXNet and Gluon can help us optimize our training loops. We have leveraged our HW (CPUs and GPUs) to address each of the steps in the training loop:

- We revisited how lazy evaluation and automatic parallelization mechanisms work together to optimize all MXNet-based flows.
- We leveraged all our CPU threads to load data and optimized that process further via preprocessing in GPU. We also compared the trade-offs between speed and memory optimizations.
- We analyzed different data types and combined the accuracy and precision of `Float32` with the speed-ups of `Float16` where possible, using AMP.
- We increased the performance of our training loops by using multiple GPUs (assuming our HW has these devices available).

We compared each of these scenarios by running two experiments, comparing the performance before a specific optimization to the performance afterward, emphasizing potential trade-offs that have to be taken into account when using these optimizations. In the recipes that follow, we will apply all these optimization techniques concurrently to optimize two familiar tasks: **image segmentation** and **text translation**.

There's more...

All the optimization features shown in this recipe have been thoroughly described in the research literature. The following are some introductory links to start understanding each of the features in depth:

- **Lazy evaluation and automatic parallelization:** `https://cljdoc.org/d/org.apache.mxnet.contrib.clojure/clojure-mxnet-linux-cpu/1.4.1/doc/ndarray-imperative-tensor-operations-on-cpu-gpu#lazy-evaluation-and-automatic-parallelization`
- **Gluon DataLoaders:** **https:**`//mxnet.apache.org/versions/master/api/python/docs/tutorials/getting-started/crash-course/5-datasets.html`
- **AMP:** `https://medium.com/apache-mxnet/simplify-mixed-precision-training-with-mxnet-amp-dc2564b1c7b0`
- **Training with multiple GPUs:** `https://mxnet.apache.org/versions/1.7/api/python/docs/tutorials/getting-started/crash-course/6-use_gpus.html`

Optimizing training for image segmentation

In the previous recipe, we saw how we could leverage MXNet and Gluon to optimize the training of our models with a variety of different techniques. We understood how we can jointly use lazy evaluation and automatic parallelization for parallel processing. We saw how to improve the performance of our DataLoaders by combining preprocessing in the CPU and GPU, and how using half-precision (`Float16`) in combination with AMP can halve our training times. Lastly, we explored how to take advantage of multiple GPUs to further reduce training times.

Now, we can revisit a problem we have been working with throughout the book: **image segmentation**. We have worked on this task in recipes from previous chapters. In the *Segmenting objects semantically with MXNet Model Zoo – PSPNet and DeepLabv3* recipe in *Chapter 5*, we learned how to use pre-trained models from GluonCV Model Zoo, and introduced the task and the datasets that we will be using in this recipe: **MS COCO** and the **Penn-Fudan Pedestrian** dataset. Furthermore, in the *Improving performance for segmenting images* recipe in *Chapter 7, Optimizing Models with Transfer Learning and Fine-Tuning* we compared the different approaches that we could take when dealing with a target dataset, training our models from scratch, or leveraging the existing knowledge of pre-trained models and adjusting it for our task using the different modalities of transfer learning and fine-tuning.

In this recipe, we will apply all these optimization techniques for the specific task of training an image segmentation model.

Getting ready

Similar to previous chapters, in this recipe, we will be using some matrix operations and linear algebra, but it will not be hard at all, as you will find lots of examples and code snippets to facilitate your learning.

How to do it...

In this recipe, we will be looking at the following steps:

1. Revisiting our current preprocessing and training pipeline
2. Applying training optimization techniques
3. Analyzing the results

Let's dive into each of these steps.

Revisiting our current preprocessing and training pipeline

In the *Improving performance for segmenting images* recipe in *Chapter 7*, we processed the data with the following approach:

1. Loaded the data from storage into the *CPU memory space*
2. Preprocessed the data using the *CPU*
3. Used the **default parameters** to process the data during training

This was a valid approach to compare the different training alternatives available to us (training from scratch, pre-trained models, transfer learning, and fine-tuning) without adding complexity to the experiments. For example, this approach worked quite well to introduce and evaluate the technique of fine-tuning directly.

Following the aforementioned approach on the dataset selected for this recipe (*Penn-Fudan Pedestrian*), the CPU-based preprocessing took the following amount of time:

```
Pre-processing time (s): 0.12470602989196777
```

Furthermore, when combined with the necessary step of reloading the data in batches and copying it to the GPU, we obtain the following performance:

```
Data-Loading in GPU time (s): 0.4085373878479004
```

After the preprocessing, the next step is the training process. As described, we will evaluate the effect of our training optimizations directly by using the technique of fine-tuning. In combination with this approach, we will use the following hyperparameters:

```
# Epochs & Batch Size
epochs = 10
batch_size = 4
# Define Optimizer and Hyper Parameters
trainer = mx.gluon.Trainer(deeplab_ft_direct_naive.collect_params(),
"sgd", {"learning_rate": 0.1})
```

In these conditions, the training process duration and performance achieved were as follows:

```
Training time for 10 epochs (s): 638.9948952198029 / Best validation
loss: 0.09416388
```

As we can see, we got an excellent validation performance (~0.09) in a little over 10 minutes.

The evolution of the training loss and the validation loss across each epoch looks as follows:

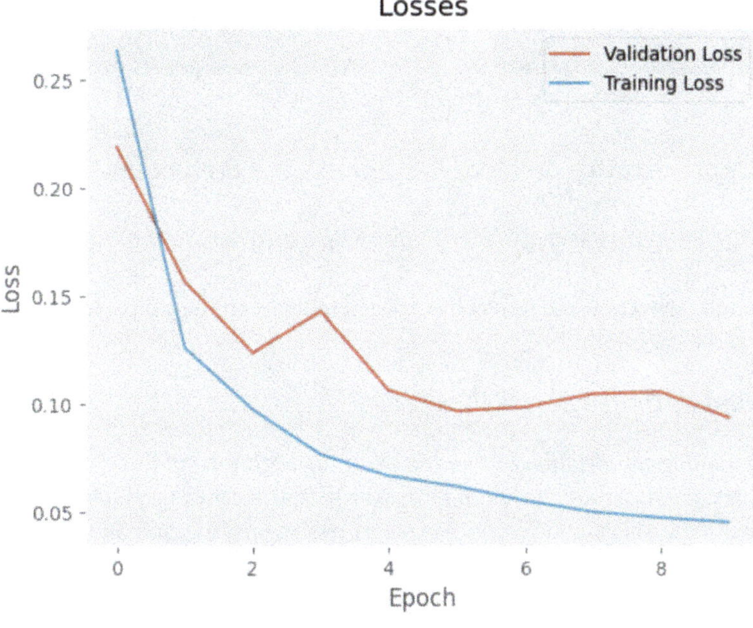

Figure 8.2 – Revisiting training: training loss versus validation loss

From *Figure 8.2*, we can see the evolution of the training and validation loss. As explored throughout the chapters, we select the model that provide the minimal validation loss (in this case, this was achieved in the last epoch, epoch 10).

After the training is completed, we can verify the overall performance in the test split of our dataset. From a quantitative point of view, these are the results we obtained:

```
PixAcc:  0.9627800347222222
mIoU  :  0.9070747450272697
```

As expected, we got excellent results by training for just a limited number of epochs (10 in this case).

From a qualitative point of view, this is what we have:

Ground Truth   Prediction

Figure 8.3 – Revisiting training: GroundTruth example and Prediction post-training

As expected, the results show how the model has learned to focus on the people in the foreground, avoiding the ones in the background.

Applying training optimization techniques

In the *Introducing training optimization features* recipe at the beginning of this chapter, we showed how different optimization techniques could improve the performance of the different steps we take when training a machine learning model, including preprocessing the data and training and evaluating the model.

In this section, we will show how, with MXNet and Gluon and just a few lines of code, we can easily apply all the techniques we've been introduced to.

As shown in the first recipe of this chapter, MXNet applies by default the best policy (`ThreadedEnginePerDevice`) to optimize lazy evaluation and automatic parallelization, taking into account the number of CPU threads available, so there is no need for us to apply any changes here (please note that this technique is also applied automatically when working with multiple GPUs).

We also showed how we could optimize our data preprocessing pipeline by combining the usage of CPU threads and GPUs, taking into account the number of devices available for each, and optimizing accordingly. For this experiment, specific HW was chosen with the following characteristics:

```
Number of CPUs: 16
Number of GPUs: 4
```

In order to use this optimization technique, we had to apply some changes to our code. Specifically, we define the GPUs available for use:

```
# Context variable is now a list,
 # with each element corresponding to a GPU device
ctx_list = [mx.gpu(0), mx.gpu(1), mx.gpu(2), mx.gpu(3)]
 num_gpus = len(ctx_list)
```

Furthermore, in our preprocessing pipeline, we now need a specific step that takes the data from CPU memory space and copies it to GPU memory space:

```
p_train_gpu = mx.gluon.data.SimpleDataset(
    [(data.as_in_context(ctx_list[idx % num_gpus]), label.as_in_
context(ctx_list[idx % num_gpus]))
     for idx, (data, label) in enumerate(pedestrian_train_dataset)])
 p_val_gpu    = mx.gluon.data.SimpleDataset(
    [(data.as_in_context(ctx_list[idx % num_gpus]), label.as_in_
context(ctx_list[idx % num_gpus]))
     for idx, (data, label) in enumerate(pedestrian_val_dataset)])
 p_test_gpu   = mx.gluon.data.SimpleDataset(
    [(data.as_in_context(ctx_list[idx % num_gpus]), label.as_in_
context(ctx_list[idx % num_gpus]))
     for idx, (data, label) in enumerate(pedestrian_test_dataset)])

p_train_opt = p_train_gpu.transform(train_val_transform, lazy=False)
 p_val_opt    = p_val_gpu.transform(train_val_transform, lazy=False)
 p_test_opt   = p_test_gpu.transform(test_transform, lazy=False)
```

As discussed in the first recipe of this chapter, in a typical production-oriented environment, we do not want to keep the data in the GPU, occupying precious GPU memory. It is usual to optimize the batch size for the GPU memory available, and to load the data from the CPU memory space into the GPU memory space in batches using *MXNet Gluon DataLoaders*. Therefore, for our GPU-based preprocessing pipeline to be complete, we need a final step to copy the data back into the CPU memory space:

```
to_cpu_fn = lambda x: x.as_in_context(mx.cpu())
```

With these code changes, our optimal preprocessing pipeline is ready, and we can continue with the next optimization technique: applying `Float16` optimizations, including AMP.

As shown in the first recipe of this chapter, in order to enable this technique, we just need a few changes in our code. First of all, we initialize the library:

```
# AMP
amp.init()
```

Secondly, we attach the trainer/optimizer to the library:

```
amp.init_trainer(trainer)
```

And lastly, due to the limitations of the `Float16` data type, there is a risk of gradients over/underflowing; therefore, we need to adjust (scale) the loss accordingly, which can be done automatically with these lines of code:

```
with amp.scale_loss(losses, trainer) as scaled_losses: mx.autograd.
backward(scaled_losses)
```

With these three simple changes, we have updated our training loop to work efficiently with the `Float16` data type (when appropriate).

Please note in the preceding code snippet how we are now working with a list of losses, instead of a single instance. This is due to our next and last training optimization technique: working with *multiple GPUs*.

As we will see, working with multiple GPUs optimally implies working with them in parallel, and therefore, computing losses and executing the training backward pass in parallel, yielding the losses list described in the previous paragraph.

In order to work with multiple GPUs in parallel, we need to define the new context as a list (seen before for preprocessing, and shown here again for convenience):

```
# Context variable is now a list,
 # with each element corresponding to a GPU device
ctx_list = [mx.gpu(0), mx.gpu(1), mx.gpu(2), mx.gpu(3)
 num_gpus = len(ctx_list)
```

As we now have multiple GPUs, we can increase our batch size to optimally use the available GPU memory space:

```
batch_size = len(ctx_list) * batch_size_per_gpu
```

Furthermore, when reading from Gluon DataLoaders, we need to split the batches of data across the GPUs. Thankfully, Gluon also provides a function that simplifies that action. We just need the following lines of code to be added (for each training and validation batch):

```
data_list  = mx.gluon.utils.split_and_load(data , ctx_list=ctx_list)
label_list = mx.gluon.utils.split_and_load(label, ctx_list=ctx_list)
```

As mentioned, this split across GPUs allows us to compute in parallel the model outputs and the losses associated with those outputs (a measure of the difference between the actual outputs and the expected outputs). This can be achieved with the following lines of code:

```
outputs = [model(data_slice) for data_slice in data_list]
losses = [loss_fn(output[0], label_slice) for output, label_slice in
zip(outputs, label_list)]
```

And lastly, we compute the backward pass used to update the weights of our model (combined with the scaled loss of AMP):

```
with amp.scale_loss(losses, trainer) as scaled_losses:
mx.autograd.backward(scaled_losses)
```

With these minimal code changes, we now have an optimal preprocessing and training pipeline and can run our experiments to analyze the performance changes.

Analyzing the results

In the previous sections, we revisited the previous performance of our preprocessing and training pipelines, and we reviewed how we had to apply the necessary changes for our training optimization techniques, specifically for our image segmentation task.

Our preprocessing pipeline steps are now the following:

1. Load the data from storage into CPU memory space.
2. Preprocess the data using the GPU.
3. Copy back the data to CPU memory space.
4. Use the optimized parameters to process the data during training.

For our experiments, we are going to use the technique of fine-tuning directly.

Applying the approach described earlier on the dataset selected for this recipe (*Penn-Fudan Pedestrian*), the preprocessing took the following amount of time:

```
Pre-processing time (s): 0.10713815689086914
```

An end-to-end preprocessing pipeline must take into account the process of batching using the *Gluon DataLoader* to load the data – in our case, into multiple GPUs as follows:

```
Data-Loading in GPU time (s): 0.18216562271118164
```

Compared to the initial section of this recipe (where the preprocessing took `0.4` seconds), we can see how, even with the added overhead of copying back the data to the CPU memory space, we have improved the preprocessing performance by >2 times.

After the preprocessing, the next step is the training process. As described, we will evaluate the effect of our training optimizations using the technique of fine-tuning directly. In combination with this approach, we use the following hyperparameters:

```
# Epochs & Batch Size
epochs = 10
batch_size_per_gpu = 4
batch_size = len(ctx_list) * batch_size_per_gpu
# Define Optimizer and Hyper Parameters
trainer = mx.gluon.Trainer(deeplab_ft_direct_opt.collect_params(),
"sgd", {"learning_rate": 0.5})
```

Please note how, by adding multiple GPUs to the training process, we can increase the batch size (multiplied by the number of GPUs), and we can also increase the learning rate (from 0.1 to 0.5). In these conditions, the training process duration and performance achieved were as follows:

```
Training time for 10 epochs: 59.86336851119995 / Best validation loss:
0.08904324161509672
```

As can be seen, we got excellent validation performance (~0.09) in less than 1 minute. When comparing with the results obtained in the recipe, we can see how there was a minimal decrease in the loss (a positive change that we will confirm with our performance analysis shortly), but the largest improvement by far was a >10x decrease in the training time. This improvement is due to all the training optimization techniques that we have applied. In a nutshell, each of the optimizations provided the following improvements:

- **Using 4 GPUs**: Provided a 4x decrease in time
- **Using Float16 and AMP**: Provided a 2x decrease (8x combined)
- **Preprocessing the datasets**: Provided a 1.25x decrease (>10x combined)

The evolution of the training loss and the validation loss across each epoch was the following:

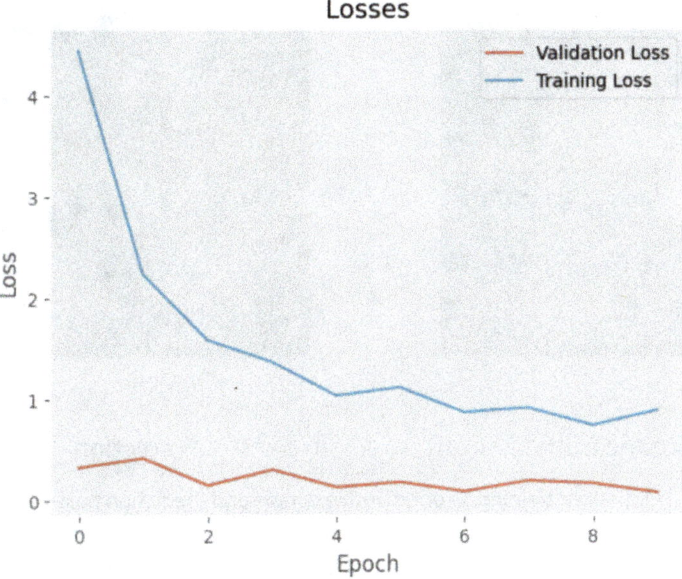

Figure 8.4 – Optimized training: training loss versus validation loss

From *Figure 8.4*, we can see the evolution of the training and validation losses. As explored throughout the chapters so far, we select the model that provided the minimal validation loss (in this case, achieved in the last epoch, epoch 10).

After training is completed, we can verify the overall performance in the test split of our dataset. From a quantitative point of view, these are the results we obtained:

```
PixAcc:  0.9679262152777778
mIoU  :  0.9176786683400912
```

As expected, we got excellent results just by training for a limited number of epochs (10 in this case). We can also confirm that the minimal improvement in the validation loss provided a minimal improvement in our test metrics (compared with 0.96/0.91 in our initial experiment).

From a qualitative point of view, we have the following:

GroundTruth Prediction

Figure 8.5 – Optimized training: GroundTruth example and Prediction post-training

As expected, the results show how the model has learned to focus on the different people in the foreground, avoiding the ones in the background.

How it works...

In this recipe, we applied the different training optimization techniques seen in the first recipe of this chapter, leveraging our HW (CPUs and GPUs) to address each of the steps in the training loop:

- We revisited how lazy evaluation and automatic parallelization mechanisms worked together to optimize all MXNet-based flows.
- We leveraged all our CPU threads to load data and optimized that process further via preprocessing in the GPU. We also compared the trade-offs between speed and memory optimizations.
- We analyzed different data types and combined the accuracy and precision of `Float32` with the speed-ups of `Float16` where possible, using AMP.
- We increased the performance of our training loops by using multiple GPUs (assuming our HW has these devices available).

We compared each of these scenarios applied specifically to the task of image segmentation, running two experiments. In the first experiment, we did not apply any of the training optimization techniques described in the previous recipe, following the approach seen in previous chapters of the book. In the second experiment, we applied all the techniques in parallel, trying to optimize as much as we could.

This proved quite useful, delivering similar algorithmic performance, with 10x improvement in training time (from 10 minutes to 1 minute). This was mostly due to using multiple GPUs (4x decrease), leveraging `Float16` AMP (2x decrease), and the optimized preprocessing (1.25x decrease).

There's more...

We have described, implemented, executed, and evaluated several training optimization techniques. However, there are even more advanced techniques that can be leveraged to achieve the optimal training loop.

One such technique is **learning rate schedules**. Throughout the book, we have been working with constant learning rates. However, there are multiple advantages of using a dynamically adjusted learning rate. Some of them are as follows:

- **Warmup**: When working with pre-trained models, it's not advisable to start with a large learning rate. The initial epochs must be used for the gradients to start adjusting. This can be thought of as a way of *adjusting the model from the source task to the target task*, retaining and leveraging the knowledge from the previous task, so smaller learning rates are recommended.
- **Decay**: In optimal training loops, as the model learns the expected representation of inputs to outputs, the objective of the training is to produce finer and finer improvements. Smaller learning rates achieve better performance at these stages (smaller and more stable weight updates). Therefore, a decaying learning rate is preferred after a few epochs.

Dive into Deep Learning provides great insights on how to implement these techniques in MXNet: `https://d2l.ai/chapter_optimization/lr-scheduler.html.`

Optimizing training for translating text from English to German

In the first recipe of this chapter, we saw how we could leverage MXNet and Gluon to optimize the training of our models, applying different techniques. We understood how to jointly use lazy evaluation and automatic parallelization for parallel processing and improved the performance of our DataLoaders by combining preprocessing in the CPU and GPU. We saw how using half-precision (`Float16`) in combination with AMP can halve our training times, and explored how to take advantage of multiple GPUs for further reduced training times.

Now, we can revisit a problem we have been working with throughout the book, that of **translating text from English to German**. We have worked with translation tasks in recipes in previous chapters. In the *Translating text from Vietnamese to English* recipe from *Chapter 6*, we introduced the task of translating text, while also learning how to use pre-trained models from GluonCV Model Zoo. Furthermore, in the *Improving performance for translating English to German* recipe from *Chapter 7*, we introduced the datasets that we will be using in this recipe: *WMT2014* and *WMT2016*, and compared the different approaches that we could take when dealing with a target dataset, training our models from scratch or leveraging past knowledge from pre-trained models and adjusting it for our task, using the different modalities of transfer learning and fine-tuning.

Therefore, in this recipe, we will apply all these optimization techniques for the specific task of training an *English-to-German text translation model.*

Getting ready

As in previous chapters, in this recipe, we will be using some matrix operations and linear algebra, but it will not be difficult to understand at all.

How to do it...

In this recipe, we will work through the following steps:

1. Revisiting our current preprocessing and training pipeline
2. Applying training optimization techniques
3. Analyzing the results

Let's dive into each of these steps.

Revisiting our current preprocessing and training pipeline

In the *Improving performance for translating English to German* recipe from *Chapter 7*, we processed the data with the following approach:

- Loaded the data from storage into CPU memory space
- Preprocessed the data using CPU
- Used the default parameters to process the data during training

This was a valid approach to compare the different training alternatives available for us (training from scratch, pre-trained models, transfer learning, and fine-tuning) without adding complexity to the experiments. For example, this approach worked quite well to introduce and evaluate the technique of fine-tuning, which is the technique that we have selected to work with in this recipe.

Applying the approach described earlier on the dataset selected for this recipe (*WMT2016*), the CPU-based preprocessing took the following amount of time:

```
Pre-processing time (s): 2.697735548019409
```

Furthermore, when combined with the necessary step of reloading the data in batches and copying it to the GPU, we obtain the following performance:

```
Data-Loading in GPU time (s): 27.328779935836792
```

After the preprocessing, the next step is the training process. As described, we will evaluate the effect of our training optimizations using the technique of fine-tuning directly. In combination with this approach, we use the following hyperparameters:

```
# Epochs & Batch Size
hparams.epochs = 5
hparams.lr = 0.00003
# hparam.batch_size = 256
```

In these conditions, the training process duration and performance achieved were as follows:

```
Training time for 5 epochs: 11406.558312892914 / Best validation loss:
1.4029905894300159
```

As we can see, we got an excellent validation performance (~1.4) for a training time of ~3 hours.

The evolution of the training loss and the validation loss across each epoch looked as follows:

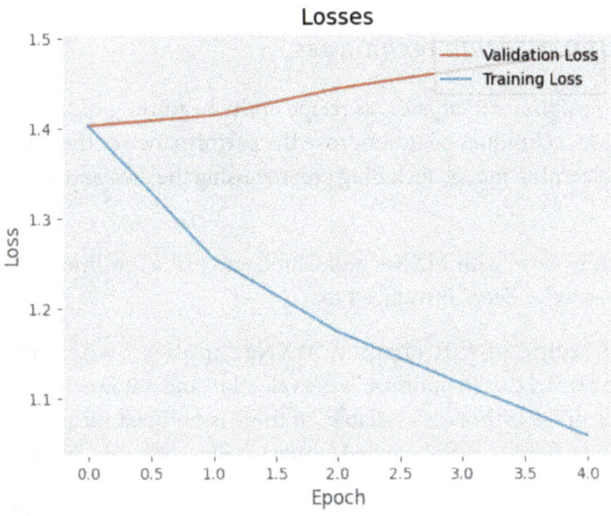

Figure 8.6 – Revisiting training: training loss versus validation loss

From *Figure 8.6*, we can see the evolution of the training and validation loss. As explored throughout the chapters, we select the model that provide the minimal validation loss (in this case, it was achieved in the first epoch, epoch 1).

After the training is completed, we can verify the overall performance in the test split of our dataset. From a quantitative point of view, these are the results we obtained:

```
WMT16 test loss: 1.28; test bleu score: 27.05
```

As expected, we got excellent results just by training for a limited number of epochs (10 in this case).

From a qualitative point of view, we can also check how well our model is performing by testing it with an example sentence. In our case, we chose `I learn new things every day`, and the output obtained is as follows:

```
Qualitative Evaluation: Translating from English to German
Expected translation:
 Ich lerne neue Dinge jeden Tag.
 In English:
 I learn new things every day.
 The German translation is:
 Immer wieder erfährt ich Neues.
```

The German sentence obtained in the output (`Immer wieder erfährt ich Neues`) means `I'm always learning new things`, and therefore, as can be seen from the results, the text has been almost perfectly translated from English to German.

Applying training optimization techniques

In the *Introducing training optimization features* recipe at the beginning of this chapter, we showed how different optimizing techniques could improve the performance of the different steps we take when training a machine learning model, including preprocessing the data and training and evaluating the model.

In this section, we will show how, with MXNet and Gluon and just a few lines of code, we can easily apply all of the techniques we've been introduced to.

As shown in the first recipe of this chapter, MXNet applies by default the best policy (`ThreadedEnginePerDevice`) to optimize lazy evaluation and automatic parallelization, taking into account the number of CPU threads available, so there is no need for us to apply any changes here (please note that this technique is also applied automatically when working with multiple GPUs).

We have shown how we could optimize our data preprocessing pipeline by combining the usage of CPU threads and GPUs, taking into account the number of devices available for each and optimizing accordingly. For this experiment, specific HW was chosen with the following characteristics:

```
Number of CPUs: 16
Number of GPUs: 4
```

In order to apply this optimization technique, we had to apply some changes to our code. Specifically, we defined the GPUs available for use:

```
# Context variable is now a list,
 # with each element corresponding to a GPU device
ctx_list = [mx.gpu(0), mx.gpu(1), mx.gpu(2), mx.gpu(3)]
 num_gpus = len(ctx_list)
```

Furthermore, in our preprocessing pipeline, we now need a specific step that takes the data from the CPU memory space and copies it to the GPU memory space:

```
wmt2016_train_data_processed_gpu = mx.gluon.data.SimpleDataset([(mx.
nd.array(data).as_in_context(ctx_list[idx % num_gpus]), mx.nd.
array(label).as_in_context(ctx_list[idx % num_gpus])) for idx, (data,
label) in enumerate(wmt2016_train_data_processed)])
wmt2016_train_data_processed_gpu = mx.gluon.data.SimpleDataset([(mx.
nd.array(data).as_in_context(ctx_list[idx % num_gpus]), mx.nd.
array(label).as_in_context(ctx_list[idx % num_gpus])) for idx, (data,
label) in enumerate(wmt2016_train_data_processed)])
wmt2016_val_data_processed_gpu = mx.gluon.data.SimpleDataset([(mx.
nd.array(data).as_in_context(ctx_list[idx % num_gpus]), mx.nd.
array(label).as_in_context(ctx_list[idx % num_gpus])) for idx, (data,
label) in enumerate(wmt2016_val_data_processed)])
wmt2016_ test _data_processed_gpu = mx.gluon.data.SimpleDataset([(mx.
nd.array(data).as_in_context(ctx_list[idx % num_gpus]), mx.nd.
array(label).as_in_context(ctx_list[idx % num_gpus])) for idx, (data,
label) in enumerate(wmt2016_ test _data_processed)])
```

As discussed in the first recipe of this chapter, in a typical production-oriented environment, we do not want to keep the data in the GPU as it occupies precious GPU memory. It is usual to optimize the batch size for the GPU memory available, and to load the data from the CPU memory space into the GPU memory space in batches using MXNet Gluon DataLoaders. Therefore, for our GPU-based preprocessing pipeline to be complete, we need a final step to copy the data back into the CPU memory space. As introduced in the *Improving performance for translating English to German* recipe from *Chapter 7*, we are using the `ShardedDataLoader` class from MXNet `GluonNLP` library. This class performs that data transfer back to the CPU memory space automatically.

However, as will be seen in our experiments, when working with multiple GPUs, performance is better when working directly with MXNet Gluon DataLoaders, as these are designed to be parallelized optimally afterward.

With these code changes, our optimal preprocessing pipeline is ready, and we can continue with the next optimization technique: applying `Float16` optimizations, including AMP.

As shown in the first recipe of this chapter, in order to enable this technique, we just need a few changes in our code. First of all, we initialize the library:

```
# AMP
amp.init()
```

Secondly, we attach the trainer/optimizer to the library:

```
amp.init_trainer(trainer)
```

In the previous recipe, when dealing with images, we described how, due to the risk of gradients over/under-flowing, there was a need to adjust (scale) the loss accordingly. This is not necessary for our use case; therefore, we do not apply **loss scaling** here.

With these two simple changes, we have updated our training loop to work efficiently with the `Float16` data type (when appropriate).

Finally, we can apply our next and last training optimization technique: working with multiple GPUs.

As we will see, working with multiple GPUs optimally implies working with them in parallel, and therefore, computing losses and executing the training backward pass in parallel, yielding the losses list described in the previous paragraph.

In order to work with multiple GPUs in parallel, we need to define the new context as a list (seen before for preprocessing, and shown here again for convenience):

```
# Context variable is now a list,
 # with each element corresponding to a GPU device
ctx_list = [mx.gpu(0), mx.gpu(1), mx.gpu(2), mx.gpu(3)]
 num_gpus = len(ctx_list)
```

As we now have multiple GPUs, we can increase our batch size to optimally use the available GPU memory space:

```
batch_size = len(ctx_list) * batch_size_per_gpu
```

Furthermore, when reading from Gluon DataLoaders, we need to split the batches of data across the GPUs. Thankfully, Gluon also provides a function that simplifies that action. We just need the following lines of code to be added (for each training and validation batch):

```
src_seq_list = mx.gluon.utils.split_and_load(src_seq, ctx_list=ctx_
list, even_split=False)
 tgt_seq_list = mx.gluon.utils.split_and_load(tgt_seq, ctx_list=ctx_
list, even_split=False)
 src_valid_length_list = mx.gluon.utils.split_and_load(src_valid_
length, ctx_list=ctx_list, even_split=False)
 tgt_valid_length_list = mx.gluon.utils.split_and_load(tgt_valid_
length, ctx_list=ctx_list, even_split=False)
```

As mentioned, this split across the GPUs allows us to compute in parallel the model outputs and the losses associated with those outputs (a measure of the difference between the actual outputs and the expected outputs). This can be achieved with the following lines of code:

```
out_slice, _ = wmt_transformer_model_ft_direct_opt(
src_seq_slice,
tgt_seq_slice[:, :-1],
src_valid_length_slice,
tgt_valid_length_slice - 1)
loss = loss_function(out_slice, tgt_seq_slice[:, 1:], tgt_valid_
length_slice - 1)
```

Typically, in order to finalize our updates to work with multiple GPUs in the training loop, we would need to apply further changes to our loss scaling. However, as discussed, for our use case, this is not necessary.

With these minimal code changes, we now have an optimal preprocessing and training pipeline, and we can run the required experiments to analyze the performance changes.

Analyzing the results

In the previous sections, we revisited the previous performance of our preprocessing and training pipelines and reviewed how we had to apply the necessary changes for our training optimization techniques, specifically for our task of translating text from English to German.

Our preprocessing pipeline steps are now the following:

1. Load the data from storage into the CPU memory space.
2. Preprocess the data using the GPU (although as we will see, we will change this to the CPU).
3. Copy back the data to the CPU memory space (won't be necessary).
4. Use the optimized parameters to process the data during training.

For our experiments, we are going to use the technique of fine-tuning directly.

Following the aforementioned approach on the dataset selected for this recipe (*WMT2016*), the GPU-based preprocessing took the following amount of time:

```
Pre-processing time (s): 50.427586793899536
```

An end-to-end preprocessing pipeline must take into account the process of batching using the Gluon DataLoader to load the data (in our case, into multiple GPUs), giving us the following performance:

```
Data-Loading in GPU time (s): 72.83465576171875
```

Compared to the initial section of this recipe (where the preprocessing took 27 seconds), we can see how, in this case, preprocessing in the GPU has not been effective. This is due to the nature of the text data, which is not as straightforward to parallelize as it is with images, for example.

In this scenario, a CPU-based preprocessing pipeline is best, avoiding the Gluon NLP `ShardedDataLoader` class and using the `Gluon DataLoader` class instead (which is better suited for parallelizing). Applying this pipeline, we get the following results:

```
Data-Loading in CPU with Gluon DataLoaders time (s):
24.988255500793457
```

This gives us a minimal edge (2 seconds), but, as mentioned, this is the best we can get with the usage of Gluon DataLoader and its parallelization capabilities.

After the preprocessing, the next step is the training process. As described, we will evaluate the effect of our training optimizations using the technique of fine-tuning directly. In combination with this approach, we use the following hyperparameters:

```
# Epochs & Batch Size
hparams.epochs = 5
hparams.lr = 0.0001
# hparams.batch_size = num_gpus * 256
```

Please note how by adding multiple GPUs to the training process, we can increase the batch size (multiplied by the number of GPUs), and we can also increase the learning rate (from 0.00003 to 0.0001). In these conditions, the training process duration and achieved performance is as follows:

```
Training time for 5 epochs: 1947.1244320869446 / Best validation loss:
1.2199710432327155
```

As we can see, we got excellent validation performance (~1.4) with training that took ~3 hours. When compared to the results obtained in the initial section of this recipe, we can see how there was a minimal decrease in the loss (a positive change, which we will confirm with our performance analysis shortly), but the largest improvement by far has been a 5.5x decrease in the training time. This improvement is due to all the training optimization techniques that we have applied. In a nutshell, each of the optimizations provided the following improvements:

- **Using 4 GPUs**: Provided a 4x decrease (as expected).
- **Using Float16 and AMP**: Provided a 1.4x decrease (5.5x combined), which is less than the 2x decrease expected, due to the small number of operations to be reduced to `Float16` without compromising algorithmic performance.
- **Preprocessing the datasets**: In this case, there were negligible improvements.

The evolution of the training loss and the validation loss across each epoch looked as follows:

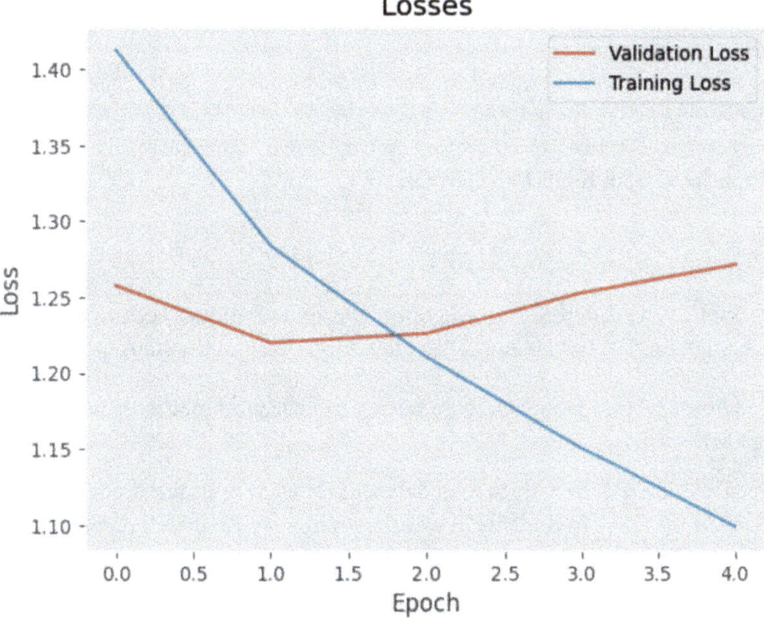

Figure 8.7 – Optimized training: training loss versus validation loss

From *Figure 8.7*, we can see the evolution of the training and validation losses. As explored throughout the chapters, we select the model that provided the minimal validation loss (in this case, achieved in the first epoch).

After training is completed, we can verify the overall performance in the test split of our dataset. From a quantitative point of view, these are the results we obtained:

```
WMT16 test loss: 1.27; test bleu score: 28.20
```

As expected, we got excellent results just by training for a limited number of epochs (5 in this case). We can also confirm how the minimal improvement in the validation loss provided a minimal improvement in our test metrics (compared to 27.05 as initially obtained).

From a qualitative point of view, we can also check how well our model is performing by testing it with an example sentence. In our case, we chose `I learn new things every day`, and the output obtained is as follows:

```
Qualitative Evaluation: Translating from English to German
Expected translation:
 Ich lerne neue Dinge.
```

```
In English:
I learn new things every day.
The German translation is:
Ich lerne jedes Mal Neues.
```

The German sentence obtained in the output (`Ich lerne jedes Mal Neues`) means `I learn something new every time`, and therefore, as can be seen from the results, the text has been almost perfectly translated from English to German.

How it works...

In this recipe, we applied the different training optimization techniques seen in the first recipe of this chapter, leveraging our HW (CPUs and GPUs) to address each of the steps in the training loop:

- We revisited how lazy evaluation and automatic parallelization mechanisms work together to optimize all MXNet-based flows.
- We leveraged all our CPU threads to load data and tested to optimize that process further via preprocessing in the GPU. In this case, it was shown how a CPU-based preprocessing pipeline in combination with Gluon DataLoader was the optimal approach.
- We analyzed different data types and combined the accuracy and precision of `Float32` with the speed-ups of `Float16`, and where possible, AMP.
- We increased the performance of our training loops by using multiple GPUs (assuming our HW has these devices available).

We compared each of these scenarios applied specifically to the task of *translating text from English to German*, running two experiments. In the first experiment, we did not apply any of the training optimization techniques described, following the approaches seen in previous chapters of the book. In the second experiment, we applied all the techniques in parallel, trying to optimize as much as we could.

This proved quite useful, delivering similar algorithmic performance, with a 5.5x improvement in training time (from 3 hours to 30 minutes). This was mostly due to using multiple GPUs (4x decrease) and leveraging `Float16` and AMP (1.4x decrease), whereas the optimized preprocessing provided negligible improvements.

There's more...

We have described, implemented, executed, and evaluated several training optimization techniques. However, there are even more advanced techniques that can be leveraged to achieve the optimal training loop.

One such technique is **Reinforcement Learning from Human Feedback** (**RLHF**), where a *human-in-the-loop* process is introduced. In this process, after a model has been trained, a person is presented with different options for output by the model (for example, different potential translations) and they rank those responses according to how they better represent the original sentence. These human inputs are then used to train a reward model that scores the output of the model and selects the one with the highest score. This technique has been proven to be extremely powerful. As an example, **OpenAI** developed **ChatGPT** on top of the **GPT-3** language model using RLHF.

To learn more about *ChatGPT* and *RLHF*, the following article is recommended: `https://huyenchip.com/2023/05/02/rlhf.html`.

9
Improving Inference Performance with MXNet

In previous chapters, we leveraged MXNet's capabilities to solve **computer vision** and **natural language processing tasks**. In those chapters, the focus was on obtaining the maximum performance out of **pre-trained models**, leveraging the **Model Zoo** API from GluonCV and GluonNLP. We trained these models using different approaches from scratch, including **transfer learning** and **fine-tuning**. In the previous chapter, we explored how some advanced techniques can be leveraged to optimize the training process. Finally, in this chapter, we will focus on improving the performance of the inference process itself, accelerating how we can obtain results from our models with several topics related to **edge AI computing**.

To achieve the objective of optimizing the performance of our inference pipeline, MXNet contains different features. We have already briefly discussed some of those features, such as the concept of **Automatic Mixed Precision** (**AMP**), which was introduced in the previous chapter to increase the training performance and can also be used to increase the inference performance. We will revisit it in this chapter, along with other features, such as **hybridization**. Moreover, we will further optimize how to use data types efficiently, leveraging the speed-ups associated with using the **INT8** data type with **quantization**.

Moreover, we will explore how our models work in terms of operations, understanding how they work internally with the help of the **MXNet profiler**. We will then take a step forward with the help of MXNet GluonCV Model Zoo and learn how to export our models to **ONNX**, which allows us to use our models in different frameworks, such as deploying our models on NVIDIA hardware platforms, such as the **NVIDIA Jetson** family of products.

Finally, we will apply all these techniques together, taking as examples problems already explored in the book. For our computer vision task, we will choose image segmentation, and for our natural language processing task, we will choose translating text from English to German.

Specifically, this chapter contains the following recipes:

- Introducing inference optimization features
- Optimizing inference for image segmentation
- Optimizing inference when translating text from English to German

Technical requirements

Apart from the technical requirements specified in the *Preface*, the following apply:

- Ensure that you have completed *Recipe 1, Installing MXNet*, from *Chapter 1, Up and Running with MXNet*.
- Ensure that you have completed *Chapter 5, Analyzing Images with Computer Vision*, and *Chapter 6, Understanding Text with Natural Language Processing*.
- Ensure that you have completed *Chapter 7, Optimizing Models with Transfer Learning and Fine-Tuning*.

The code for this chapter can be found at the following GitHub URL: `https://github.com/PacktPublishing/Deep-Learning-with-MXNet-Cookbook/tree/main/ch09`.

Furthermore, you can access each recipe directly from Google Colab, for example, for the first recipe of this chapter: `https://github.com/PacktPublishing/Deep-Learning-with-MXNet-Cookbook/blob/main/ch09/9_1_Introducing_inference_optimization_features.ipynb`.

Introducing inference optimization features

In the previous chapters, we have seen how we can leverage MXNet, GluonCV, and GluonNLP to retrieve pre-trained models in certain datasets (such as ImageNet, MS COCO, or IWSLT2015) and use them for our specific tasks and datasets. Furthermore, we used transfer learning and fine-tuning techniques to improve the algorithmic performance of those tasks/datasets.

In this recipe, we will introduce (and revisit) several concepts and features that will optimize our inference loops to improve our runtime performance, and we will analyze the trade-offs involved.

Getting ready

As in previous chapters, in this recipe, we will be using some matrix operations and linear algebra, but it will not be hard at all.

How to do it...

In this recipe, we will be carrying out the following steps:

1. Hybridizing our models
2. Applying float16 and AMP for inference
3. Applying quantization by using INT8
4. Profiling our models

Let's dive into each of these steps.

Hybridizing our models

In the initial chapters where we were exploring the features of MXNet, we focused on **imperative programming**. If you have coded in the past with languages such as Java, C/C++, or Python, it is very likely you used imperative programming. It is the usual way of coding, as it is more flexible.

With imperative programming, a step-by-step sequential execution of the statements set in the code is expected. For example, typically in our evaluation paths, we run these statements step-by-step inside a loop:

1. Load new samples from our data loader.
2. Transform the input and expected output so that it can be consumed by our model and our metrics computations.
3. Pass the input through the model to compute the output.
4. Compare the model output with the expected output and update the corresponding metrics.

In this programming paradigm, each of the statements is executed in sequence, and the output can be checked or debugged for each step if we wait for its completion (as MXNet uses **lazy evaluation**).

With a different programming paradigm, called **symbolic programming**, symbols are used instead, which are basically abstractions for operations, and no actual computation happens until a defined point (typically known as the compile step). This is especially useful for **deep learning**, as all models can be defined as graphs, use this graph as a symbol, optimize the operation paths in the underlying graph, and only run the optimized computation when needed.

However, as the computation hasn't happened yet, the output for each step cannot be checked or debugged, making the finding and fixing of issues much more difficult. On the other hand, due to the capabilities of graph optimization, symbolic programming requires less memory and is faster.

Thankfully, with MXNet, we can leverage the best of both worlds. We can define our model with imperative programming, test it, debug it, and fix it with the usual mechanisms (*print* statements, tests, debugging, and so on). When we are ready for optimization, we just need to call the `hybridize` function, and it will take care of everything under the hood, working with our graph in symbolic programming. This approach is called hybrid programming and is one of the best advantages of MXNet. Moreover, there is no hardware limitation for this feature, and it can be used for both CPU and GPU computations.

As a toy example, we can run some experiments with the inference of a model and compare the different results for different configurations. Specifically, these are the configurations we will test:

- CPU:
  - With imperative execution
  - With symbolic execution and default parameters
  - With symbolic execution with a specific backend
  - With symbolic execution, specific backend, and static allocation of memory
  - With symbolic execution, specific backend, static allocation of memory, and invariant input shapes
- GPU:
  - With imperative execution
  - With symbolic execution and default parameters
  - With symbolic execution and static allocation of memory
  - With symbolic execution, static allocation of memory and invariant input shapes

Please note that in order to verify the computation time properly, we are adding calls to the `mx.nd.waitall()` function. The method chosen is to use the **ADE20K** validation split (dataset available from MXNet GluonCV) and process it with a **DeepLabv3** model. We will be using a batch size of 4:

1. For the initial CPU computing configuration, with imperative execution, the processing of the dataset by the model took the following time:

   ```
   Time (s): 115.22693085670471
   ```

2. For the second CPU computing configuration, we just need to leverage the MXNet hybrid programming model and transform our model with the following:

   ```
   deeplab_pt_cpu_hybrid.hybridize()
   ```

   With this configuration, the processing of the dataset by the model took the following time:

   ```
   Time (s): 64.75840330123901
   ```

   As we can see, the optimizations performed reduced to almost half the computation time.

3. For the third CPU computing configuration, we just need to slightly modify our hybridization call to define a specific backend. We will leverage our Intel CPU architecture, use the `MKLDNN` backend, and transform our model with the following:

```
deeplab_pt_cpu_hybrid.hybridize(backend = "MKLDNN")
```

With this configuration, the processing of the dataset by the model took the following time:

```
Time (s): 55.860424757003784
```

As we can see, the specific backend further reduced the computation time by ~20%.

4. For the fourth CPU computing configuration, we just need to slightly modify our hybridization call to define that we want to use the static memory allocation. We can update our call with the following:

```
deeplab_pt_cpu_hybrid.hybridize(backend = "MKLDNN", static_
alloc=True)
```

With this configuration, the processing of the dataset by the model took the following time:

```
Time (s): 53.905478715896606
```

As we can see, the static memory allocation allowed us to reduce the computation time by another ~4%.

5. For the fifth CPU computing configuration, we just need to slightly modify our hybridization call to define that we want to leverage our invariant input shapes (we have already preprocessed our data to have the same input shape, `480x480`). We can update our call with the following:

```
deeplab_pt_cpu_hybrid.hybridize(backend = "MKLDNN", static_
alloc=True, static_shape=True)
```

With this configuration, the processing of the dataset by the model took the following time:

```
Time (s): 52.464826822280884
```

As we can see, the invariant input shape constraint allowed us to reduce the computation time by another ~2%.

6. For the initial GPU computing configuration, with imperative execution, the processing of the dataset by the model took the following time:

```
Time (s): 13.315197944641113
```

7. For the second GPU computing configuration, we just need to leverage the MXNet hybrid programming model and transform our model with the following:

```
deeplab_pt_gpu_hybrid.hybridize()
```

With this configuration, the processing of the dataset by the model took the following time:

```
Time (s): 12.873461246490479
```

As we can see, when the optimizations are performed in the GPU, they yield almost no improvement in the computation time as GPUs are already optimized internally for these types of computations.

8. For GPU computing, there are no specific backends to be selected. Therefore, for the third GPU computing configuration, we just need to slightly modify our hybridization call to define that we want to use static memory allocation. We can update our call with the following:

```
deeplab_pt_gpu_hybrid.hybridize(static_alloc=True)
```

With this configuration, the processing of the dataset by the model took the following time:

```
Time (s): 12.752988815307617
```

As we can see, the static memory allocation produced another negligible improvement on the GPU.

9. For the fourth GPU computing configuration, we just need to slightly modify our hybridization call to define that we want to leverage our invariant input shapes (we already preprocessed our data to have the same input shape, 480x480). We can update our call with the following:

```
deeplab_pt_gpu_hybrid.hybridize(static_alloc=True, static_
shape=True)
```

With this configuration, the processing of the dataset by the model took the following:

```
Time (s): 12.583650827407837
```

As we can see, the invariant input shape constraint produced another negligible improvement on the GPU.

The results show that when using the CPU, we can reduce the inference time to half the original time, which is a significant improvement. With the GPU, the improvements are negligible due to the internal optimizations.

> **Important Note**
>
> Please note in the code how we used the `mx.nd.waitall()` function to verify that all computations have been strictly completed before computing the time these operations took.

Applying float16 and AMP for inference

In the previous chapter, *Chapter 8, Improving Training Performance with MXNet,* we introduced the `float16` data type and AMP optimization, an extremely simple way to leverage this half-precision data type only when it was most useful.

In *Recipe 1, Introducing training optimization features,* from the previous chapter, we compared single-precision (`float32`) and half-precision (`float16`) data types, understanding their characteristics and memory/speed trade-offs. You are encouraged to review the recipe if you haven't already done so as it is very relevant to this topic.

As most concepts were introduced previously, in this section, we will focus on how to apply AMP to the inference process. As usual, MXNet provides a very simple interface for this operation, just requiring a call to the `amp.convert_hybrid_block()` function.

This optimization can be applied to both CPU and GPU environments, so let's run these experiments.

To modify our CPU model to use AMP, we just need the following line of code:

```
deeplab_pt_cpu_hybrid_amp = amp.convert_hybrid_block(deeplab_pt_cpu_
hybrid, ctx=mx.cpu())
```

With this modified model, the processing of the dataset took the following time:

```
Time (s): 56.16465926170349
```

As we can see, AMP produced negligible improvements on the CPU. This is due to the largest gains being achieved on the backward pass required for training, but not necessary during inference. Furthermore, CPUs do not typically have specific circuitry to work directly with float16, limiting the improvements.

To modify the GPU model to use AMP, we just need the following line of code:

```
deeplab_pt_gpu_hybrid_amp = amp.convert_hybrid_block(deeplab_pt_gpu_
hybrid, ctx=mx.gpu())
```

With this modified model, the processing of the dataset took the following time:

```
Time (s): 3.371366024017334
```

As we can see, AMP produced excellent results on the GPU, reducing the inference time to almost ~25%. This is due to GPUs having specific circuitry to work directly with float16, improving the results massively.

> **Important Note**
>
> The `amp.convert_hybrid_block()` function accepts different parameters. You are encouraged to try different options (such as `cast_optional_params`) to find the optimal configuration.

Applying quantization by using Int8

In the previous sections, we saw how to optimize our inference loops by using different approaches optimizing how to use the CPU and GPU for maximum performance, given a model. We also explored how to leverage single-precision (float32) and half-precision (float16) data types. In this section, we will explore how our data inputs, our model parameters, and the different arithmetic calculations among them can be optimized with a new data type, Int8.

This data type modification has larger implications than a change in precision. We are also modifying the underlying representation from a floating-point number to an integer, which yields a reduction in both memory and computing requirements. Let's analyze this data type.

Int8 indicates two things: that it is a data type that only supports integer numbers (no floating radix point), and that the amount of bits used to store a single number in this format is 8 bits. The most important features of this format are the following:

- Capability of representing integer numbers from -128 to 127, or from 0 to 255 (depending on whether it is of the signed or unsigned type)
- Constant precision (each consecutive number differs by exactly 1)

To explain the core idea behind `Int8` quantization, and also to show the loss of precision, we can display the approximated value of the number 1/3 (one third-th) in both formats (Float32 and Int8) with this code excerpt:

```
a = mx.nd.array([1/3], dtype=mx.np.float32)
 int_value = 85
scaling_factor = 255
b = int_value / scaling_factor
print("1/3 as 0.333... (Float32): {0:.30f}".format(a.asscalar())))
print("1/3 as 85/255   (Int8)    : {0:.30f}".format(b))
```

This yields the following:

```
1/3 as 0.333... (Float32): 0.333333343267440795898437500000
1/3 as 85/255   (Int8)   : 0.333333333333333314829616256247
```

As we can see, none of the representations are exact, with `float32` yielding a very high precision as expected. With `Int8`, we did a small shortcut; we used two 8-bit integers, `85` and `255`, and used one as the scaling factor. This scaling factor is typically applied for several sets of numbers at the same time. It can be the same scaling factor for the whole model (unlikely), per layer, and so on. The scaling factor does not need to be represented in `Int8`; it can be a `float32`.

> **Important Note**
>
> For this particular example, the chosen Int8 representation is more exact to the intended number, but this is a coincidence. In common scenarios, there is a loss of precision that translates to a loss of performance.

To minimize this loss of performance, typically quantization tuning techniques ask for a **calibration dataset**. This dataset is then used to compute the parameters that minimize the mentioned performance loss.

In addition to using a calibration dataset, there are several techniques to optimize the computation of the most accurate Int8 values, and MXNet provides a very simple API to facilitate the optimization of our networks. With a simple call to the `mx.contrib.quantization.quantize_net_v2()` function, we will update our network to Int8.

Specifically, for our experiments, this is the call that we used:

```
deeplab_pt_cpu_q_hybrid = quantization.quantize_net_v2(
deeplab_pt_cpu,
 quantized_dtype='auto',
 exclude_layers=None,
 exclude_layers_match=None,
 calib_data=ade20k_cal_loader_gpu_cpu,
 calib_mode='entropy',
 logger=logger,
 ctx=mx.cpu())
```

> **Important Note**
>
> `Int8` quantization is a nuanced process, and tailoring it for a specific application requires in-depth analysis and some trial-and-error experiments. For more information regarding the parameters involved, you are encouraged to read the function documentation: `https://github.com/apache/mxnet/blob/v1.9.1/python/mxnet/contrib/quantization.py#L825`.

With this modified CPU-based model, the processing of the dataset took the following time:

```
Time (s): 36.10324692726135
```

As we can see, `Int8` produced a strong improvement in the CPU, yielding almost another ~50% reduction in runtime.

Unfortunately, for GPUs, this feature cannot be introduced. Although recent GPUs have dedicated `Int8` circuitry, this is quite a new development and MXNet does not support these operators yet.

Profiling our models

In this recipe, we have seen how to use different techniques to optimize our inference loops. However, sometimes, even after introducing these optimization techniques, our models might not reach the runtime performance we are targeting. This could be due to a number of reasons:

- Architecture is not optimal for edge computing.
- Operators have not been optimized adequately.
- Data transfers among components.
- Memory leaks.

In order to verify how our model is working internally, to check where to optimize further and/or investigate the possible reasons why our models might not be performing well, MXNet provides us with a tool for low-level analysis called the MXNet profiler.

The MXNet profiler runs in the background and records all operations and data transfers happening on our models in real time. It is also very lightweight, consuming a minimal amount of resources. Best of all, it is extremely easy to configure and use.

In order to profile a set of statements, we need to take two steps:

1. Configure the profiler.
2. Start and stop the profiler before and after the statements to be profiled.

To configure the profiler, we just need one line of code, as follows:

```
mx.profiler.set_config(
profile_all=True,
 aggregate_stats=True,
 continuous_dump=True,
 filename='profile_output_cpu.json')
```

To start and stop the profiler, we need to add the following lines at the beginning and end of the statements to be analyzed:

```
mx.profiler.set_state('run')
[... code statements to analyze ...]
# Wait until all operations have completed
mx.nd.waitall()
# Stop recording
mx.profiler.set_state('stop')
# Log results
mx.profiler.dump()
```

Please note how we need three statements to stop recording: finalize all instructions, stop recording, and dump the information of the file configured in the first step.

> **Important Note**
>
> Please note in the code how we used the `mx.nd.waitall()` function to verify that all computations have been strictly completed before computing the time these operations took.

The instructions described previously generate a JSON file, which can then be analyzed with tracing applications. I recommend the Tracing app included with Google Chrome as it is very easy to use and access. Simply type the following in the address bar: `chrome://tracing`.

In order to verify the functionality of the MXNet profiler, let's take the example of **ResNet** architectures, which are used extensively, such as in our image segmentation task, being the backbone of the DeepLabv3 network we use.

The typical architecture of a ResNet network is the following:

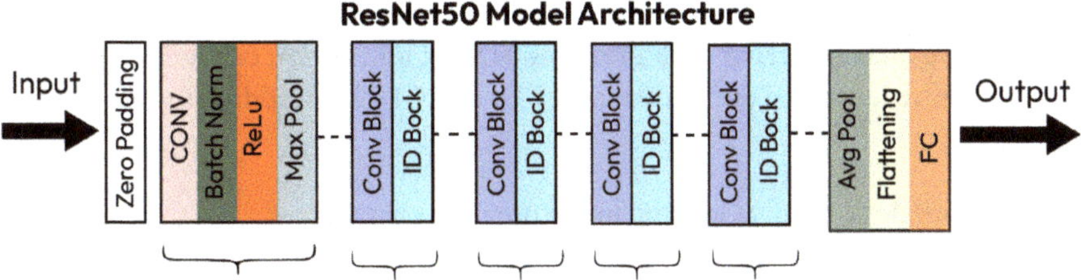

Figure 9.1 – ResNet50 model architecture

Please note that in *Figure 9.1*, the initial steps (stage 1) are convolution, batch normalization, and activation.

From our profiled model, the Google Chrome Tracing app provides the following screen:

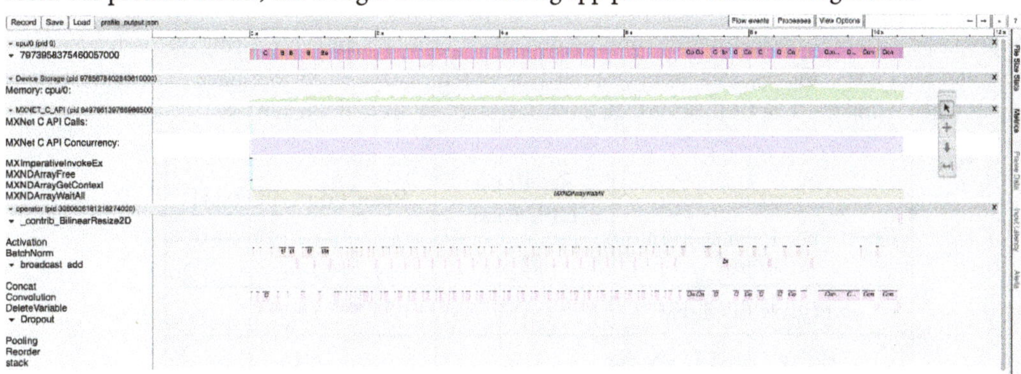

Figure 9.2 – Profiling ResNet

In *Figure 9.2*, we can see the general execution of the model. Zooming on the earlier layers, we can see it as follows:

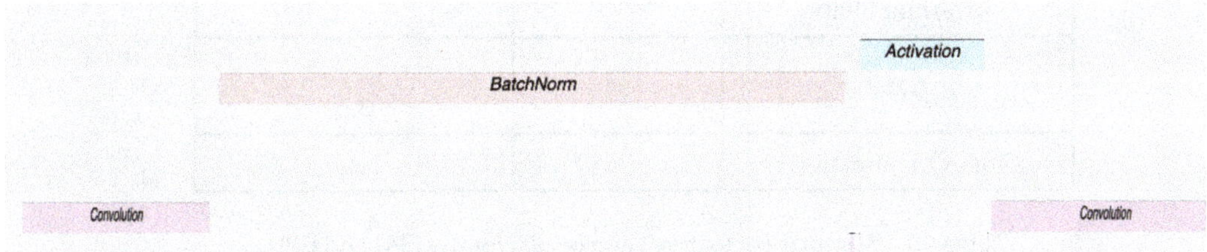

Figure 9.3 – Profiling ResNet: Zoom

In *Figure 9.3*, we can see how the stage 1 steps of convolution, batch normalization, and activation are clearly displayed. We can also now very clearly see how the batch normalization operation takes about 4x longer than the convolution and activation steps, potentially indicating an avenue of improvement.

How it works...

In this recipe, we have taken a deeper look into how MXNet and Gluon can help us optimize our inference loops. We have leveraged our hardware (CPUs and GPUs) by addressing each of the steps in the inference loop:

- Reused the work done for data loading in the previous chapter
- Optimized graph computation via hybridization
- Analyzed different data types and combined the accuracy and precision of float32 with the speed-ups of `float16` (leveraging the specific circuitry of GPUs) where possible, using AMP
- Taken another step forward by using Int8 quantization
- Analyzed low-level performance using the MXNet profiler

We compared each of these features by running several experiments and comparing the performance before and after a specific optimization, emphasizing potential trade-offs that have to be taken into account when using these optimizations. To summarize, these were the results:

Feature	Result on CPU (ms)	Result on GPU (ms)
Standard	115	13.3
Hybridize / Default	65	12.9
Hybridize / MKLDNN	56	N/A
Hybridize / MKLDNN + Static Alloc	54	12.8
Hybridize / MKLDNN + Static Alloc + Invariant Shape	52	12.6
AMP	54	3.5
Int8 Quantization	36	N/A

Table 9.1 – Summary of features and results for the CPU and GPU

In the next recipes, we will apply all these optimization techniques concurrently for the best cases for the CPU (MKL-DNN + static allocation + invariant shape + Int8 quantization) and GPU (static allocation + invariant shape + automatic mixed precision) to optimize two familiar tasks: image segmentation and text translation.

There's more...

All the optimization features shown in this recipe have been thoroughly described in the literature. In this section, we share some introductory links to start understanding each of the features in depth:

- **Hybridization**: `https://mxnet.apache.org/versions/1.9.1/api/python/docs/tutorials/packages/gluon/blocks/hybridize.html`
- **Automatic Mixed Precision (AMP)**: `https://mxnet.apache.org/versions/1.9.1/api/python/docs/tutorials/packages/gluon/blocks/hybridize.html`
- **Int8 quantization**: `https://mxnet.apache.org/versions/1.9.1/api/python/docs/tutorials/packages/gluon/blocks/hybridize.html`
- **MXNet profiler**: `https://mxnet.apache.org/versions/1.9.1/api/python/docs/tutorials/packages/gluon/blocks/hybridize.html`

Optimizing inference for image segmentation

In the previous recipe, we saw how we can leverage MXNet and Gluon to optimize the inference of our models, applying different techniques, such as improving the runtime performance using hybridization; how using half-precision (float16) in combination with AMP can strongly reduce our inference times; and how to take advantage of further optimizations with data types such as Int8 quantization.

Now, we can revisit a problem we have been working with throughout the book, image segmentation. We have worked with this task in recipes from previous chapters. In *Recipe 4, Segmenting objects semantically with MXNet Model Zoo – PSPNet and DeepLabv3*, from *Chapter 5, Analyzing Images with Computer Vision*, we introduced the task and the datasets that we will be using in this recipe, *MS COCO and Penn-Fudan Pedestrian*, and learned how to use pre-trained models from GluonCV Model Zoo.

Furthermore, in *Recipe 3, Improving performance for segmenting images*, from *Chapter 7, Optimizing Models with Transfer Learning and Fine-Tuning*, we compared the different approaches that we could take when dealing with a target dataset, training our models from scratch or leveraging past knowledge from pre-trained models and adjust them for our task, using the different modalities of transfer learning and fine-tuning. Lastly, in *Recipe 2, Optimizing training for image segmentation*, from *Chapter 8, Improving Training Performance with MXNet*, we applied different techniques to improve the runtime performance of our training loops.

Therefore, in this recipe, we will apply all the introduced optimization techniques for the specific task of optimizing the inference of an image segmentation model.

Getting ready

As in previous chapters, in this recipe, we will be using some matrix operations and linear algebra, but it will not be hard at all.

How to do it...

In this recipe, we will be using the following steps:

1. Applying inference optimization techniques
2. Visualizing and profiling our models
3. Exporting our models to ONNX and TensorRT

Let's dive into each of these steps.

Applying inference optimization techniques

In *Recipe 1, Introducing inference optimization features*, at the beginning of this chapter, we showed how different optimization techniques could improve the performance of the different steps we take in the inference of a machine learning model, including hybridization, AMP, and Int8 quantization.

In this section, we will show how, with MXNet and Gluon, just with a few lines of code, we can easily apply each and every technique we've introduced and verify the results of each technique.

Without applying these optimization techniques, as a baseline, these are the quantitative results obtained with the CPU:

```
PixAcc:  0.9602144097222223
mIoU  :  0.4742364603465315
Time (s): 27.573920726776123
```

We can display an image for qualitative results:

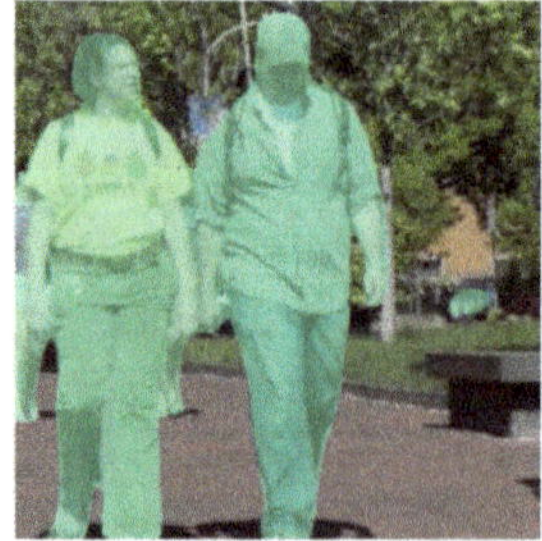

Figure 9.4 – Qualitative results: CPU baseline

As expected from the quantitative metrics, *Figure 9.4* shows excellent results as well.

As concluded in the previous recipe, for maximum performance on the CPU, the best approach is the following:

- Use hybridization: Using the Intel MKL-DNN backend, combined with static memory allocation and invariant input shapes.
- Do not use AMP.
- Use Int8 quantization.

Let's apply each of these techniques for our current specific task, image segmentation.

For hybridization, we just need one line of code (which includes the necessary parameters for the Intel `MKLDNN` backend, combined with static memory allocation and invariant input shapes):

```
deeplab_pt_cpu_q_hybrid.hybridize(backend="MKLDNN", static_alloc=True,
static_shape=True)
```

We do not need to add an AMP step as it was shown not to add benefits to CPU-based workloads.

For `Int8` quantization, we need two separate steps. On one hand, we need to define the calibration dataset. This can be achieved with a small number of lines:

```
# Dataset Loading & Transforming
# Limit to 10 samples (last ones)
 max_samples = 10
samples = range(0, max_samples)
 p_cal_cpu_pre = mx.gluon.data.SimpleDataset([(pedestrian_
val_dataset[-i][0], pedestrian_val_dataset[-i][1]) for i in
tqdm(samples)])
 p_cal_gpu_cpu = p_cal_cpu_pre.transform_first(input_transform_fn_gpu_
cpu, lazy=False)
# DataLoader for Calibration
# For CPU, Pre-processed in GPU, copied back to CPU memory space)
 num_workers = 0
batch_size = 4
p_cal_loader_gpu_cpu = mx.gluon.data.DataLoader(
    p_cal_gpu_cpu,    batch_size=batch_size,
    num_workers=num_workers,
    last_batch="discard")
```

Then, to apply `Int8` quantization, optimized using the calibration dataset, just another line of code is required:

```
deeplab_pt_cpu_q_hybrid = quantization.quantize_net_v2(
    deeplab_pt_cpu,
    quantized_dtype='auto',
    exclude_layers=None,
    exclude_layers_match=None,
    calib_data=p_cal_loader_gpu_cpu,
    calib_mode='entropy',
    logger=logger,
    ctx=mx.cpu())
```

Applying these optimization techniques, these are the quantitative results obtained for an optimized CPU inference:

```
PixAcc:   0.9595597222222222
mIoU  :   0.47379941937958425
Time (s): 8.355125904083252
```

As we can see, the differences in performance (`0.959` versus `0.960` and `0.473` versus `0.474`) are negligible. However, with these inference optimization techniques, we have been able to reduce the inference runtime by 4x (8.4 seconds versus 27.6 seconds), which is an impressive result.

We can also display an image for qualitative results:

GroundTruth

Prediction

Figure 9.5 – Qualitative results: CPU-optimized inference

As expected from the quantitative metrics, *Figure 9.5* shows excellent results as well, with negligible differences (if any).

What about GPU-based inference? Let's follow the same steps:

1. Without applying these optimization techniques, as a baseline, these are the quantitative results obtained with the GPU:

```
PixAcc:  0.9602144097222223
mIoU  :  0.4742364603465315
Time (s): 13.068315982818604
```

   As expected, there is no change relative to algorithmic performance from the CPU baseline. The runtime inference is indeed twice as fast in the GPU (`13.1` seconds versus `27.6` seconds).

2. We can display an image for qualitative results:

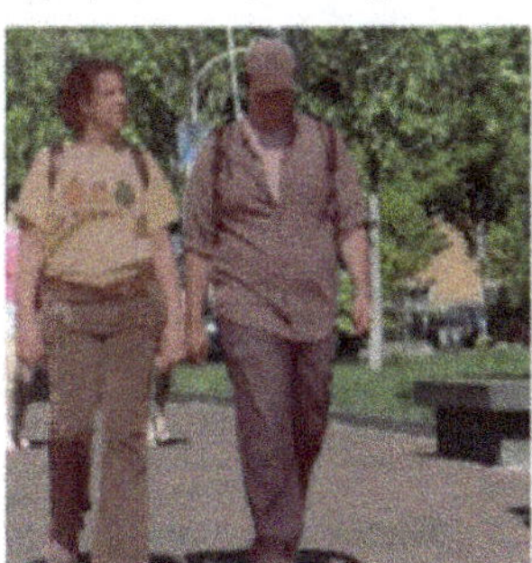

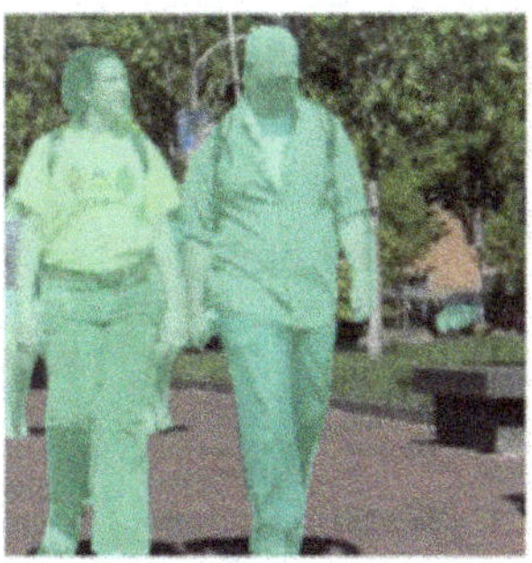

GroundTruth Prediction

Figure 9.6 – Qualitative results: GPU baseline

As expected from the quantitative metrics, *Figure 9.6* shows excellent results as well.

As concluded in the previous recipe, for maximum performance on the GPU, the best approach is the following:

- **Use hybridization**: Using static memory allocation and invariant input shapes. Do not use the Intel MKL-DNN backend.
- Use AMP.
- Do not use Int8 quantization (not supported).

Let's apply each of these techniques to our current specific task, image segmentation.

For hybridization, we just need one line of code (which includes the necessary parameters for static memory allocation and invariant input shapes):

```
deeplab_pt_gpu_hybrid.hybridize(static_alloc=True, static_shape=True)
```

For AMP, we need to follow two simple steps, a forward pass and the conversion of the model, as follows:

```
deeplab_pt_gpu_hybrid(single_sample_gpu);
deeplab_pt_gpu_hybrid_amp = amp.convert_hybrid_block(deeplab_pt_gpu_
hybrid, ctx=mx.gpu())
```

No further steps are required.

By applying these optimization techniques, these are the quantitative results obtained for an optimized GPU inference:

```
PixAcc:   0.9602565972222222
mIoU   :  0.4742640561133744
Time (s): 0.8551054000854492
```

As we can see, the differences in performance (`0.960` versus `0.960` and `0.474` versus `0.474`) are non-existent. Furthermore, with these inference optimization techniques, we have been able to reduce the inference runtime by 15x (0.85 seconds versus 13.1 seconds), which is an impressive result.

We can also display an image for qualitative results:

GroundTruth

Prediction

Figure 9.7 – Qualitative results: GPU-optimized inference

As expected from the quantitative metrics, *Figure 9.7* shows excellent results as well, with negligible differences (if any) from the results in *Figure 9.6*.

Visualizing and profiling our models

In the previous sections, we saw the different techniques that we could apply to optimize our inference loops, and the results these techniques achieved. However, how exactly do these techniques work? Why are they faster?

We are going to use two tools that MXNet provides for exactly this purpose:

- **Model visualization**
- **Model profiling**

Model visualization provides us with an intuitive way to see how the different layers interact with each other. This is particularly interesting for networks that use ResNet backbones (such as `DeepLabv3`, which we use for image segmentation in this recipe) because of the **residuals** being transferred through layers.

Visualizing our model architecture with MXNet is very easy. When working with symbolic models, just one line of code is necessary. In our case, as we work with Gluon models, these are the lines of code necessary:

```
deeplab_pt_cpu_q_hybrid.export('deeplab_pt_cpu_q_hybrid_sym')
 sym, arg_params, aux_params = mx.model.load_checkpoint('deeplab_pt_
cpu_q_hybrid_sym', 0)
 mx.visualization.plot_network(sym)
```

As shown in the previous recipe, ResNet-based networks are composed of ResNet blocks, which include the convolution, batch normalization, and activation steps.

For our CPU-optimized model (hybridized and `Int8`-quantized), this is what some of the connections among those blocks look like:

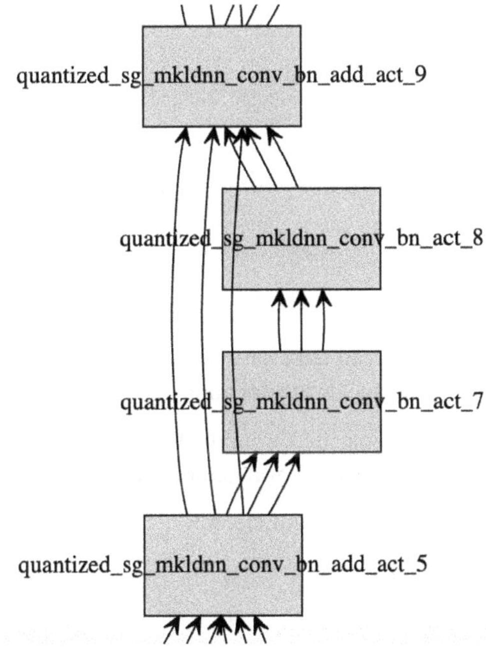

Figure 9.8 – GraphViz of ResNet blocks (CPU-optimized)

As we can see in *Figure 9.8*, there are no individual blocks for each of the expected ResNet block operations; they are all part of single blocks that perform all computations. This combination of operations is aptly called operator fusion, where as many operations as possible are fused together, instead of computing an operation and then the next one (with the typical data transfers occurring). The largest benefit is that fused operations can happen in the same memory space. This is one of the optimizations performed by hybridization, as once the graph for the network is finished, it is quite straightforward to find the operations that are candidates to be fused.

OK, so the model visualization tells us those optimizations will happen, but how can we verify they are actually happening? This is what model profiling is good at, and can also help us understand issues happening during runtime.

As mentioned in the recipe, *Introducing inference optimization features*, and the section Profiling our models, the output of model profiling is a JSON file that can be visualized with tools such as the Google Chrome Tracing app. For a non-optimized CPU workload, our `DeepLabv3` model shows the following timing profile:

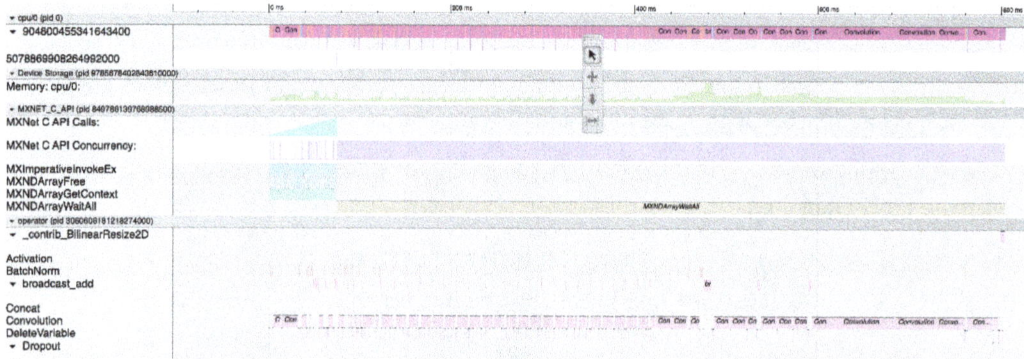

Figure 9.9 – Profiling DeepLabv3: Non-optimized CPU workload

In *Figure 9.9*, we can see the following characteristics:

- Almost all of the tasks are handled by a single process.
- Around 80 ms into the operation, all tasks have been sent to be dispatched, and the control is returned for operations to continue (lazy evaluation and `mx.nd.waitall`).
- All tasks have been sent to be dispatched around 80ms into the operation, and the control is returned for operations to continue (lazy evaluation and mx.nd.waitall).
- All operations are atomic and executed individually.
- The full operation takes around 800 ms.

For a CPU-optimized workload, our DeepLabv3 model shows the following timing profile:

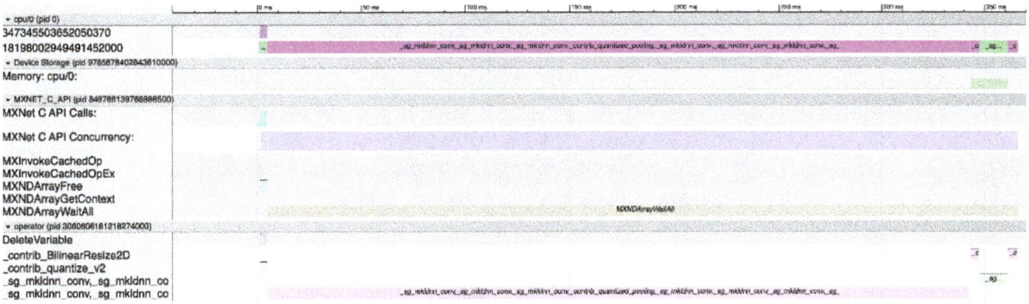

Figure 9.10 – Profiling DeepLabv3: Optimized CPU workload

In *Figure 9.10*, we can see the following characteristics:

- Almost all of the tasks are handled by a single process, similar to the non-optimized counterpart.
- Around 5 ms into the operation, all tasks have been sent to be dispatched, and the control is returned for operations to continue (lazy evaluation and `mx.nd.waitall`), much faster than the non-optimized counterpart.
- Memory is used in a synchronous/structured way, in stark contrast to the non-optimized counterpart.
- All operations are fused together, again in stark contrast to the non-optimized counterpart.
- The full operation takes around 370 ms.

In summary, for CPU-based optimizations, we can clearly see the effects:

- Hybridization has fused all operators together, basically executing almost the full workload in a single operation.
- The MKL-DNN backend and Int8 quantization have improved those operations with accelerated operators.

For our GPU-optimized model (hybridized and AMP-ed), this is what some of the connections among ResNet blocks look like:

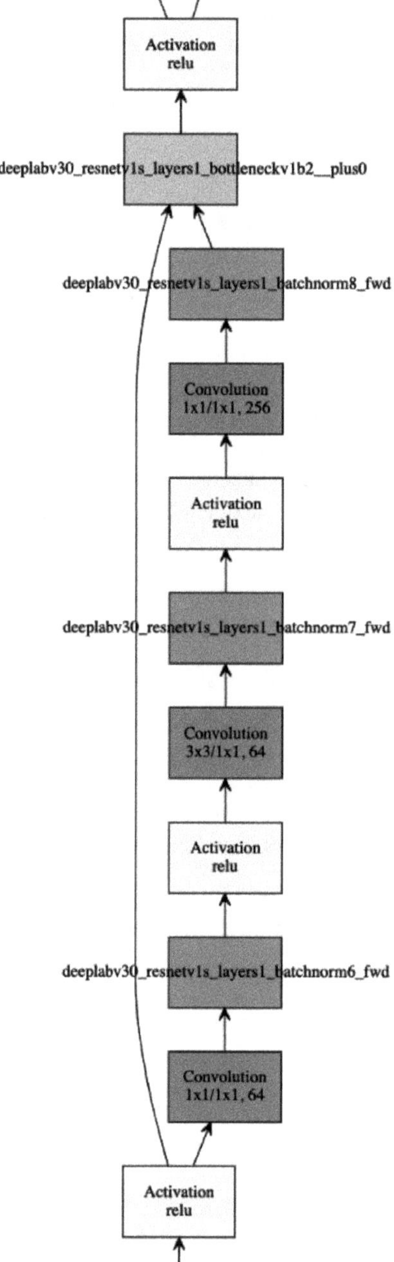

Figure 9.11 – GraphViz of ResNet blocks (GPU-optimized)

As we can see in *Figure 9.11*, this is a very different visualization than the CPU-optimized one as all the individual blocks of the expected ResNet block operations can be identified. As we saw in the first recipe of this chapter, hybridization had a very limited effect on GPUs.

So, where do the accelerations come from? Let's get some help from model profiling.

For a non-optimized GPU workload, our DeepLabv3 model shows the following timing profile:

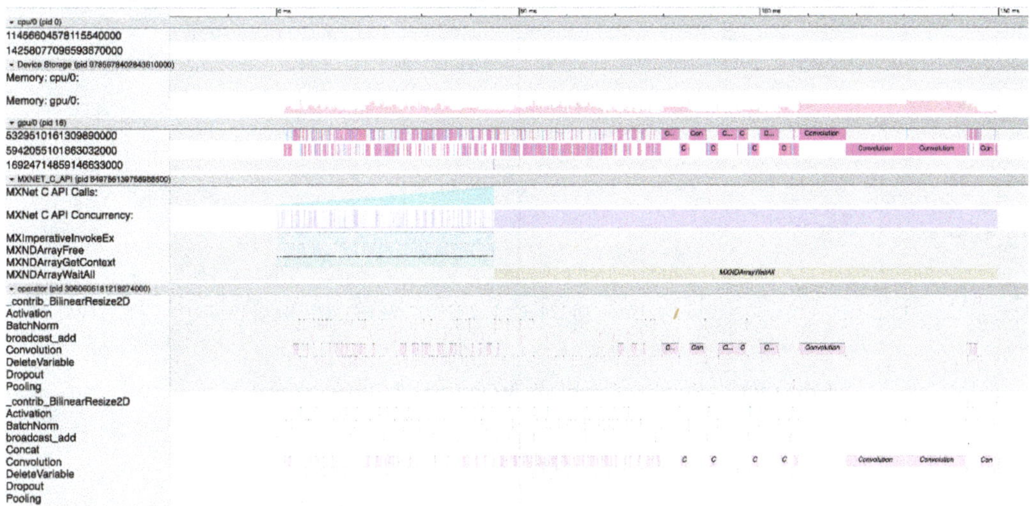

Figure 9.12 – Profiling DeepLabv3: Non-optimized GPU workload

In *Figure 9.12*, we can see the following characteristics:

- Almost all of the tasks are handled by two GPU processes.
- Around 40 ms into the operation, all tasks have been sent to be dispatched, and the control is returned for operations to continue (lazy evaluation and `mx.nd.waitall`).
- Asynchronous/unstructured usage of memory.
- All operations are atomic and executed individually.
- The full operation takes around 150 ms.

For a GPU-optimized workload, our DeepLabv3 model shows the following timing profile:

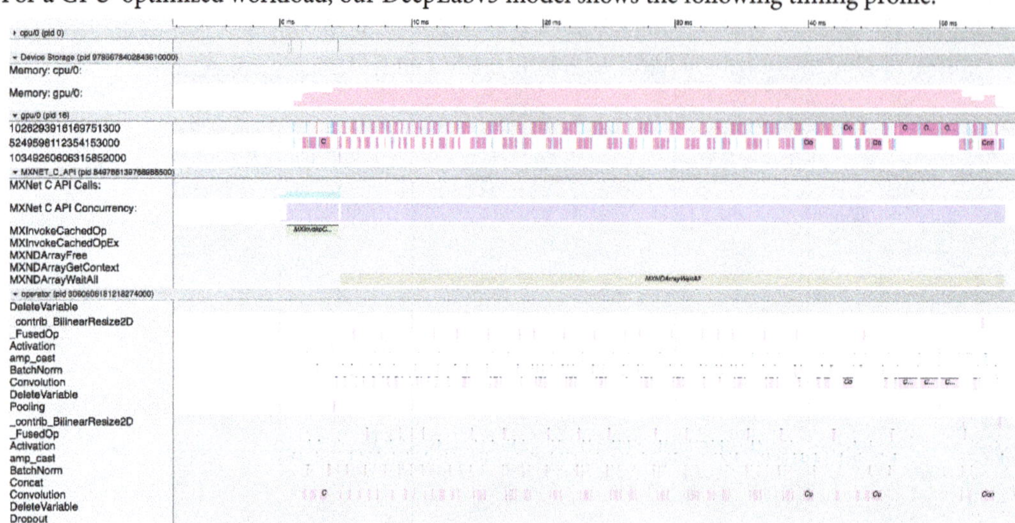

Figure 9.13 – Profiling DeepLabv3: Optimized GPU workload

In *Figure 9.13*, we can see the following characteristics:

- Almost all of the tasks are handled by two processes, similar to the non-optimized counterpart.
- Around 4 ms into the operation, all tasks have been sent to be dispatched, and the control is returned for operations to continue (lazy evaluation and `mx.nd.waitall`), much faster than the non-optimized counter-part.
- Synchronous/structured usage of memory, in stark contrast to the non-optimized counterpart.
- All operations are atomic and executed individually; similar to the non-optimized counterpart, they are just much faster. For example, large convolution operations take ~1 ms in the GPU non-optimized case, whereas they take one-third of that time (~0.34 ms) in the GPU-optimized case.
- The full operation takes around 55 ms (one-third of the non-optimized time).

In summary, for GPU-based optimizations, we can clearly see the effects:

- Hybridization, as expected, has no effect and no operator fusion can be identified.
- AMP makes operations run much faster if the GPU has float16-dedicated circuitry.

Exporting our models to ONNX and TensorRT

MXNet and GluonCV also provide tools to export our models externally. This makes the most sense for optimizing runtime computation times (inference) might need:

- Specific algorithms that MXNet/GluonCV might not support
- Deployment and optimizations on specific hardware platforms

In this section, we are going to study one example of each category.

For specific algorithms, we are going to export our models in ONNX format. **ONNX** stands for **Open Neural Network eXchange** and is an open format that describes how deep learning models can be stored and shared. This is extremely useful to leverage specific tools for highly specialized tasks. For example, **ONNX Runtime** has really powerful inference tools, including quantization (for example, ONNX Runtime has support for GPU-based INT8 quantization). Therefore, we can export our model in ONNX format and start working directly with ONNX Runtime.

As usual, MXNet will allow us to accomplish this with just a few lines of code. We will need to carry out two steps. Firstly, we need to transform our model from Gluon to symbolic format (hybridizing and then exporting):

```
deeplab_pt_gpu_hybrid.hybridize(static_alloc=True, static_shape=True)
 deeplab_pt_gpu_hybrid(single_sample_gpu)
 # Need to be exported externally for the symbols to be loaded
deeplab_pt_gpu_hybrid_filename = "deeplab_resnet101_coco_pt_gpu_
hybrid"
deeplab_pt_gpu_hybrid.export(deeplab_pt_gpu_hybrid_filename)
```

Next, we can transform the symbolic model into ONNX:

```
# Files exported
sym_filename = deeplab_pt_gpu_hybrid_filename + "-symbol.json"
params_filename = deeplab_pt_gpu_hybrid_filename + "-0000.params"
in_shapes = [single_sample_gpu.shape]
 in_types = [mx.np.float32]
 onnx_model_path = mx.onnx.export_model(
    sym_filename,
    params_filename,
    in_shapes,
    in_types,
    onnx_file_name)
```

ONNX also provides a checker to verify our model has been exported correctly. This can be done with the following lines of code:

```
# Model Verification
import onnx
# Load the ONNX model
onnx_model = onnx.load_model(onnx_model_path)
 # Check the ONNX graph
onnx.checker.check_graph(onnx_model.graph)
```

And that's it! Following these instructions, we will have our ONNX model stored in a file (in our example, `'deeplab_resnet101_coco_pt_gpu_hybrid.onnx'`), ready to be used with any tool that accepts ONNX models as input.

On the other hand, sometimes we would like to deploy and/or optimize our models on specific hardware platforms, such as the NVIDIA family of products (for example, Nvidia Jetson platforms). Specifically, Nvidia works with a specific machine learning framework designed to run inference on their own hardware. This framework is called **TensorRT**.

Although MXNet features direct TensorRT integration, it's not enabled by default, requiring building MXNet directly from the source, with specific parameters enabling TensorRT integration. Much more straightforwardly, we can leverage our recently described ONNX export to generate a TensorRT-capable model.

To achieve this, it is enough to write a few lines of code:

```
import tensorrt as trt
trt_file_name = "deeplab_resnet101_coco_pt_gpu_hybrid.trt"
TRT_LOGGER = trt.Logger(trt.Logger.INFO)
 builder = trt.Builder(TRT_LOGGER)
 config = builder.create_builder_config()
explicit_batch = 1 << (int) (trt.NetworkDefinitionCreationFlag.
EXPLICIT_BATCH)
 deeplab_pt_gpu_hybrid_trt = builder.create_network(explicit_batch)
with open(onnx_file_name, 'rb') as model:
    with trt.OnnxParser(deeplab_pt_gpu_hybrid_trt, TRT_LOGGER) as
parser:
        assert parser.parse(model.read()) == True
    deeplab_pt_gpu_hybrid_engine_serialized = builder.build_
serialized_network(deeplab_pt_gpu_hybrid_trt, config=config)
with open(trt_file_name, 'wb') as f:
    f.write(bytearray(deeplab_pt_gpu_hybrid_engine_serialized))
```

With this, we will write a serialized TensorRT-capable model.

We can verify the model can be read by deserializing and reading it. We can do so with the following lines of code:

```
# Check it can be read back
runtime = trt.Runtime(TRT_LOGGER)
 with open(trt_file_name, 'rb') as f:
    deeplab_pt_gpu_hybrid_engine_deserialized = runtime.deserialize_
cuda_engine(f.read())
```

And we are done! In this section, we have been able to successfully write ONNX and TensorRT models. Congratulations!

How it works...

In this recipe, we have applied the different inference optimization techniques seen in the first recipe of this chapter, leveraging our hardware (CPUs and GPUs) to optimize our model runtime performance by doing the following:

- Hybridizing the model
- Leveraging AMP
- Quantizing with the INT8 data type for accelerated inference

Moreover, we have learned how to use model visualizations (powered by GraphViz) and the MXNet profiler and have used these tools to analyze the inference optimizations from a low-level perspective.

Finally, we have learned how to export our models for specific scenarios and purposes, using the ONNX and TensorRT libraries.

There's more...

In this recipe, we have presented the inference optimization problem from a post-training perspective. We were given a (pre-)trained model and tried to squeeze as much performance as we could from it.

However, there is another avenue that can be explored, which starts thinking about maximizing inference performance from a machine learning model design perspective. This is known as **model compression** and is an active area of research, with lots of improvements published periodically. Recently active research topics include the following:

- **Knowledge distillation**: `https://arxiv.org/pdf/1503.02531.pdf`
- **Pruning**: `https://arxiv.org/pdf/1510.00149.pdf`
- **Quantization-aware training**: `https://arxiv.org/pdf/1712.05877.pdf`

Optimizing inference when translating text from English to German

In the initial recipe, we saw how we can leverage MXNet and Gluon to optimize the inference of our models, applying different techniques: improving the runtime performance using hybridization; how using half-precision (float16) in combination with AMP can strongly reduce our inference times; and how to take advantage of further optimizations with data types such as Int8 quantization.

Now, we can revisit a problem we have been working with throughout the book: translating English to German. We have worked with translation tasks in recipes from previous chapters. In *Recipe 4, Translating text from Vietnamese to English*, from *Chapter 6, Understanding Text with Natural Language Processing*, we introduced the task of translating text, while also learning how to use pre-trained models from GluonCV Model Zoo.

Furthermore, in *Recipe 4, Improving performance for translating English to German*, from *Chapter 7, Optimizing Models with Transfer Learning and Fine-Tuning*, we introduced the datasets that we will be using in this recipe: **WMT 2014** and **WMT 2016**. We also compared the different approaches that we could take when dealing with a target dataset: training our models from scratch or leveraging past knowledge from pre-trained models and adjusting them for our task, using the different modalities of transfer learning and fine-tuning. Lastly, in *Recipe 3, Optimizing training for translating English to German*, from *Chapter 8, Improving Training Performance with MXNet*, we applied different techniques to improve the runtime performance of our training loops.

Therefore, in this recipe, we will apply all the introduced optimization techniques for the specific task of optimizing the inference for translating English to German.

Getting ready

As in previous chapters, in this recipe, we will be using some matrix operations and linear algebra, but it will not be hard at all.

How to do it...

In this recipe, we will be carrying out the following steps:

1. Applying inference optimization techniques
2. Profiling our models
3. Exporting our models

Let's dive into each of these steps.

Applying inference optimization techniques

In *Recipe 1, Introducing inference optimization features*, at the beginning of this chapter, we showed how different optimization techniques could improve the performance of the different steps we take in the inference of a machine learning model, including hybridization, AMP, and Int8 quantization.

In this section, we will show how, with MXNet and Gluon, just with a few lines of code, we can easily apply each and every technique introduced and verify the results of each technique.

Without applying these optimization techniques, as a baseline, these are the quantitative results obtained with the CPU:

```
WMT16 test loss: 1.53; test bleu score: 26.40
Time (s): 373.5446252822876
```

From a qualitative point of view, we can also check how well our model is performing with a sentence example. In our case, we chose *I learn new things every day*, and the output obtained is the following:

```
Qualitative Evaluation: Translating from English to German
Expected translation:
 Ich lerne neue Dinge.
 In English:
 I learn new things every day.
 The German translation is:
 Ich lerne neue Dinge, die in jedem Fall auftreten.
```

The German sentence obtained in the output (*Ich lerne neue Dinge, die in jedem Fall auftreten*) means *I learn new things that arise in every case*, and therefore, as can be seen from the results, the text has been almost perfectly translated from English to German.

As concluded in the previous recipe, for maximum performance on the CPU, the best approach is the following:

- Use hybridization: Using the Intel MKL-DNN backend, combined with static memory allocation and invariant input shapes.
- Do not use AMP.
- Use Int8 quantization.

Unfortunately, we won't be able to use Int8 quantization, as this is not supported for GluonNLP models.

Let's apply each of these techniques for our current specific task, translating from English to German.

For hybridization, we just need a couple of lines of code (which include the necessary parameters for the Intel MKL-DNN backend, combined with static memory allocation and invariant input shapes, and the hybridization of the loss function as well):

```
wmt_transformer_pt_cpu_hybrid.hybridize(backend="MKLDNN", static_
alloc=True, static_shape=True)
loss_function = nlp.loss.MaskedSoftmaxCELoss()
loss_function.hybridize(backend="MKLDNN", static_alloc=True, static_
shape=True)
```

We do not need to add any steps connected to AMP as it was shown not to add benefits to CPU-based workloads. Similarly, GluonNLP does not support Int8 quantization, and therefore, we don't need to make any further changes to our code.

Applying these optimization techniques, these are the quantitative results obtained for an optimized CPU inference:

```
WMT16 test loss: 1.53; test bleu score: 26.40
Time (s): 312.5660226345062
```

As we can see, the differences in performance (1.53 versus 1.53 for the loss and 26.40 versus 26.40 for the BLEU score) are negligible. However, with these inference optimization techniques, we have been able to reduce the inference runtime by 20% (313 seconds versus 374 seconds), which is a very good result.

From a qualitative point of view, we can also check how well our model is performing with a sentence example. In our case, we chose *I learn new things every day*, and the output obtained is the following:

```
Qualitative Evaluation: Translating from English to German
Expected translation:
 Ich lerne neue Dinge.
 In English:
 I learn new things every day.
 The German translation is:
 Ich lerne neue Dinge, die in jedem Fall auftreten.
```

The German sentence obtained in the output (*Ich lerne neue Dinge, die in jedem Fall auftreten*) means *I learn new things that arise in every case*, and therefore, as can be seen from the results, the text has been almost perfectly translated from English to German. Moreover, the results are equivalent to the non-optimized case (as expected).

What about GPU-based inference? Let's follow the same steps:

1. Without applying these optimization techniques, as a baseline, these are the quantitative results obtained with the GPU:

   ```
   WMT16 test loss: 1.53; test bleu score: 26.40
   Time (s): 61.67868137359619
   ```

   As expected, there is no change relative to algorithmic performance from the CPU baseline. Runtime inference is indeed six times as fast in the GPU (61.7 seconds versus 374 seconds).

2. From a qualitative point of view, we can also check how well our model is performing with a sentence example. In our case, we chose *I learn new things every day*, and the output obtained is the following:

   ```
   Qualitative Evaluation: Translating from English to German
   Expected translation:
    Ich lerne neue Dinge.
    In English:
    I learn new things every day.
    The German translation is:
    Ich lerne neue Dinge, die in jedem Fall auftreten.
   ```

   The German sentence obtained in the output (*Ich lerne neue Dinge, die in jedem Fall auftreten*) means *I learn new things that arise in every case*, and therefore, as can be seen from the results, the text has been almost perfectly translated from English to German (and is equivalent to both CPU cases).

As concluded in the previous recipe, for maximum performance on the GPU, the best approach is the following:

- Use hybridization: Using static memory allocation and invariant input shapes. Do not use the Intel MKL-DNN backend.
- Use AMP.
- Do not use Int8 quantization (not supported).

Unfortunately, we won't be able to use AMP, as this is not supported for GluonNLP models.

Let's apply each of these techniques for our current specific task, translating from English to German.

For hybridization, we just need a couple of lines of code (which include the necessary parameters for static memory allocation and invariant input shapes, and the hybridization of the loss function as well):

```
wmt_transformer_pt_gpu_hybrid.hybridize(static_alloc=True, static_
shape=True)
loss_function = nlp.loss.MaskedSoftmaxCELoss()
loss_function.hybridize(static_alloc=True, static_shape=True)
```

We do not need to add any steps connected to AMP or Int8 quantization, as GluonNLP does not support these features. Therefore, no further steps are required.

By applying these optimization techniques, these are the quantitative results obtained for an optimized GPU inference:

```
WMT16 test loss: 1.53; test bleu score: 26.40
Time (s): 56.29795598983765
```

As we can see, the differences in performance (1.53 versus 1.53 for the loss and 26.40 versus 26.40 for the BLEU score) are negligible. However, with these inference optimization techniques, we have been able to reduce the inference runtime by 10% (56.3 seconds versus 61.7 seconds), which is a very good result.

From a qualitative point of view, we can also check how well our model is performing with a sentence example. In our case, we chose *I learn new things every day*, and the output obtained is the following:

```
Qualitative Evaluation: Translating from English to German
Expected translation:
 Ich lerne neue Dinge.
 In English:
 I learn new things every day.
 The German translation is:
 Ich lerne neue Dinge, die in jedem Fall auftreten.
```

The German sentence obtained in the output (*Ich lerne neue Dinge, die in jedem Fall auftreten*) means *I learn new things that arise in every case*, and therefore, as can be seen from the results, the text has been almost perfectly translated from English to German. Moreover, the results are equivalent to the non-optimized case (as expected).

Profiling our models

In the previous sections, we saw the different techniques that we could apply to optimize our inference loops, and the results these techniques achieved. However, how exactly do these techniques work? Why are they faster?

In this section, we are going to use the MXNet profiler, which can help us understand issues happening during runtime.

As mentioned in the initial section, the output of model profiling is a JSON file that can be visualized with tools such as the Google Chrome Tracing app. For a non-optimized CPU workload, our transformer model shows the following timing profile:

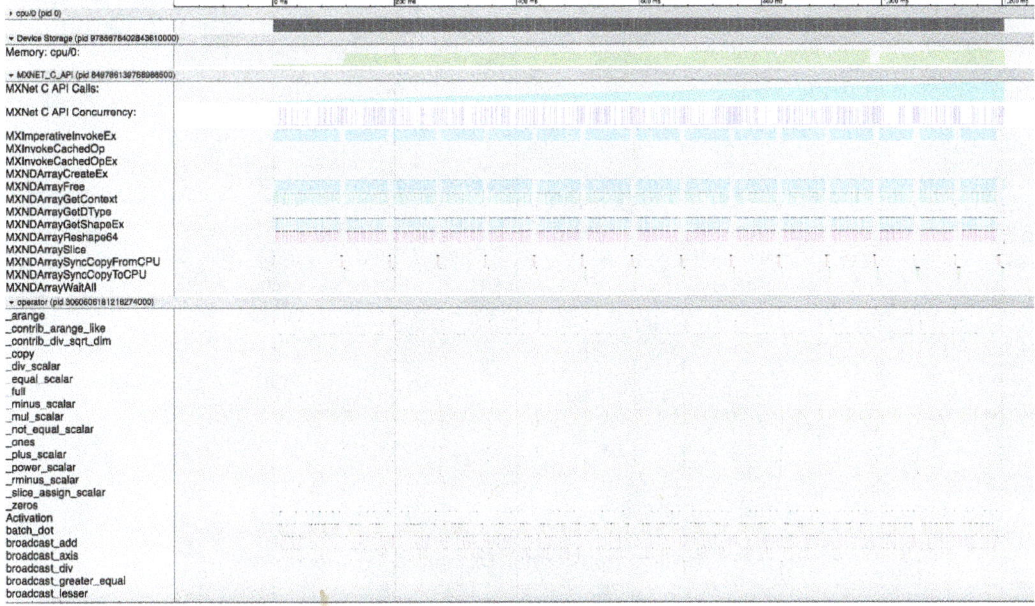

Figure 9.14 – Profiling Transformer: Non-optimized CPU workload

In *Figure 9.14*, we can see the following characteristics:

- Almost all of the tasks are handled by two processes.
- There is almost no waiting time (lazy evaluation and `mx.nd.waitall`).
- Synchronous/structured usage of memory.
- All operations are atomic and executed individually.
- The full operation takes around 1,200 ms.

For a CPU-optimized workload, our Transformer model shows the following timing profile:

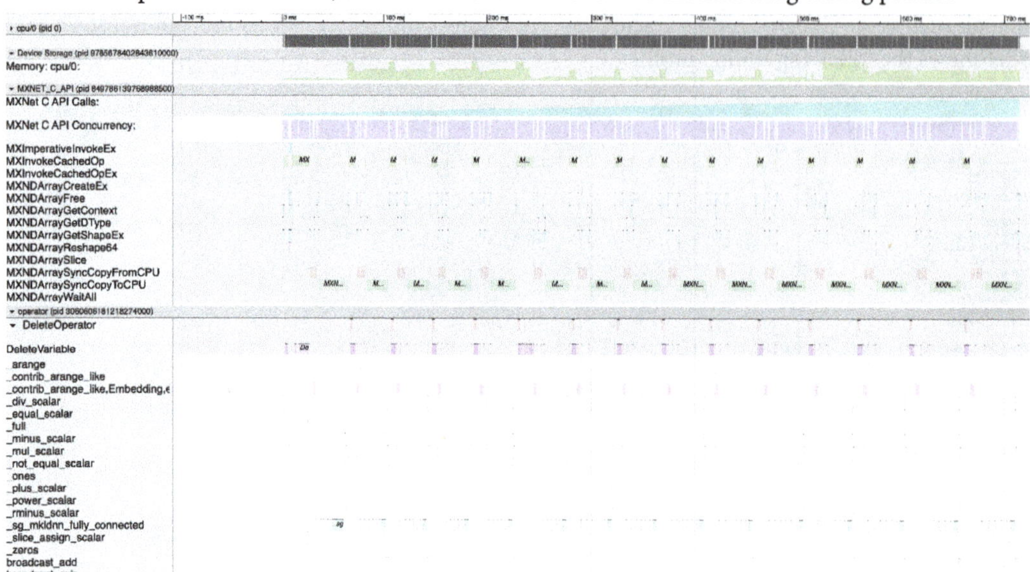

Figure 9.15 – Profiling Transformer: Optimized CPU workload

In *Figure 9.15*, we can see the following characteristics:

- Almost all of the tasks are handled by two processes, similar to the non-optimized counterpart.
- There is almost no waiting time (lazy evaluation and `mx.nd.waitall`), similar to the non-optimized counterpart.
- Memory is used in a more asynchronous/structured way, in comparison to the non-optimized counterpart.
- Some operations are fused together. Although the visualizations are not very clear, operator fusion (hybridization) seems to be working, with most of the time spent on fused operations.
- The full operation takes around 720 ms.

Let's take a zoomed look into one of the operator fusion steps:

DeleteVariable
elemwise_add
elemwise_mul
elemwise_sub
Embedding
Embedding,expand_dims,broadca
expand_dims
expand_dims,broadcast_axis,Resh

Figure 9.16 – Profiling Transformer: Optimized CPU workload (zoom into OperatorFusion)

In *Figure 9.16* we can see how operator fusion has fused together several different operations, including embeddings, layer normalization, fully connected layers, and MKL-DNN-accelerated layers.

In summary, for CPU-based optimizations, we can clearly see the effects:

- Hybridization has fused most of the operators together, although the visualization is difficult to see, and this happens many times.
- The MKL-DNN backend has improved those operations with accelerated operators.

Let's discuss the GPU case now.

For a non-optimized GPU workload, our Transformer model shows the following timing profile:

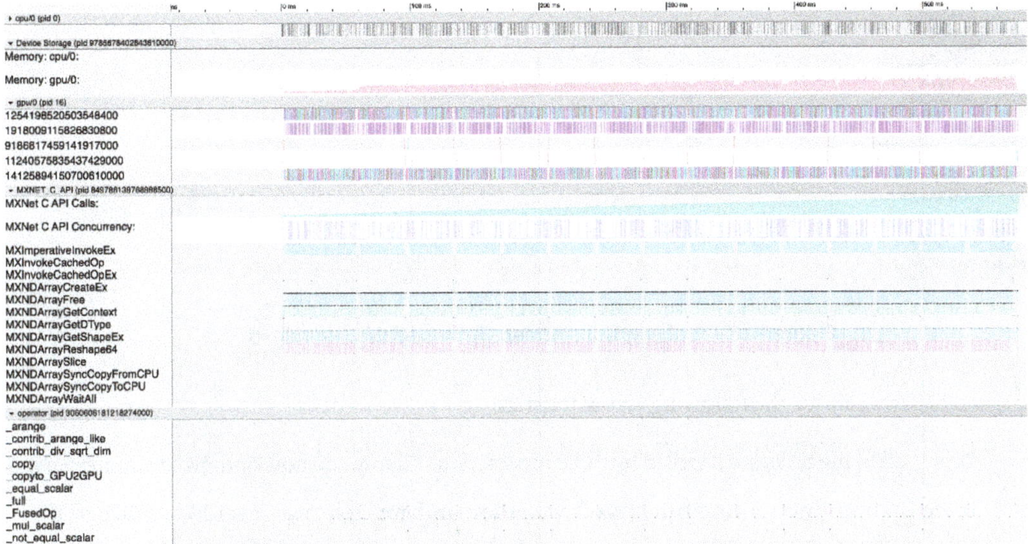

Figure 9.17 – Profiling Transformer: Non-optimized GPU workload

In *Figure 9.17*, we can see the following characteristics:

- Tasks are mostly handled by several (three) GPU processes.
- There is almost no waiting time (lazy evaluation and `mx.nd.waitall`).
- Memory is gradually increasing.
- All operations are atomic and executed individually.
- Several copies from/to CPU, which do not seem to degrade performance.
- The full operation takes around 580 ms.

For a GPU-optimized workload, our Transformer model shows the following timing profile:

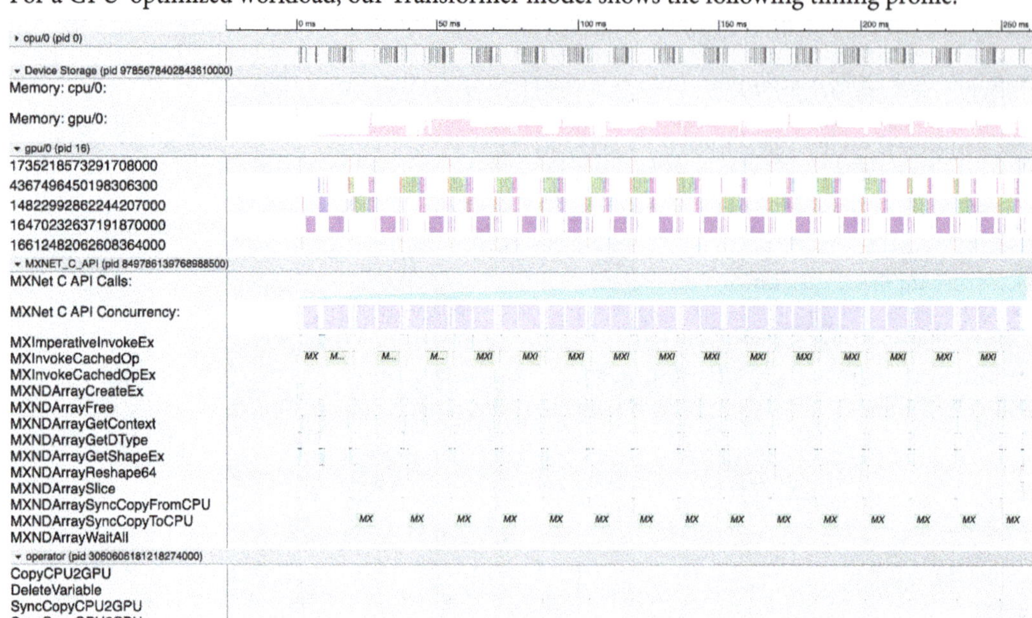

Figure 9.18 – Profiling Transformer: Optimized GPU workload

In *Figure 9.18*, we can see the following characteristics:

- Almost all of the tasks are handled by three processes, similar to the non-optimized counterpart.
- There is almost no waiting time (lazy evaluation and `mx.nd.waitall`), similar to the non-optimized counterpart.
- More asynchronous/unstructured usage of memory, in comparison to the non-optimized counterpart.
- Some operations are fused together. Although the visualizations are not very clear, operator fusion (hybridization) seems to be working, spending most of the time in fused operations.
- Data copy operations from/to CPU do not seem to degrade performance, although there are several.
- The full operation takes around 260 ms.

Let's take a zoomed look into one of the operator fusion steps:

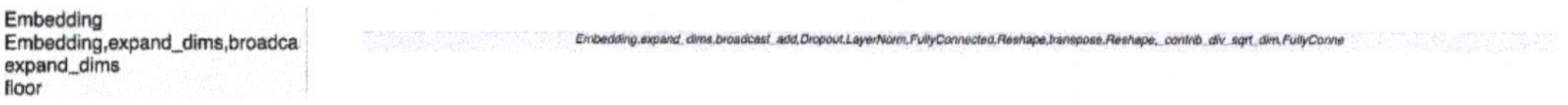

Embedding
Embedding,expand_dims,broadca
expand_dims
floor

Figure 9.19 – Profiling Transformer: Optimized GPU workload (zoom into OperatorFusion)

In *Figure 9.19*, we can see how operator fusion has fused together several different operations, including embeddings, layer normalization, and fully connected layers.

In summary, for GPU-based optimizations, we can clearly see the effect of hybridization, where all operations have been fused together, although the visualization is difficult to interpret, and this happens many times.

Exporting our models

MXNet and GluonNLP also provide tools to export our models. However, these tools are mostly for internal usage of MXNet/Gluon. The reason for this is that GluonNLP mostly deals with **Large Language Models** (**LLMs**) and these workloads are not typically optimized for inference (edge AI computing). Therefore, the only method available for model export is the `save()` function.

This function can be easily called:

```
wmt_transformer_pt_gpu_hybrid.save('transformer_pt_gpu_hybrid')
```

We can verify the files associated with the model, the parameters (the `.params` extension), and the architecture (the `.json` extension) have been saved with these commands:

```
assert os.path.exists("transformer_pt_gpu_hybrid-model.params")
assert os.path.exists("transformer_pt_gpu_hybrid-model.json")
```

And we are done! In this section, we have been able to successfully export our Transformer model. Congratulations!

How it works...

In this recipe, we have applied the different inference optimization techniques seen in the first recipe of this chapter, leveraging our hardware (CPUs and GPUs) to optimize our model runtime performance by hybridizing the model.

Moreover, we have learned how to use the MXNet profiler to analyze the inference optimizations from a low-level perspective.

Finally, we have learned how to export our models using internal MXNet libraries.

There's more...

In this recipe, we presented the inference optimization problem from a post-training perspective. We were given a (pre-)trained model and we tried to squeeze as much performance as we could from it.

However, there is another avenue that can be explored, which starts thinking about maximizing inference performance from a machine learning model design perspective. Several improvements to how LLMs can be used without large compute workloads have been published, such as the following:

- **Low Ranking Adaptation (LORA)**: `https://arxiv.org/pdf/2012.13255.pdf`
- **LORA meets pruning**: `https://arxiv.org/pdf/2305.18403.pdf`
- **GPT4All (quantization)**: `https://gpt4all.io`
- **Int4 quantization**: `https://arxiv.org/pdf/2301.12017.pdf`

Index

A

absolute value 97
activation function 68, 70
linear activation function 70
Rectified Linear Unit activation function 70
Adadelta 83
Adagrad 83
Adam 83
AdaMax 83
ADE20K 300
AlexNet 133
AlexNet custom model
creating, from scratch 135, 136
evaluating 137
training 136, 137
AlexNet pre-trained model
evaluating, from Model Zoo 140
loading, from Model Zoo 139, 140
Area Under the Curve (AUC) 147
Atrous Spatial Pyramid Pooling (ASPP) 170
attention 184
attention heads 185
cross-attention head 186
self-attention head 186
Automatic Mixed Precision (AMP) 271
applying, for inference 303
enabling, for training optimization 271
using 313
average pooling layer 130

B

Back-Propagation Through Time (BPTT) 180
Bag-Of-Words (BOW) 58
batches 42
batch normalization 81
batch size 42
Beginning-of-Sentence (BOS) tokens 207
BERT 184, 199
BERT base model 200
BERT large model 200
masked language model 199
next-sentence prediction 199
bigrams 62
BiLingual Evaluation Understudy (BLEU) 208
binary classification model
defining 103, 104
binary segmentation mask 161
biological neuron
Axon/Axon terminals 67
cell body 67
dendrites 67
modeling 67
soma 67

C

calibration dataset 304
Categorial Cross-Entropy (CCE) loss function 115
ChatGPT 295
CIFAR10 144
classification datasets 33
  Iris dataset 33
classification model evaluation 120
  qualitative evaluation 120, 122
  quantitative evaluation 123, 124
  training performance, measuring 121, 122
classification models 102
  Area Under the Curve (AUC) 112, 113
  binary classification model 103, 104
  confusion matrix 109, 110
  cross-entropy loss function 108, 109
  evaluating 107
  evaluation metrics 107, 108, 111
  features, defining 105, 106
  initializing 106
  loss functions 107, 108
  multi-label classification model 104, 105
  working 107
classification model training 114
  batch size 118
  dataset, splitting 115, 116
  epochs 118, 119
  loss function, defining 115
  model, improving 114, 115
  optimizer and learning rate 116-118
  optimizer, defining 115
  training loop 116
clustering algorithms
  expectation maximization clustering 191
  mean shift clustering 191
Computer Vision (CV) 125, 215
Conditional Random Fields (CRFs) 170
confusion matrix 109, 228
context modules 164
Convolutional Neural Network (CNN) 126
  convolutional layer equations 127
  convolutional layer example, running with MXNet 130
  convolution parameters 128
  pooling layer equations 130
  pooling layer example, running with MXNet 131
  receptive field 129
  summarizing 131, 132
convolution parameters 128
corpus 55, 62
CPUs 3
cross-attention head 186
cross-entropy loss function 108, 109
CUDA (Compute Unified Device Architecture) 3
custom LSTM networks 182, 183

D

DataFrames 34
DataLoader 83
data parallelization
  applying 272, 273
  training with 272
Data Scientist (DS) 272
dataset split, regression model training 84
  test set 84
  training set 84
  validation set 84
decay 285
DeepLab-v1 architecture 170
DeepLab-v2 architecture 169, 170
DeepLab-v3 architecture 164, 170

DeepLab-v3+ architecture 170
DeepLab-v3 model 239, 300
training, with Penn-Fudan Pedestrian dataset 241-243
DeepLab-v3 pre-trained model
evaluating, from Model Zoo 171
loading, from Model Zoo 168-171
deep learning 299
deep learning network 69
dilation 129
dimensionality reduction techniques 45
Doc2Vec model 196
Dogs vs Cats dataset 226-228
pre-trained ResNet model, fine-tuning on 234-237
used, for training ResNet model 228-230
domain adaptation problem 143
domain gap 143
dropout method 81

E

encoder-decoder 164
End-of-Sentence (EOS) tokens 207
English to German translation
performance, improving 251, 252
pre-trained Transformer model, used for performance optimization via transfer learning WMT2014 to WMT2016 255-258
pre-trained Transformer model, fine-tuning on WMT2016 258-261
transfer learning and fine-tuning techniques, applying 261, 262
Transformer model, training in WMT2016 dataset 254, 255
WMT2014 dataset 252, 253
WMT2016 dataset 252, 253
Enron Email dataset 62
analyzing 50
content analysis 52-54
data cleaning 55, 56
data structure 51
examples per class 51, 52
N-grams 56, 57
PCA, applying 60
t-SNE, applying 61, 62
word clouds 58
word embeddings 59, 60
word processing 58
evaluation and loss functions
defining, for regression 74-76
mean absolute error 76, 77
mean squared error 77, 78
properties 75, 76
smooth mean absolute error/ smooth L1 loss 78, 79
evaluation metrics, classification models 111
accuracy 111
F1 score 111
precision 111
recall 111
specificity 111
exploding gradients problem 180
Exploratory Data Analysis (EDA) 134

F

Fashion-MNIST dataset 40
analyzing 41, 42
data structure 42
dimensionality reduction techniques 44, 45
examples per class 43, 44
PCA, visualizing 45
PyMDE, visualizing 48
t-SNE, visualizing 46
UMAP, visualizing 47

Faster R-CNN 154
Faster R-CNN pre-trained model
evaluating, from Model Zoo 154, 155
loading, from Model Zoo 152, 153
Fast R-CNN 148, 153
feature pyramid network 154
fine-tuning 216, 217, 222
concepts 225
Float16 270
used, for training optimization 270
Float32 269
used, for training optimization 269, 270
Fully Connected Networks (FCNs) 163

G

GloVe algorithm 59
GluonCV 215
GluonCV Model Zoo 133, 139, 223
Gluon DataLoader 266
GluonNLP 215
GluonNLP Model Zoo 176, 183
Google Bard 205
Google Colab 3
Google Colab notebook
setting up 4, 5
Google Neural Machine Translator (GNMT) model 208
architecture 209
GPT-3 295
GPU-based preprocessing pipeline 279
GPUs 3
gradient descent 83

H

House Sales dataset 22
analyzing 23, 24
correlation study 25-27
data features 22, 23
data structure 24, 25
exploratory data analysis (EDA) 24
grade analysis 29
location analysis 32
room analysis 29, 30
square feet analysis 28
view analysis 30
year-built and year-renovated analysis 31
hybridization
using 313
hyperparameters 222

I

ILSVRC Scene Parsing Challenge 167
image classification 40, 131
classifier 131
Dogs vs Cats dataset 227, 228
feature extraction 131
ImageNet-1k dataset 227, 228
performance, improving 226
pre-trained ResNet model, fine-tuning on Dogs vs Cats 234-237
pre-trained ResNet model, used for performance optimization via transfer learning from ImageNet-1 to Dogs vs Cats 230-234
ResNet model, training with Dogs vs Cats 228-230
transfer learning and fine-tuning techniques, applying 238

image classification, with MXNet 133
AlexNet custom model, creating from scratch 135, 136
AlexNet custom model, evaluating 137
AlexNet custom model, training 136, 137
AlexNet pre-trained model, evaluating from Model Zoo 140, 141
AlexNet pre-trained model, loading from Model Zoo 139, 140
Dogs vs. Cats dataset reduced version, exploring 134, 135
ImageNet pre-trained models 139
Model Zoo 138, 139
ResNet pre-trained model, evaluating from Model Zoo 142, 143
ResNet pre-trained model, loading from Model Zoo 141, 142

image datasets 40
Fashion-MNIST dataset 40
working 49

ImageNet-1k dataset 226-228

ImageNet-1k, to Dogs vs Cats
pre-trained ResNet model, used for optimizing performance via transfer learning 230-234

ImageNet Large Scale Visual Recognition Challenge (ILSVRC) 135

ImageNet pre-trained models 139

image segmentation
DeepLab-v3 model, training with Penn-Fudan Pedestrian 241-243
inference optimization 309
MS COCO dataset 239-241
Penn-Fudan Pedestrian dataset 239-241
performance, improving 239
pre-trained DeepLab-v3 model, fine-tuning on Penn-Fudan Pedestrian 246-250
pre-trained DeepLab-v3 model, used for performance optimization via transfer learning from MS COCO to Penn-Fudan Pedestrian 243-246
transfer learning and fine-tuning techniques, applying 250, 251

imperative programming 299

inductive transfer learning 220

inference optimization features 298
AMP, applying 303
float16, applying 302
models, hybridizing 299-302
models, profiling 305-309
quantization, applying with Int8 303-305
working 308, 309

inference optimization, for image segmentation
inference optimization techniques, applying 310-314
models, exporting to ONNX and TensorRT 321-323
models, profiling 314-320
models, visualizing 314-320
performing 309
working 323

inference optimization, when translating text from English to German
inference optimization techniques, applying 325-328
models, exporting 333
models, profiling 328-332
performing 324
working 333

Int8 304

Intel CPUs 3

Intersection over Union (IoU) 146

Iris dataset 33
one-versus-one comparison (pair plots) 37-38
analyzing 34
correlation study 35-37
data structure 35
violin plot 39
IWSLT2015 dataset 264
exploring 206

J

Jupyter 2

K

Kaiming initialization method 82
kernel 127
K-means 191

L

Large Language Models (LLMs) 184, 333
lazy evaluation 265
learned embeddings 185
learning rate (LR) 80, 116
learning rate schedules 285
libraries
installing 6-8
verifying 6-8
linear activation function 70
linear regression model 66, 102
Long Short-Term Memory (LSTM) 176
cell state (ct) 181
equations 181, 182
forget gate (ft) 182
hidden state (ht) 181
input gate (it) 182
memory cell candidate (gt) 182
memory cell (ct) 182
output gate (ot) 182
output state (ht) 182
used, for improving RNNs 181-183
loss scaling 271

M

Machine Learning Engineer (MLE) 272
machine learning (ML)
reinforcement learning (RL) 22
supervised learning (SL) 22
unsupervised learning (UL) 22
machine translation 252
machine translators (BLEU)
evaluating 208
Math Kernel Library (MKL) 3, 10
matrices 10
max-over-time pooling layers 177
max pooling layer 130
Mean Absolute Error (MAE) 77
Mean Absolute Percentage Error (MAPE) 77
mean Average Precision (mAP) 147
mean Intersection over Union (mIoU) 162, 241
Mean Squared Error (MSE) 77, 78
Mini-Batch Gradient Descent 83
MNIST database 40
model compression 323
model parallelization 272
model profiling 315
model visualization 315
Model Zoo 133, 138
Momentum/Nesterov accelerated gradient 83
MS COCO dataset 239-241
inference optimization, for image segmentation 309

MS COCO pre-trained models
working with 151
MS COCO, to Penn-Fudan Pedestrian
pre-trained DeepLab-v3 model, used for performance optimization via transfer learning 243-246
MSRA PReLU initialization method 82
multi-label classification model
defining 104, 105
Multi-Layer Perceptron (MLP) network architecture 81, 114
multiple GPUs 291
training with 272
MXBoard 225
MXNet 1, 215
MXNet installation
performing 3
verifying 9, 10
MXNet ND arrays 10-12
conclusions, drawing 16-19
data structures, timing 13, 14
matrix multiplication 15
operations, performing 12
working 19
MXNet version 3

N

Nadam 83
Natural Language Processing (NLP) 50, 175, 215
text translation, from Vietnamese to English 205, 206
topic modeling 189
Natural Language Toolkit (NLTK) 58
N-dimensional arrays (ND arrays) 10
Neural Machine Translation (NMT) 205
NLLB-200 model 214
NLP networks 176
GluonNLP Model Zoo 176, 183
Recurrent Neural Networks (RNNs) 176-180
TextCNNs 176
TextCNNs, applying 177
transformers 183
working 187-189
Non-Maximum Suppression (NMS) 148, 154
NumPy 10
NVIDIAGPUs 3

O

object classifier 148
object detection Model Zoo 150, 151
object detection, with MXNet 144-146
Faster R-CNN pre-trained model, evaluating from Model Zoo 154
Faster R-CNN pre-trained model, loading from Model Zoo 152
MS COCO pre-trained models, working with 151
object detection Model Zoo 150, 151
object detectors, evaluating 146-148
Penn-Fudan Pedestrian dataset, exploring 149
single-stage detectors, versus two-stage object detectors 148
YOLOv3 pre-trained model, evaluating from Model Zoo 157, 158
YOLOv3 pre-trained model, loading from Model Zoo 155-157
object segmentation, with MXNet 160
DeepLab-v3 pre-trained model, evaluating from Model Zoo 171
DeepLab-v3 pre-trained model, loading from Model Zoo 168-171

network architectures, comparing for semantic segmentation 163, 164
Penn-Fudan Pedestrian dataset, exploring with segmentation ground-truth 165
PSPNet pre-trained model, evaluating from Model Zoo 167, 168
PSPNet pre-trained model, loading from Model Zoo 167
segmentation models, evaluating 162, 163
semantic segmentation 161, 162
Semantic Segmentation Model Zoo 165, 166
one-hot encoding scheme 71
One Million News Headlines dataset
exploring 190, 191
ONNX Runtime 321
OpenAI 295
OpenAI GPT 205
Open Neural Network eXchange (ONNX) 321
overfitting 94

P

padding 128
Penn-Fudan Pedestrian dataset 239, 241, 309
exploring 149
exploring, with segmentation ground-truth 165
pre-trained DeepLab-v3 model, fine-tuning 246-250
used, for training DeepLab-v3 model 241-243
Perceptron 66-68
bias 68
output 68
sum 68
weights 68
Perplexity 208
Pixel accuracy 162
positional encodings 185
PR curve 147
pre-trained DeepLab-v3 model
fine-tune, on Penn-Fudan Pedestrian 246-250
pre-trained ResNet model
fine-tuning, on Dogs vs Cats 234-237
used, for optimizing performance via transfer learning from ImageNet-1k to Dogs vs Cats 230-234
pre-trained Transformer model
fine-tuning, on WMT2016 258-261
Principal Component Analysis (PCA) 135
visualizing 45
PSPNet 164
PSPNet pre-trained model
evaluating, from Model Zoo 167, 168
loading, from Model Zoo 167
PyMDE
visualizing 48
Pyramid Scene Parsing Network (PSPNet) 167
Python 2

Q

qualitative evaluation, regression model 94, 96
quantitative evaluation, regression model 94
mean absolute error 97
mean absolute percentage error 97
thresholds and percentage 97, 98
quantization
applying for inference, with Int8 303, 304

R

R-CNN 153
receiver operating characteristic curves 113
Rectified Linear Unit (ReLU) 131
Recurrent Neural Networks (RNNs) 176, 178
architecture 179
custom RNN 180
improving, with LSTM 181-183
training 180
vanilla RNN 178
Region Proposal Network (RPN) 148, 154
Regions Of Interest (ROIs) 154
regression datasets 22
House Sales dataset 22
regression model evaluation
performing 94
qualitative evaluation 94-96
quantitative evaluation 94-98
training performance measuring 94, 95
regression models 66
activation functions 70
biological neuron, modeling 67, 68
defining 68, 69
evaluating 73
evaluation and loss functions 74
features, defining 71, 72
initializing 72, 73
working 74
regression models, training 79, 80
batch size 80, 89
dataset split 80-85
epochs 80, 88
fairness and diversity, analyzing 85-88
loss function 80, 83
model, improving 81, 82
optimizer 80, 83
training loop 89
regression problems 66
Reinforcement Learning from Human Feedback (RLHF) 295
relative error rate 97
ReLU activation function 70, 81, 114
representation learning
focusing, on practical applications 222-225
fundamentals 220-222
representations 218
ResNet 133
ResNet-101 Model Zoo version 143
ResNet architectures 307
ResNet model 226
training, with Dogs vs Cats dataset 228-230
ResNet pre-trained model
loading, from Model Zoo 141, 142
RMSprop 83
Root Mean Squared Error (RMSE) 78

S

SacreBLEU 213
scalars 10
scikit-learn 84, 191
segmentation models
evaluating 162, 163
self-attention 133
self-attention head 186
semantic segmentation 161, 162, 239
network architectures, comparing for 163, 164
working 172, 173
Semantic Segmentation Model Zoo 165, 166
Sentence-BERT model 196
sentiment analysis 196
sentiment analysis, in movie reviews
BERT 199, 200
IMDb Movie Reviews dataset, exploring 197, 198

implementing 200-204
TextCNNs, combining with word embeddings 198
working 204, 205
setosa iris 38
shuffling 42
Single Shot Detector (SSD) 148
single-stage object detectors 148
softmax cross-entropy 228
spatial pyramid pooling 164
architectures 168
State-of-the-Art (SOTA) models 254
Stochastic Gradient Descent 83
Support Vector Machine (SVM) classifier 153
symbolic programming 299

T

t-distributed stochastic neighbor embedding (t-SNE) 135
TensorRT 322, 323
text classification 50
TextCNNs 176
applying 176, 177
text corpus 184
text dataset
corpus 62
Enron Email dataset 50
text translation, from Vietnamese to English 205, 206
BiLingual Evaluation Understudy (BLEU) 208
GNMT model 208, 209
implementing 210-212
IWSLT2015 dataset 206, 207
working 213
tokenizing 58
tokens 58
topic modeling 189
1 Million News Headlines dataset, exploring 190, 191
implementing 192-195
topics, clustering with K-means 191, 192
word embeddings, applying 191
working 195
training loop, regression model
batch size 91
epochs 91, 92
learning rate 89, 90
optimizer 89
working 92, 93
training optimization features 264
Automatic Mixed Precision (AMP), enabling 271, 272
Float16, using 269-271
Float32, using 269
lazy evaluation and automatic parallelization, working with 265-268
training, with multiple GPUs and data parallelization 272-274
working 274, 275
training optimization, for image segmentation
current preprocessing and training pipeline 276-278
results, analyzing 281-283
training optimization techniques, applying 278-281
working 284
training optimization, for translating text from English to German
current preprocessing and training pipeline 286-288
performing 285, 286
results, analyzing 291-294

training optimization techniques, applying 288-291
working 294
transfer learning 216-218
concepts 225
usage, advantages 218, 220
Transformer model
training, in WMT2016 254, 255
transformers 176, 183
architecture 184
attention heads 185, 186
encoder-decoder architecture 184, 185
implementing, in GluonNLP 187
input and output preprocessing 184, 185
key and value 186
query 186
translation invariance 127
trigrams 62
t-SNE
visualizing 46
two-stage object detection 148

U

U-Net 164
Uniform Manifold Approximation and Projection (UMAP) 47, 135
visualizing 47
Universal Approximation Theorem 107

V

vanishing gradient problem 141, 180
vectors 10
versicolor 38
virginica 38

W

warmup 285
weakly SL 22
WMT2014 dataset 252, 253
WMT2016 dataset 252, 253
pre-trained Transformer model, fine-tuning on 258-261
Transformer model, training 254, 255
word2vec 199
word2vec algorithm 59
Workshop on Statistical Machine Translation (WMT) 262
references 262

X

Xavier initialization method 82

Y

YOLO 157
architecture 155
YOLOv2 157
architecture 156
YOLOv3 155, 157
architecture 156
model 149
YOLOv3 pre-trained model
evaluating, from Model Zoo 157, 158
loading, from Model Zoo 155-157
You Only Look Once (YOLO) 148

Z

Zenodo 134

Packtpub.com

Subscribe to our online digital library for full access to over 7,000 books and videos, as well as industry leading tools to help you plan your personal development and advance your career. For more information, please visit our website.

Why subscribe?

- Spend less time learning and more time coding with practical eBooks and Videos from over 4,000 industry professionals
- Improve your learning with Skill Plans built especially for you
- Get a free eBook or video every month
- Fully searchable for easy access to vital information
- Copy and paste, print, and bookmark content

Did you know that Packt offers eBook versions of every book published, with PDF and ePub files available? You can upgrade to the eBook version at packtpub.com and as a print book customer, you are entitled to a discount on the eBook copy. Get in touch with us at customercare@packtpub.com for more details.

At www.packtpub.com, you can also read a collection of free technical articles, sign up for a range of free newsletters, and receive exclusive discounts and offers on Packt books and eBooks.

Other Books You May Enjoy

If you enjoyed this book, you may be interested in these other books by Packt:

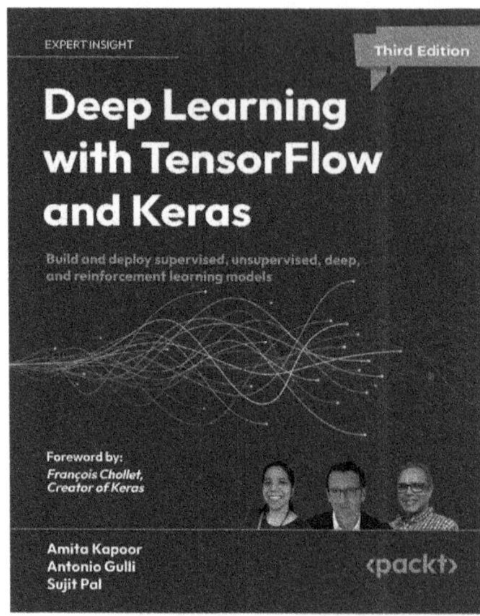

Deep Learning with TensorFlow and Keras - Third Edition

Amita Kapoor, Antonio Gulli, Sujit Pal

ISBN: 978-1-80323-291-1

- Learn how to use the popular GNNs with TensorFlow to carry out graph mining tasks
- Discover the world of transformers, from pretraining to fine-tuning to evaluating them
- Apply self-supervised learning to natural language processing, computer vision, and audio signal processing
- Combine probabilistic and deep learning models using TensorFlow Probability
- Train your models on the cloud and put TF to work in real environments
- Build machine learning and deep learning systems with TensorFlow 2.x and the Keras API

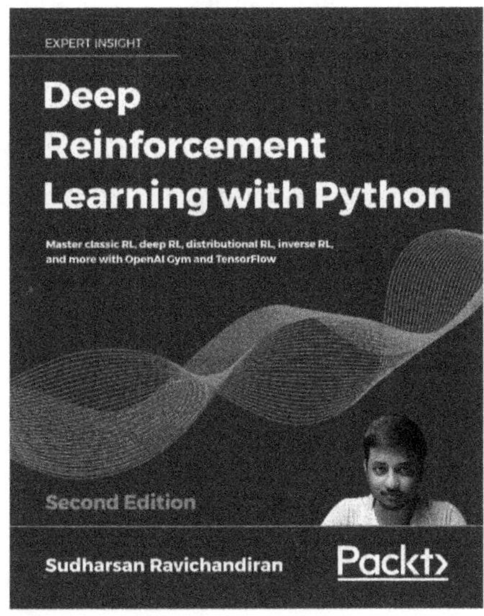

Deep Reinforcement Learning with Python - Second Edition

Sudharsan Ravichandiran

ISBN: 978-1-83921-068-6

- Understand core RL concepts including the methodologies, math, and code
- Train an agent to solve Blackjack, FrozenLake, and many other problems using OpenAI Gym
- Train an agent to play Ms Pac-Man using a Deep Q Network
- Learn policy-based, value-based, and actor-critic methods
- Master the math behind DDPG, TD3, TRPO, PPO, and many others
- Explore new avenues such as the distributional RL, meta RL, and inverse RL
- Use Stable Baselines to train an agent to walk and play Atari games

Packt is searching for authors like you

If you're interested in becoming an author for Packt, please visit `authors.packtpub.com` and apply today. We have worked with thousands of developers and tech professionals, just like you, to help them share their insight with the global tech community. You can make a general application, apply for a specific hot topic that we are recruiting an author for, or submit your own idea.

Share Your Thoughts

Now you've finished *Deep Learning with MXNet Cookbook*, we'd love to hear your thoughts! Scan the QR code below to go straight to the Amazon review page for this book and share your feedback or leave a review on the site that you purchased it from.

`https://packt.link/r/1-800-56960-2`

Your review is important to us and the tech community and will help us make sure we're delivering excellent quality content.

Download a free PDF copy of this book

Thanks for purchasing this book!

Do you like to read on the go but are unable to carry your print books everywhere?

Is your eBook purchase not compatible with the device of your choice?

Don't worry, now with every Packt book you get a DRM-free PDF version of that book at no cost.

Read anywhere, any place, on any device. Search, copy, and paste code from your favorite technical books directly into your application.

The perks don't stop there, you can get exclusive access to discounts, newsletters, and great free content in your inbox daily

Follow these simple steps to get the benefits:

1. Scan the QR code or visit the link below

`https://packt.link/free-ebook/9781800569607`

2. Submit your proof of purchase
3. That's it! We'll send your free PDF and other benefits to your email directly

www.ingramcontent.com/pod-product-compliance
Lightning Source LLC
Chambersburg PA
CBHW081045220326
41598CB00038B/6995

* 9 7 8 1 8 0 0 5 6 9 6 0 7 *